Air Pollution Control,   Part I

ENVIRONMENTAL SCIENCE AND TECHNOLOGY

A Wiley-Interscience Series of Texts and Monographs

Edited by ROBERT L. METCALF, *University of Illinois*
JAMES N. PITTS, Jr., *University of California*

PRINCIPLES AND PRACTICES OF INCINERATION
Richard C. Corey

AN INTRODUCTION TO EXPERIMENTAL AEROBIOLOGY
Robert L. Dimmick

AIR POLLUTION CONTROL, Part I
Werner Strauss

APPLIED STREAM SANITATION
Clarence J. Velz

AIR POLLUTION CONTROL

CONTROL

Part I

EDITED BY

WERNER STRAUSS

University of Melbourne
Victoria, Australia

Wiley-Interscience

A DIVISION OF JOHN WILEY & SONS, INC.

NEW YORK · LONDON · SYDNEY · TORONTO

Library of Congress Catalogue Card Number: 69-18013

ISBN 0-471-83320-7

Printed in the United States of America

10 9 8 7 6 5 4 3 2

SERIES PREFACE

Environmental Science and Technology

The Environmental Science and Technology Series of Monographs, Textbooks, and Advances is devoted to the study of the quality of the environment and to the technology of its conservation. Environmental science therefore relates to the chemical, physical, and biological changes in the environment through contamination or modification, to the physical nature and biological behavior of air, water, soil, food, and waste as they are affected by man's agricultural, industrial, and social activities, and to the application of science and technology to the control and improvement of environmental quality.

The deterioration of environmental quality, which began when man first collected into villages and utilized fire, has existed as a serious problem under the ever-increasing impacts of exponentially increasing population and of industrializing society, environmental contamination of air, water, soil, and food has become a threat to the continued existence of many plant and animal communities of the ecosystem and may ultimately threaten the very survival of the human race.

It seems clear that if we are to preserve for future generations some semblance of the biological order of the world of the past and hope to improve on the deteriorating standards of urban public health environmental science and technology must quickly come to play a dominant role in designing our social and industrial structure for tomorrow. Scientifically rigorous criteria of environmental quality must be developed. Based in part on these criteria, realistic standards must be established and our technological progress must be tailored to meet them. It is obvious that civilization will continue to require increasing amounts of fuel, transportation, industrial chemicals, fertilizers, pesticides, and countless other products and that it will continue to produce waste products of all descriptions. What is urgently needed is a total systems approach to modern civilization through which the pooled talents of scientists and engineers, in cooperation with social scientists and the medical profession, can be focused on the development of order and equilibrium to the presently disparate segments of the human environment. Most of the skills and tools that are needed are already in existence. Surely a technology that has created such manifold environmental problems is also capable of solving them. It is our hope that this Series in Environmental Sciences and Technology will not only serve to make this challenge more explicit to the established professional but that it also will help to stimulate the student toward the career opportunities in this vital area.

Robert L. Metcalf
James N. Pitts, Jr.

PREFACE

The increasing civilization of man unfortunately brings with it increasing pollution of his environment. In recent years we have become aware of this pollution, and studies of the control of solid, liquid, and gaseous wastes are being undertaken by a large number of universities, research institutes, and industrial research organizations. This has led to a rapid growth in the literature on pollution and its control, and it is becoming difficult to remain aware of current work in some of the critical areas.

The main aim of this volume on air pollution control is to present authoritative reviews of specific fields currently of major importance. With this in mind, the authors of the individual chapters have been selected because of their current active participation in the subject.

The reviews attempt to be not only comprehensive but also selective and critical in the constructive sense. Whenever possible, worked examples that illustrate methods of calculation have been included.

It is my hope that the papers will provide an adequate background in their specific fields for engineers and scientists who are faced with particular air pollution control problems. Furthermore, they should be helpful to those who are working in the field application of processes, methods, or techniques as well as to those who are engaged in research for better control methods.

I should like to thank the authors for their cooperation in following the rather stringent guidelines laid down, which have resulted in the patterns of the different chapters aimed at presenting the critical "state of the art." I also wish to express my appreciation for the help I received from my colleagues at the University of Melbourne, Australia, and at the Westinghouse Research and Development Center, Pittsburgh, Pennsylvania—in particular, Dr. J. Bagg, Dr. E. V. Somers, Dr. B. W. Lancaster, and Mr. J. R. Hamm. Much of the typing of the manuscript and the compilation of tables and diagrams was done by Mrs. Tracee Tang, without whose assistance this volume would not have been produced so speedily and efficiently and to whom I wish to express my gratitude.

Melbourne, Australia   WERNER STRAUSS

March 1971

CONTENTS

Dispersion of Pollutants Emitted into the Atmosphere

E. V. Somers

Westinghouse Electric Corporation, Research Laboratories,
Pittsburgh, Pennsylvania

I. INTRODUCTION

Atmospheric contamination that accompanies industrial society arises from the exhausts generated by industrial plants, power plants, refuse disposal plants, domestic and commercial heating, and transportation. These pollutants, which are in the form of dirt, smog, odors, and undesirable gases, arise mostly from combustion processes and in addition to particulates contain trace amounts of gases such as oxides of sulfur, oxides of nitrogen, hydrocarbons, and carbon monoxide. The expanding needs by society for more energy, and advancing transportation technology, coupled with the rapid growth of urban areas has led to ever increasing amounts and concentrations of pollutants in the atmosphere. The emitted pollutants, picked up and diffused by the wind as it flows past the emission sources, are dispersed into the atmosphere or settle to the ground. The speed and diffusion of the wind affect the ground-level concentration of the pollutants to which men, animals, and plants are exposed.

Emitting process gases into the atmosphere at great heights through stacks disperses the pollutants over much larger area and greatly reduces the ground-level concentration near the source. The meteorological environment into which the stack discharges its effluent is important as is the height of the stack relative to the size and position of obstructions associated with either the natural topography or buildings.

Procedures to estimate pollution concentrations at different points downwind of the emission sources must account as best as possible for the above features. If in design procedures, all the complex interchanges of the above factors are lacking, resort must be made to either field observations or scale-model testing. Frequently acceptable answers may be obtained by using simplified models that neglect complexities. Within the framework of simple topographic and meteorological models, assumptions are made to give procedures that supply preliminary design answers.

II. METEOROLOGY

The dispersion of pollutants in the atmosphere is related to the complex flow behavior within the atmospheric "heat engine." Motions within the atmosphere arise from the applied pressure gradients and the Coriolis-force and centrifugal-force per unit volume of the large scale circulation, and these lead to dispersion by forced convection and by free convection established by local temperature differences. Most important in the analysis of dispersion of pollutants is the free-convective contribution, since the diurnal heating-and-cooling cycle associated with the sun markedly affects the near-ground layer of air within which most pollutants are emitted and dispersed.

The variation of temperature with height above the earth's surface is referred to as the temperature profile. During a warm sunny day the temperature falls with height above the surface. After sunset, during a clear night, the temperature of the surface falls rapidly and cools the adjacent layer of air so that the temperature of the air increases with height above the surface, to establish an inversion.

Of particular interest is the vertical temperature gradient that occurs under adiabatic vertical air movement; this gradient is referred to as the dry adiabatic lapse rate. With inertial and frictional terms negligible due to slow vertical motion:

$$-\frac{\partial p}{\partial z} - g\rho = 0 \tag{1}$$

where $\partial p/\partial z$ is the change in pressure p with height, z, g is the gravity acceleration and ρ is the density. Combining an adiabatic change in pressure of a perfect gas by the energy equation with the equation of state yields the expression for temperature as a function of pressure:

$$T = Cp^{(\gamma-1)/\gamma} \tag{2}$$

where T is the temperature, γ is the adiabatic process exponent and C is a constant.   Differentiating eq. 2 with respect to height and combining it and eq. 1 with the equation of state for an ideal gas, $p = \rho RT$ (R = gas constant) gives the change in temperature with height:

$$\frac{dT}{dz} = -\frac{g}{R}\left(\frac{\gamma - 1}{\gamma}\right) \tag{3}$$

For air with $\gamma = 1.41$, $R = 159.6$ m^2/sec^2 °R, and $g = 9.8$ m/sec^2

$$\frac{dT}{dz} = -1.78°\text{F}/100\text{ m} = -5.43°\text{F}/1000\text{ ft}$$

(or, in metric units 1°C/100 m).

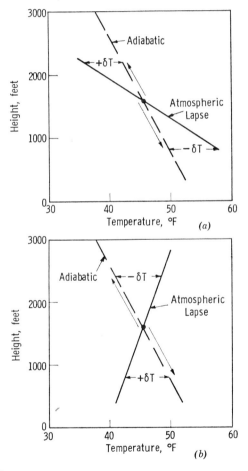

Fig. 1. (a and b) Unstable and stable atmospheric lapse rate.

If the temperature of the atmosphere has a vertical gradient equal to the adiabatic lapse rate of $-5.4°F/1000$ ft, the atmosphere is said to be "neutral." If the atmosphere has a vertical gradient less than $-5.4°F/100$ ft, it is unstable, and if more than $-5.4°F/1000$ ft, including positive gradients characteristic of inversion, it is stable. As shown in Figure 1a, adiabatic displacement of an air packet upward in an unstable atmosphere will produce a higher temperature of the air packet than that of the surrounding air; this positive temperature difference produces a buoyant force upward which produces further upward displacement. Similarly, downward displacement of the air packet will produce a downward buoyant force on it, so that the air packet will continue to sink. Similar reasoning for a stable atmosphere (Fig. 1b) will establish that the buoyant force developed upon displacement either upward or downward is a restoring and stabilizing force.

The flow of the atmosphere across the earth's surface is generally turbulent in nature. Vertical thermal gradients in the atmosphere accent the vertical turbulence if the temperature profile is unstable, and depress the turbulence if the temperature profile is stable. Accordingly, dispersion of pollutants exhausted into the atmosphere are affected not only by the average wind flow but by the lapse rate existing within that flow.

III. DISPERSION OF POLLUTANTS IN THE ATMOSPHERE

A. General Description

Dispersion of pollutants exhausted into the atmosphere is accomplished by diffusion. In laminar flow, diffusion occurs by molecular transport, and in turbulent flow by eddy transport. Large differences between the two types exist in the mechanism and in the magnitude of the diffusion. Dispersion occurs by turbulent transport in neutral and unstable atmospheres and approaches laminar diffusion under highly stable conditions.

The diffusion equation describes the transport of a contaminant through a solid medium (1)

$$\frac{\partial \chi}{\partial t} = \nabla \cdot (k \nabla \chi) \tag{4}$$

where χ is the concentration of pollutants, t the time and k is the diffusion coefficient. Extension of the principle to fluid systems is approximated by replacing the partial derivative of time of eq. 4 with the total derivative with respect to time (2).

$$\frac{d\chi}{dt} = \nabla \cdot (k \nabla \chi) \tag{5}$$

Solutions to eq. 4 with many types of initial and boundary conditions are detailed by Carslaw and Jaeger (3). Solutions to eq. 5 are abundant in boundary-layer literature (4).

If a pollutant is released continuously from a point source at a rate Q, into a steady wind of uniform velocity u, the plume of pollutant will expand by diffusion downwind of the source. This is represented in Figure 2, where Q is released at a steady rate at $x = y = z = 0$. Here x is the downwind distance and y the crosswind distance. The total time derivative is approximated by $d/dt \simeq u(\partial/\partial x)$, and if a constant value of the diffusion coefficient is assumed eq. 5 becomes

$$u \frac{\partial \chi}{\partial x} = k \nabla^2 \chi \tag{6}$$

Solutions to this problem have been formulated by Roberts (5).

$$\chi = \frac{Q}{4\pi kr} \exp\left[-\frac{u}{2k}(r - x) \right] \tag{7}$$

where $r = (x^2 + y^2 + z^2)^{1/2}$ (radial coordinate). Since $r = x[1 + (y^2 + z^2)/x^2)]^{1/2}$, one can use the following form of eq. 7 for the usual case of $y^2 + z^2 \ll x^2$

$$\chi = \frac{Q}{4\pi kx} \exp\left[-\frac{u(y^2 + z^2)}{4kx} \right] \tag{8}$$

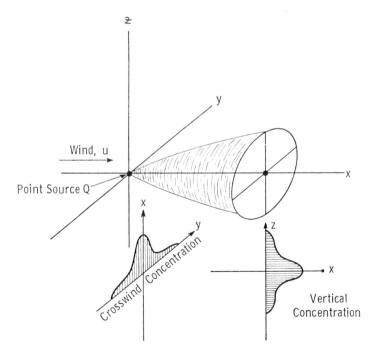

Figure 2.

Two other emission sources are commonly used: (*a*) the instantaneous point source in which Q in mass units is released at $x = y = z = t = 0$ and (*b*) the continuous line source in which Q in mass per unit time distance is released uniformly along a crosswind line referred in Figure 2 to the line $x = z = 0$.

Further, Roberts published solutions for the three sources with an anisotropic diffusion coefficient,

$$k = \begin{bmatrix} k_x & 0 & 0 \\ 0 & k_y & 0 \\ 0 & 0 & k_z \end{bmatrix} \tag{9}$$

These solutions are reproduced below.

1. Instantaneous Point Source

$$\frac{\chi}{Q} = \frac{1}{8(\pi^3 t^3 k_x k_y k_z)^{1/2}} \exp\left[-\frac{1}{4t}\left(\frac{x^2}{k_x} + \frac{y^2}{k_y} + \frac{z^2}{k_z}\right)\right] \tag{10}$$

2. Continuous Crosswind Line Source

$$\frac{\chi}{Q} \sim \frac{1}{(2\pi k_z x u)^{1/2}} \exp\left(-\frac{u z^2}{4 k_z x}\right) \tag{11}$$

3. Continuous Point Source

$$\frac{\chi}{Q} \sim \frac{1}{4\pi x (k_y k_z)^{1/2}} \exp\left[-\frac{u}{4x}\left(\frac{y^2}{k_y} + \frac{z^2}{k_z}\right)\right] \tag{12}$$

These solutions assume u to be constant with respect to z and k_x, k_y, and k_z to be constant with respect to x, y, and z. That this is not the case is well known in that friction affects the flow near the ground such that $u = 0$ at the ground and a velocity gradient in either laminar or turbulent flow is produced. It is well known that boundary-layer velocity profiles can be approximated in the form:

$$u = u_0 \left(\frac{z}{z_0}\right)^m \tag{13}$$

where m is a dimensionless exponent. Here m, u_0, and z_0 depend on the turbulence level developed by the surface roughness and by any free-convective forces established by non-neutral lapse rates (6). The turbulent diffusivity is related to the vertical velocity gradient $\partial u/\partial z$ such that with a neutral atmosphere, where the shear stress is constant throughout the layer, one obtains

$$k = k_0 \left(\frac{z}{z_0} \right)^{1-m} \tag{14}$$

As such, the values of k_x, k_y, and k_z are dependent on z. Accordingly, eqs. 10, 11, and 12 approximately conform to observed data on plume diffusion. More accurately,

$$u(z) \frac{\partial \chi}{\partial x} \sim \frac{\partial}{\partial y} \left[k_y(z) \frac{\partial \chi}{\partial y} \right] + \frac{\partial}{\partial z} \left[k_z(z) \frac{\partial \chi}{\partial z} \right] \tag{15}$$

Solutions to this equation with use of eqs. 13 and 14 and with reasonable (for example, equal) values assigned to m for both k_y and k_z are not available (7).

Available solutions for application to field problems revert back to the form of eq. 12 integrated from eq. 5 with the assumptions that the wind velocity is invariable with height and that the diffusion coefficient is dependent on height and on the temperature gradient of the atmosphere.

Several integrated forms are available from other investigators: the ASME Standard No. APS-1 (8) for estimation of dust emissions for combustion chambers uses the work of Bosanquet and Pearson (9); much use is made of Pasquill's work (10,11); the Sutton formulas, which are modified forms of eqs. 10, 11, and 12, are in widespread use (12); a new guide for predicting dispersion has been recently issued by the A.S.M.E. (53), and it uses the Sutton formulas.

If the plane $z = 0$ is taken to be the surface of the earth and if the pollutant is reflected at the ground, as is characteristic of gaseous pollutants, the source strength Q is dispersed completely in the upper half space $(z > 0)$ rather than in both the upper $(z > 0)$ and lower $(z < 0)$ half spaces. The concentration expressed in eq. 12 is accordingly doubled so that

$$\frac{\chi u}{Q} = \frac{1}{\pi \sigma_y \sigma_z} \exp \left[-\frac{1}{2} \left(\frac{y^2}{\sigma_y^2} + \frac{z^2}{\sigma_z^2} \right) \right] \tag{16}$$

with the transformations

$$\sigma_y^2 = \frac{2k_y x}{u} \qquad \sigma_z^2 = \frac{2k_z x}{u} \tag{17}$$

The quantities σ_y and σ_z are referred to as spreading coefficients, since they measure the spreading of the plume in the y and z directions as it blows downwind. It is immediately recognized that $\chi u / 2Q$ is the probability density function of two Gaussian-distributed variables y and z with zero means and with standard deviations of σ_y and σ_z, respectively. This leads

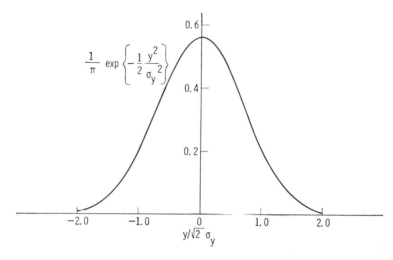

Fig. 3. Gaussian distribution of concentration function.

to the familiar bell-shaped curve for variation of concentration in the y and z direction at any distance x downwind of the source (13), as shown in Figures 2 and 3.

Sutton (14–16) developed expressions for σ_y and σ_z based on the Taylor statistical theory of turbulence and with an average value of u. He found that

$$\sigma_y = \frac{1}{\sqrt{2}} C_y x^{(2-n)/2}$$

$$\sigma_z = \frac{1}{\sqrt{2}} C_z x^{(2-n)/2} \tag{18}$$

where C_y and C_z are the constants for the spreading coefficients and n is the spreading-coefficients exponent for downwind variation. He fitted them into equations of the form of eqs. 16, 10, 11, and 12. Walters (17) used a mixing-length development for k_y and k_z unequal but having the same power variation α, and with u constant

$$k_y = k_{0y} z^{\alpha} \qquad k_z = k_{0z} z^{\alpha} \tag{19}$$

and his derived equation compares with Sutton's if α of eq. 19 equals $1 - m$ of eq. 14. Both treatments partly circumvent the lack of a solution for eq. 15; Sutton states that the error introduced with the assumption of constant wind velocity is small (18).

Downwind concentrations of pollutant emitted by continuous point source can be evaluated with the use of Sutton's eqs. 16 and 18 under meterological and topographic conditions that reasonably fit the simple model used to derive the equations.

B. Extension to Elevated Sources and Inversion Traps

Most pollutants are emitted from stacks above ground so that the ground-level expression, eq. 16, must be modified to account for this. Consider a source located at $x = y = 0$ and at a height $z = h$, as shown in Figure 4. The maximum concentration at ground level near the source of Figure 2 is moved aloft to $z = h$, and the pollutant plume spreads out as it is transported downwind, as approximately described by eq. 16 without the factor of two until ground reflection becomes important.

Accurate description of the function $\chi u/Q$ to include ground reflection is easily obtained by use of the method of images (19). This consists of establishing an image source of equal strength at $x = y = 0$ and $z = -h$ and adding the solutions for both the real and image source together, pictorially illustrated in Figure 4.

$$\frac{\chi u}{Q} = \frac{1}{2\pi\sigma_y\sigma_z}\left(\exp\left\{-\frac{1}{2}\left[\frac{y^2}{\sigma_y^2} + \frac{(z-h)^2}{\sigma_z^2}\right]\right\}\right.$$

$$\left. + \exp\left\{-\frac{1}{2}\left[\frac{y^2}{\sigma_y^2} + \frac{(z+h)^2}{\sigma_z^2}\right]\right\}\right) \quad (20)$$

For $h = 0$, this reduces to eq. 16.

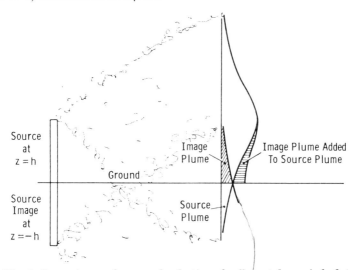

Fig. 4. Source images for ground reflection of pollutant downwind of stack.

The concentration at ground level $z = 0$, is given by

$$\frac{\chi u}{Q} = \frac{1}{\pi \sigma_y \sigma_z} \exp\left[-\frac{1}{2}\left(\frac{y^2}{\sigma_y^2} + \frac{h^2}{\sigma_z^2} \right) \right] \tag{21}$$

The maximum ground-level concentration and the point from the foot of the stack of maximum ground-level concentration can easily be obtained using eqs. 18 and 21 together,

$$\frac{\chi_{\max} u}{Q} = \frac{2}{e\pi h^2} \frac{C_z}{C_y}$$

$$x(\text{at } \chi_{\max}) = \left(\frac{h^2}{C_z^2} \right)^{1/(2-n)} \tag{22}$$

Another condition of interest is the "inversion trap," in which an inversion layer present at some height above the ground, at $z = H$, serves as a reflecting boundary such that pollution emissions into the turbulent layer between the ground and the inversion layer are pocketed there. Treatments of this problem are discussed by Bierly and Hewson (20), Pooler (21), and Miller and Holzwarth (22). Discussion of this meteorological phenomenon and its air-pollution effect will be deferred to a later section, but the simple calculation of such a model is indicated by Morse and Feshbach (19) using the method of images for multiple reflections.

As shown in Figure 5, let the source be located at $x = y = 0$, $z = h$ with the ground reflecting at $z = 0$, and an inversion layer reflecting at $z = H$. Images are established in pairs after the first image is matched to the source. For geometrical clarity it can be interpreted loosely in terms of wave phenomena rather than in terms of diffusion phenomena.

Basic pair solution—eq. 20:

$$\frac{\chi_1 u}{Q} = \frac{1}{2\pi \sigma_y \sigma_z}\left(\exp\left\{ -\frac{1}{2}\left[\frac{y^2}{\sigma_y^2} + \frac{(z-h)^2}{\sigma_z^2} \right] \right\} + \exp\left\{ -\frac{1}{2}\left[\frac{y^2}{\sigma_y^2} + \frac{(z+h)^2}{\sigma_z^2} \right] \right\} \right)$$

Source at $z = h$, plus source-image at $z = -h$.

First inversion reflection:

$$\frac{\chi_2 u}{Q} = \frac{1}{2\pi \sigma_y \sigma_z}\left(\exp\left\{ -\frac{1}{2}\left[\frac{y^2}{\sigma_y^2} + \frac{(z - 2H + h)^2}{\sigma_z^2} \right] \right\} \right.$$

$$\left. + \exp\left\{ -\frac{1}{2}\left[\frac{y^2}{\sigma_y^2} + \frac{(z - 2H - h)^2}{\sigma_z^2} \right] \right\} \right)$$

Second image of source at $z = 2H - h$, plus first image of source image at $z = 2H + h$.

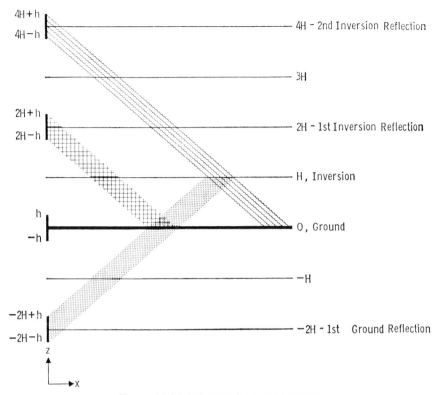

Fig. 5. Multiple images for inversion trap.

First ground reflection:

$$\frac{\chi_3 u}{Q} = \frac{1}{2\pi\sigma_y\sigma_z}\left(\exp\left\{-\frac{1}{2}\left[\frac{y^2}{\sigma_y{}^2} + \frac{(z + 2H - h)^2}{\sigma_z{}^2}\right]\right\}\right.$$

$$\left. + \exp\left\{-\frac{1}{2}\left[\frac{y^2}{\sigma_y{}^2} + \frac{(z + 2H + h)^2}{\sigma_z{}^2}\right]\right\}\right)$$

Third image of source at $z = -2H + h$, plus second image of source image at $z = -2H - h$.

Second inversion reflection:

$$\frac{\chi_4 u}{Q} = \frac{1}{2\pi\sigma_y\sigma_z} \text{ (Fourth image of source at } z = 4H - h \text{ plus}$$

$$\text{third image of source image at } z = 4H + h)$$

The solution is usually given as:

$$\frac{\chi u}{Q} = \frac{u}{Q}\left(\sum_{i=1}^{\infty}\chi_{2i-1} + \sum_{i=1}^{\infty}\chi_{2i}\right) \tag{23}$$

or by Bierly and Hewson (20)

$$\frac{\chi u}{Q} = \frac{\exp\left[-\frac{1}{2}(y^2/\sigma_y{}^2)\right]}{2\pi\sigma_y\sigma_z} \sum_{j=-\infty}^{\infty} \exp\left\{-\frac{1}{2}\left[\frac{(z - h + 2jH)^2}{\sigma_z{}^2}\right]\right\}$$

$$+ \exp\left\{-\frac{1}{2}\left[\frac{(z + h + 2jH)^2}{\sigma_z{}^2}\right]\right\} \tag{23a}$$

Since the series converges rapidly, only the first few pairs of terms (either $i = 1, 2, 3$ in eq. 23 or $j = 0, \pm 1, \pm 2$ in eq. 23a) need be considered.

C. Deposition of Particulates

Dispersion of particulates emitted from a stack differs from that of pollutant gases contained in the plume, in that the particles settle toward the earth under gravity and generally, when they reach the earth's surface, they are not re-entrained in the atmosphere by the wind. Quantitative descriptions of particulate dispersion do not correspond to the Sutton formulas developed earlier, eqs. 16, 20, and 23, since they must account for the gravitational settling and the ground-level absorption.

Particulates are characterized by their free settling velocity, w_p and in air, this combines particle shape, size, and density into an approximation (23)

$$w_p \sim \left(\frac{2gm_p}{\rho_p A_p C_D}\right)^{1/2} \tag{24}$$

where m_p = particle mass, ρ_p = particle density, A_p = cross-sectional area of particle, and C_D = drag coefficient of particle. The drag coefficient can be estimated from

$$\mathrm{Re} = \frac{w_p}{\nu}\sqrt[3]{\frac{m_p}{\rho_p}}$$

where ν = kinematic viscosity

$$C_D = \begin{cases} 24/\mathrm{Re} & \text{for} \quad \mathrm{Re} < 2, \\ 0.4 + 40\,\mathrm{Re} & \text{for} \quad 2 < \mathrm{Re} < 500 \\ 0.44 & \text{for} \quad 500 < \mathrm{Re} \end{cases} \tag{24a}$$

Although there are other approaches to the settling problem (24), a procedure discussed in an AEC article (25) employs a simple modification of

Sutton's equation.   Here the factor of two in eqs. 16 and 21 is removed to approximate nonreflection from the ground, and the axis of the plume is tilted downward by replacing h with $h - xw_p/u$.   The ground-level concentration multiplied by the particle-settling velocity gives the deposition rate per unit area-time:

$$G = \frac{Q_p w_p}{2\pi\sigma_y\sigma_z u} \exp\left\{-\frac{1}{2}\left[\frac{y^2}{\sigma_y^2} + \frac{(h - xw_p/u)^2}{\sigma_z^2}\right]\right\} \tag{25}$$

Another scheme is suggested by considering a source plus sink-image form of eq. 20, so that the absorbing (Dirichlet) condition, $\chi = 0$, exists along the ground plane rather than the reflecting (Neumann) condition $\partial\chi/\partial z = 0$, formed by a source plus source image (19) used in eq. 20.

Here no tilting of the plume is assumed and all transport to the ground occurs by eddy diffusion.

$$\frac{\chi u}{Q_p} = \frac{1}{2\pi\sigma_y\sigma_z}\left(\exp\left\{-\frac{1}{2}\left[\frac{y^2}{\sigma_y^2} + \frac{(z - h)^2}{\sigma_z^2}\right]\right\}\right.$$
$$\left. - \exp\left\{-\frac{1}{2}\left[\frac{y^2}{\sigma_y^2} + \frac{(z + h)^2}{\sigma_z^2}\right]\right\}\right) \tag{26}$$

At ground level, $z = 0$, $\chi u/Q_p = 0$, that is, all pollutants reaching the ground are absorbed by it.   See Fig. 6.

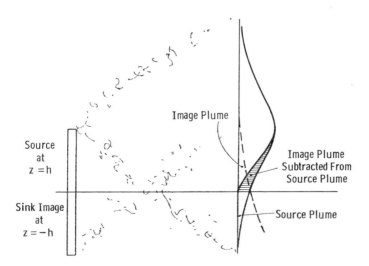

Image Plume

Source
at
z = h

Sink Image
at
z = −h

Image Plume
Subtracted From
Source Plume

Source Plume

Fig. 6. Sink image for ground absorption of pollutant downwind of stack.

The rate of absorption of particulate-pollutant by the ground is given by:

$$G = k_z \frac{\partial \chi}{\partial z}\Big|_{z=0} = \frac{k_z Q_p h}{\pi \sigma_y \sigma_2^3 u} \exp\left[-\frac{1}{2}\left(\frac{y^2}{\sigma_y^2} + \frac{h^2}{\sigma_z^2}\right)\right] \tag{27}$$

With the approximation that at ground level the anisotropic diffusion coefficient, $k_z = \sigma_z^2 u / 2x$, according to eq. 17, and with integration over the y coordinate $(-\infty < y < \infty)$, the ground settling rate per unit downwind distance is obtained:

$$L = \frac{Q_p \xi}{\sqrt{\pi}\, x} \exp\left(-\xi^2\right) \tag{28}$$

where

$$\xi = \frac{h}{\sqrt{2}\,\sigma_z}$$

Equation 25, describing the particle-settling procedure, and with integration over the y coordinate $(-\infty < y < \infty)$, is given by

$$L = \frac{Q_p \beta}{\sqrt{\pi}\, x} \exp\left[-(\xi - \beta)^2\right]$$

$$\beta = \frac{w_p}{u} \cdot \frac{x}{\sqrt{2}\,\sigma_z} \tag{29}$$

Since two-thirds of the range, $4\sigma_y$, contains about 95% of the settled material, a rate per unit area averaged over the range can be obtained by dividing L by $4\sigma_y$. Deposition rates are of importance over time periods of days and months, and more accurate representation would account for changes in the direction and speed of the wind over the period of concern. Averages over a time period can be composed of sums over subintervals of time using prevailing wind direction and speed during the subinterval to compute the deposition rate per unit area, as follows

$$\bar{G} = \frac{L}{4\sigma_y} = \frac{Q_p \xi}{4\sqrt{\pi}\,\sigma_y x} \exp\left(-\xi^2\right) \tag{28a}$$

and

$$\bar{G} = \frac{L}{4\sigma_y} = \frac{Q_p \beta}{4\sqrt{\pi}\,\sigma_y x} \exp\left[-(\xi - \beta)^2\right] \tag{29a}$$

A comparison of the two procedures for two meteorological conditions and for a 20-μm particle with a specific gravity of 3.0, falling in air is shown in the following example:

Stack: A stack of height 100 m (330 ft) is used, that is, $h = 100$ m.

Particle (26): The particle is represented by the particle free-fall velocity, $w_p = 0.04$ m/sec.

Meteorology (27): The meteorological conditions are approximated by

$$u = 2 \text{ m/sec}, \quad \text{and} \quad u = 10 \text{ m/sec}$$

$$\sigma_z = 0.51x^{0.86}, \quad \text{and} \quad \sigma_z = 0.22x^{0.86}$$

$$\sigma_z = 0.7\sigma_y$$

(The exponent used above for describing downwind variation is higher than that used in deriving eq. 28; no great error is introduced). The results of the computation are given in Table I.

The maximum fallout value in $\sqrt{\pi}\,\bar{G}/Q_p$ using the particle-settling procedure occurs at approximately 300 m at a value of $2.06 \times 10^{-7}/\text{m}^2$, for $u = 2$ m/sec, and at approximately 800 m, at a value of $3.89 \times 10^{-8}/\text{m}^2$ for $u = 10$ m/sec.

TABLE I

Calculated Deposition Rates for a Representative Particle under Specified Meteorological Conditions

(a) $u = 2$ m/sec

Distance downwind (meters)	Deposition Rate					
	200	300	500	1000	3000	5000
σ_z (meters)	49	69	107	194	502	777
$\dfrac{\sqrt{\pi}\bar{G}}{Q_p}$ Eddy diffusion (eq. 27) ($\times 10^7$), (meters^{-2})	32.70	30.30	14.00	2.90	0.16	0.04
$\dfrac{\sqrt{\pi}\bar{G}}{Q_p}$ Particle settling (eq. 25), ($\times 10^7$), (meters^{-2})	1.49	2.06	1.52	0.59	0.10	0.04
Ratio	21.9	14.7	9.2	4.9	1.6	1.0

(b) $u = 10$ m/sec

Distance downwind (meters)						
	500	800	1000	3000	10,000	25,000
σ_z (meters)	46	69	84	215	611	1345
$\dfrac{\sqrt{\pi}\bar{G}}{Q_p}$ Eddy diffusion (eq. 27) ($\times 10^8$), (meters^{-2})	110.7	113.7	86.5	8.00	0.34	0.03
$\dfrac{\sqrt{\pi}\bar{G}}{Q_p}$ Particle settling (eq. 25) ($\times 10^8$), (meters^{-2})	2.69	3.89	3.66	0.98	0.13	0.03
Ratio	41.1	29.2	23.6	8.2	2.6	1.0

The eddy diffusion is independent of particle size and is 10–30 times larger than the particle-settling method near the peak settling rates. The two methods agree best in the vicinity of the downwind distance where the tilted axis of the plume touches ground; before that point, the eddy diffusion method gives the higher settling rates and beyond it, the lower settling rates. The ratios of distance of plume–axis interception of the ground to peak-settling distances are of the order of 10:1 to 20:1, for the 20-μm particles. If 10-μm particles were used, this ratio will become 40:1 to 80:1 and, if 1μm particles are considered, 4000:1 to 8000:1.

The ASME procedure (8) for estimating maximum dust concentration at ground level approximates the particle-settling procedure with two exceptions: one, it uses the Bosanquet and Pearson formula for estimating the dispersion, instead of the "one-half value" Sutton formula; and two, it corresponds to the reflecting plume formula of Sutton, eq. 21. The maximum ground-level dust concentration differs by a ratio of 0.215/0.232, for the Bosanquet and Pearson formula and the reflecting Sutton formula, respectively (11). The maximum ground-level dust concentrations for 30 min to one hour of Figure 1 and Figure 2 of the ASME standard correspond to about 2 times the calculated value above for the 10 m/sec result at 1600 m. Since this result is about 5–10 times lower than that for the eddy diffusion it corresponds to settling of the particles from a reflecting plume, that is, the settling has little effect on the plume reflection and re-entrainment is a major factor; for such a case, the factor of two should be included in the particle-settling formulas, eqs. 25, 29, and 29a.

Re-entrainment of settled dust particles is high on smooth surfaces, but low on rough surfaces characterized by open fields, etc. Experimental detail for comparison on the accuracy of the methods is lacking.

Calculation of dust settling by either the Particle-Settling and ASME methods, or by the absorbing ground model proposed above can be extended to an inversion trap. In the Particle-Settling and ASME methods, use is made directly of eq. 23 to calculate the ground-level concentration. In the ground absorption method with a reflecting lid due to the inversion, all pairs of equations $\chi_i u / Q_p$ are formed by replacing the sum of source and image, for example, eq. 20, with a difference of source and image, for example, eq. 26, and the additional image contributions for the ground and inversion lid are summed by the following:

$$\chi = \sum_{i=0}^{\infty} [(\chi_{4i+1} + \chi_{4i+2}) - (\chi_{4i+3} + \chi_{4i+4})] \qquad (23a)$$

In forming χ, χ_{4i+1} and χ_{4i+2} correspond to reflections from the inversion and χ_{4i+3} and χ_{4i+4} correspond to absorption at the ground.

D. Formation of Fog Plumes

Stack effluents containing water vapor will form a fog or mist whenever the exhaust plume cools by mixing with ambient air to provide a local temperature and vapor-concentration condition such that the concentration of the water in the plume exceeds its dry and saturated value at the local temperature. The design of plant effluents today must account for possible formation and elimination of foggy plumes that arise from cooling towers and industrial, utility, and chemical plants (28–30). Dispersion of water vapor can be handled in the same design way as the dispersion of gaseous pollutants; by Sutton's eqs. 20–23.

Design estimates of fog plumes are obtained by calculating the atmospheric dispersion of the water and coupling it with the temperature-versus-density condition accompanying the dilution process of the water concentration in the plume. The increase above ambient of water concentration downwind of the source can be calculated by Sutton's equations realizing that the near-zone dispersion is in error because of the point assumption in the mathematical derivation. This error becomes smaller as the ratio of exhaust gas to ambient air entrained in the plume becomes smaller at farther downwind locations. The temperature rise above ambient can be calculated with the same equation by replacing the pollution source, Q, with an equivalent thermal source, $Q_h/\rho C_p$ (C_p = specific heat at constant pressure of the gas)

$$\frac{\chi u}{Q} = \frac{1}{\pi \sigma_y \sigma_z} \exp\left[-\frac{1}{2}\left(\frac{y^2}{\sigma_y{}^2} + \frac{z^2}{\sigma_z{}^2} \right) \right]$$

$$\frac{\chi u}{Q} = \frac{T u \rho c_p}{Q_h}$$

(30)

(Equation 30 corresponds to a ground-level source with ground reflection. Dispersion of fog plumes from stacks requires the use of eq. 20.) The near-zone approximation for T is in error by the same relative amount as is the concentration, χ, aside from any latent heat effects.

A typical calculation method is illustrated with the help of Figure 7; here air or gas at 90°F with a water content of 32 g/m³ issues from a stack at a given efflux rate in m³/sec, into an ambient at 50°F with a water content of 7.5 g/m³ (relative humidity of 75%). The continuous source of contaminant, Q, is calculated as a product of $(32 - 7.5)$ g/m³ by the efflux rate, and from this, with known meteorology and stack height, the value of $\chi u/Q$ can be calculated for downwind locations. The value of the thermal source, $Q_n/\rho C_p$, can be calculated as a product of the temperature rise above ambient, $(90-50)$°F, by the efflux rate, and from this coupled with the estimate of Q, the approximate slope of the χ versus T process line can be obtained, as

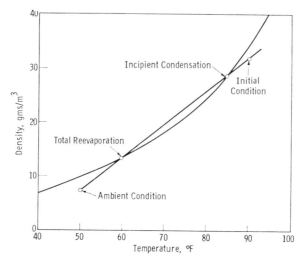

Fig. 7. Dilution process curve.

shown in Figure 7. The slope of the dilution process line is approximate in this example, since it disregards change with temperature of the specific-heat-density product of the water–air mixture; more accurate calculations including this will produce a dilution process line that deviates from the linear relationship of Figure 7 and eq. 30. From the dilution process line with χ known at downwind locations, T can be estimated, and more important, the locations of incipient condensation and total re-evaporation within the plume can be estimated. The calculated dilution process is valid so long as no drizzle falls from the plume, from the exhaust point down to the incipient condensation point and from the total re-evaporation point to the ambient point.

From the point of incipient condensation to the point of total evaporation, the plume process tends toward the saturation line, but based on the following reasonable assumptions, the above method of estimating is valid: first, the density change within the water–air mixture due to condensation is negligible and it has little effect on the dispersion of the water in the plume, especially since the increase in density is usually less than 1%; second, the plume dispersion of the water is independent of both the local phase and energy conditions of the air–water mixture, i.e., the spreading coefficients, σ_y and σ_z, are not affected by wet adiabatic stability within the foggy part of the plume as compared to dry adiabatic stability in the clear part of the plume; and third, drizzle falling out of the plume is negligibly small.

Drizzle is estimated to occur (30) whenever the *liquid* water density exceeds 1 g/m³. Handling a plume with drizzle requires estimates of the rainfall to be factored into the dilution process line, as indicated in Figure 8.

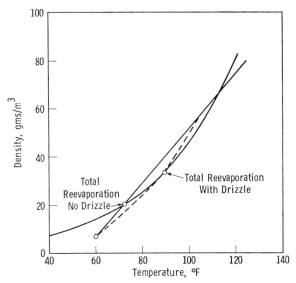

Fig. 8. Dilution process curve with drizzle.   Drizzle produces the dotted process line
by water extraction from the plume.

Drizzle from the foggy part of the plume increases the density and temperature at which total re-evaporation takes place, and it shortens the dimensions
of the foggy plume.   Blum (30) calculated precipitation from a simple plume
model; estimating the local drizzle rate from the above plume description is
more involved.

IV. EVALUATION OF SPREADING COEFFICIENTS σ_y AND σ_z

The values of the spreading coefficients σ_y and σ_z appearing in Sutton's
equation, eqs. 17–20, have been the subject of much field experimentation
which is not yet sufficient to establish undebatably accurate estimates of
σ_y and σ_z.   Despite this, approximations are cited, with caution suggested
in their use.

Singer and Smith (31) of the Meteorology Group at Brookhaven report
results of σ_y and σ_z versus travel time and/or distance downwind of the source,
based on their gustiness classification.   They cite data for four classes of
gustiness:

Class	Stability	Velocity, m/sec
B_2	Very unstable	3
B_1	Unstable	7
C	Neutral	13
D	Stable	6.4

These data are reproduced in Figure 9 for σ_y and in Figure 10 for σ_z.

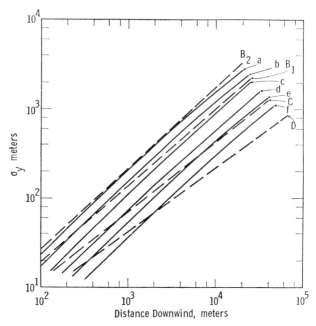

Fig. 9. Horizontal spreading coefficient versus downwind distance.

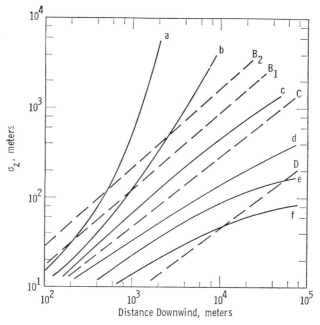

Fig. 10. Vertical spreading coefficient versus downwind distance.

Also contained in Figures 9 and 10 are data evaluated by Gifford (32). Gifford's work is based primarily on diffusion data cited by Pasquill (10,11). Gifford cites six stability categories, as follows:

Class	Stability
a	Extremely unstable
b	Moderately unstable
c	Slightly unstable
d	Neutral
e	Slightly stable
f	Moderately stable

There is good agreement between the Gifford and the Singer-Smith data for σ_y and poorer, but still general, agreement for σ_z.

Also available in another form are the data of Pasquill and Gifford reworked by Turner (33). Turner presents his data as spreading coefficient versus time-of-travel. Comparison of the Turner data with that of Singer and Smith follows in Figures 11 and 12 with Turner's seven categories of stability somewhat similar to Gifford's cited as:

Class	Stability
1	Extremely unstable
2	Unstable
3	Slightly unstable
4	Neutral
5	Slightly stable
6	Stable
7	Extremely stable.

Turner's paper deletes data on classes 6 and 7. Singer and Smith's data of Figures 9 and 10 are converted to travel time using velocities derived from their curves for each of the four classes of gustiness. Again σ_y data are in better agreement than the σ_z data.

Much of the illustrated data is adjusted where possible to correspond to the same instrument height using the correction scheme suggested by Singer, Frizzola, and Smith (27).

There appears to be generally good agreement of the discussed data in the unstable to stable categories. It is more difficult to assess good agreement in the stable areas.

Extensive data have been reported by TVA (34); these data do not display as good agreement with the different data shown in Figures 9–12. They are generally lower in value for σ_y and σ_z than the Singer and Smith data, because they are for higher heights.

Recommendations for σ_y and σ_z have been suggested by Singer, Frizzola,

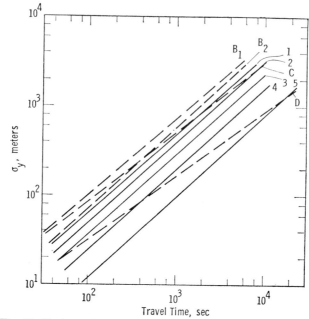

Fig. 11. Horizontal spreading coefficient versus downwind travel time.

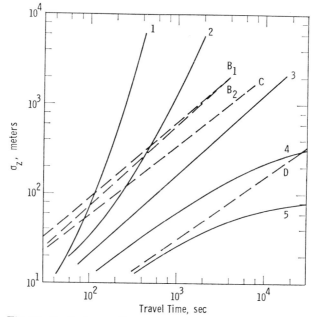

Fig. 12. Vertical spreading coefficient versus downwind travel time.

and Smith (27) for use in obtaining rough estimates of downwind concentration. They caution the user of the data and point out that these should not be substituted for a thorough analysis, including preliminary meteorological investigation, on the site of interest.

Two cases are developed:

Stable: Inversion

$$\sigma_y = 0.15\sigma_a x^{0.71}$$
$$\sigma_z = 0.15\sigma_e x^{0.71}$$

(31)

Unstable: Isothermal or adiabatic

$$\sigma_y = 0.045\sigma_a x^{0.86}$$
$$\sigma_z = 0.045\sigma_e x^{0.86}$$

(32)

If the azimuth angle is available at stack height,

$$\sigma_e = 0.2\sigma_a, \quad \text{stable}$$
$$\sigma_e = 0.7\sigma_a, \quad \text{unstable}$$

(33)

If only an anemometer is available

$$\sigma_a = 2, \quad \text{stable}$$
$$\sigma_a = \frac{23}{u} + 4.75, \quad \text{unstable, at 100 m}$$

(34)

Since the values of σ_y and σ_z are suggested to correspond to the height of the emission source, correction of wind velocity measured at one elevation can be made to another height by

$$\frac{u}{u_0} = \left(\frac{z}{z_0}\right)^{1/2}, \quad \text{stable}$$
$$\frac{u}{u_0} = \left(\frac{z}{z_0}\right)^{1/4}, \quad \text{unstable}$$

(35)

At any height

$$\sigma_a u = \sigma_{a0} u_0$$
$$\sigma_e u = \sigma_{e0} u_0$$

(36)

Methods of evaluating meteorological data such as wind speed, azimuthal and elevation angles, and the standard deviation for these quantities have been summarized by Slade (35) and are applicable to the method of estimating dispersion suggested per eqs. 31–36.

V. STACK DESIGN FOR DISPERSION OF POLLUTANTS

Effluents from industrial, chemical-processing, and electric power plants are usually emitted to the atmosphere from stacks in order to spread the pollutant over a much wider downwind area and so reduce the ground-level concentrations. Considerable attention has been given to the process design of stack emissions that are both heated and unheated. Two types of emissions, at present strongly noticed by the general public, are those from stacks of the large electric power plants and those from water-cooling towers, particularly the large water-cooling towers associated with heat rejection from electric power plants. Some modern electric power plants, e.g., the Cardinal Plant of American Electric Power system, have stacks at the 800 feet and higher level above ground (36). The new large mine-mouth plants in Western Pennsylvania will also have stacks of this height.

The estimate of dispersion of pollutants from stacks includes consideration of many effects including: local and average meteorological conditions, the thermal and momentum rises of the plume from the stack, topography and nearby buildings, and multiple stacks close to each other.

A. Local and Average Meteorology

The effect of short-time meteorology within the diurnal cycle, of the existence of inversion layers and of large scale thermal convection is most important in estimating gaseous concentrations of short duration, while the long-time meteorology (over a month or a year) is important in dustfall.

Types of plumes of concern to stack designers were discussed by Bierly and Hewson (20), Thomas et al. (37), and Scorer and Barrett (38); these are shown in Figure 13.

The *coning plume* occurs when the plume is dispersed by the wind in a neutral atmosphere, a condition characteristic of afternoon meteorology. Plumes under this condition can be calculated by Sutton's formulas, eqs. 20–23, for gaseous pollutants and eqs. 24–29 for particulates, using appropriate neutral or unstable values for σ_y and σ_z, for example, either eqs. 32 or the values for corresponding stability from Figures 9–12.

The *fanning plume* occurs when the plume is dispersed in at atmospheric inversion, during the evening, night, or early morning. Approximations of plume dispersion under this condition can be obtained from the same Sutton equations for gaseous pollutants and with less accuracy, due to particle settling, from the corresponding equations for particulates. The σ_y and σ_z values characteristic of a stable atmosphere must be used here, for example, either eqs. 31 or values selected from Figures 9–12 with the spreading coefficients matched to the estimated degree of stability.

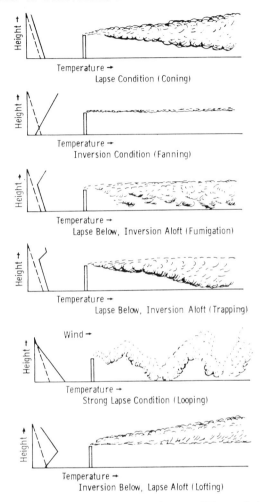

Fig. 13. Six types of plume behavior under various conditions of stability.   Broken
lines, dry adiabatic lapse rate; full lines, existing lapse rates.

The *fumigating plume* occurs when an inversion is breaking up during the
day, i.e., when the turbulent layer rising from the heated ground reaches a
fanning plume emitted into and trapped at effective stack height in the
inversion the night before and downwashes it to ground level.   Estimates
of the fumigation effect can be obtained from the simple procedure (21,39),
wherein the pollutant trapped in the fanning plume is calculated to be
dispersed in the unstable or neutral air mass between the inversion and the
ground.   The pollutant is assumed to be dispersed uniformly in the vertical
direction and with a Gaussian distribution in the y-direction.

Thus, if the *fanning plume* extends on its top side to a height equal to the effective stack height plus a distance equal to two spreading coefficients in the z-direction for the fanning plume, $h + 2\sigma_z$, (98% of the plume is contained within $0 \leqslant z \leqslant h + 2\sigma_z$) the *fumigation* concentration at ground level is estimated to be (21):

$$\frac{\chi u}{Q} = \frac{1}{\sqrt{2\pi}\,\sigma_y(h + 2\sigma_z)}\,\exp\left(-\frac{1}{2}\frac{y^2}{\sigma_y{}^2}\right) \tag{37}$$

where σ_y accounts for the original spread of fanning plume in the y-direction, as given by either eqs. 31 or the stable values as contained in Figures 9 and 11.

Pooler (21) suggests a simple method of estimating time of the inversion fumigation by matching the solar heating of a layer per unit time to that required to heat the air layer moving between the ground and $h + 2\sigma_z$, from its original inversion condition to its final adiabatic lapse condition. Mixing depth forecasts for different times of day, when available, can be used equally well (40). (The mixing depth is the height to which the dry adiabatic lapse rate extends into the inversion.) The vertical temperature profile and wind condition, if known from Weather Bureau ascents, can be coupled with daily synoptic forecasts of the surface temperature to estimate the time that the adiabatic lapse rate extends to any point in the inversion. This effectively is the heat balance estimate of Pooler used in a synoptic manner. Such predictions are available from meteorological forecasting companies.

The existence of "fumigation" conditions from high stacks has been observed on the U.K. Central Electricity Generating Board High Marnham Power Station (41,42), but in no case did the detected fumigation of SO_2 from the power plant exceed in concentration that occurring under neutral lapse conditions. TVA has produced similar findings (37) and has indicated that results calculated on the basis of eq. 37 are higher than observed values by two to three times (43).

The *trapped plume* occurs when the pollutant is emitted into an unstable layer of air trapped between an inversion layer and the ground. Approximations to the pollutant concentration at ground level are obtained from the Sutton formulas for multiple reflections developed in eq. 23, or by the method of Pooler (21). In eq. 23 fast convergence of the terms will show that but a few pairs of terms need be calculated. If Pooler's method is to be used the term $h + 2\sigma_z$ must be replaced with H, the height of the inversion layer, and σ_y is referred to the unstable values, given by either eqs. 32 or contained in Figures 9 and 11. Miller and Holzworth (22) have calculated estimates of the ground-level concentration of pollutant versus inversion depth for continuous line-source emission.

The *looping plume* occurs during the unstable condition of a light wind on a hot summer afternoon, when the large-scale eddying carries portions of the plume to the ground in a corkscrew pattern. The plume touching the ground for short duration will yield a high pollutant concentration during that period. This type of plume is cited by Thomas (39) to give ground-level concentrations for 30-min concentration averages, about 40% of those as calculated by the method for coning plumes. Either eqs. 32 or the extremely unstable values of Figures 9–12 can be used to estimate σ_y and σ_z. Concentrations during a few minutes or less consisting of whiffs of the threadlike part of the plume when it touches the ground can, however, rise to very high values (38,42).

The *lofting plume* occurs when the stack exhausts above an inversion, or when the plume buoyancy carries a stack emission through an inversion layer into a lapse layer aloft. The plume disperses above the inversion, since the top of the inversion layer acts as a barrier that prevents all gaseous and small-particulate emissions from reaching the ground. This type of plume is one of the major goals of tall-stack operation of electric-utility and industrial plants.

B. Thermal and Momentum Contributions to Plume Rise

The Sutton eqs. 20–23 and eqs. 25–29 contain an emission height, h, that accounts for emission of the pollutant from an elevated point source. Determination of this point is difficult and has been the subject of much discussion (44). Since it is greater than the physical height of the stack, h_s, the equivalent stack height includes the stack height plus additional rise due to the thermal buoyancy and to vertical discharge momentum from the stack. The plume leaving the stack is not only influenced by its thermal buoyancy and discharge momentum, but it immediately comes under the influence of the atmospheric dispersive forces. Coupling the plume-rise and dispersive forces into a design scheme that can be easily calculated is afforded by separating the effects assuming, first, the calculation of the plume rise to an equivalent stack height independent of the atmospheric dispersion and, second, the calculation of the atmospheric dispersion of the plume issuing from the virtual point source at the equivalent stack height, h. This is illustrated in Figure 14. The virtual point source can be located by shifting the plume upwind of the stack so that the calculated plume width, $-3\sigma_y$ to $+3\sigma_y$, at the stack corresponds to the width of the buoyant plume, e.g., the stack width can be used as a first approximation. Use of a virtual point source is only of importance for rough estimates in the near-zone of the stack and for fanning plumes.

Recent publications have documented much experimental data on CEGB

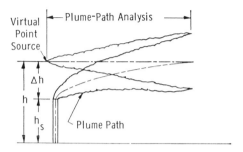

Fig. 14. Two-step plume analysis.

and TVA installations (41,42,45), and comparisons of calculated results with the data have indicated several acceptable formulas.

Carpenter et al. (45) cite the work of Carson and Moses, referring to it as the A.S.M.E. formulation

$$\Delta h = A \frac{4.12 V_s d + 1.47 (Q_h)^{\frac{1}{2}}}{u} \tag{38}$$

where V_s = emission velocity from stack (m/sec),

d = diameter of top of stack (m),

Q_h = heat rate (continuous point source),

and A is a numeric, such that

$$A = 1 \qquad \frac{d\theta}{dz} \geqslant -0.4°F/100 \text{ m}$$

$$A = 2.43 \qquad \frac{d\theta}{dz} < -0.4°F/100 \text{ m}$$

Here $d\theta/dz$ is the change in potential temperature with height from the top of the stack to the top of the plume,

$$\frac{d\theta}{dz} \sim \left(\frac{d\Gamma}{dz} + \Gamma\right)\frac{\theta}{\Gamma} \sim \frac{dT}{dz} + \Gamma$$

$$\Gamma \sim 1.8°F/100 \text{ m} \tag{39}$$

(negative dry adiabatic lapse rate, eq. 3). The A.S.M.E. formulation (53) finally adopted for their standard differs from the Carson and Moses formulation; results calculated using that of Carson and Moses are in substantial agreement with those calculated by the A.S.M.E. standard.

The CEGB uses the Lucas-Moore-Spurr formula (41,42,46) as does the present ASME standard on particulate emissions (8),

$$\Delta h = \frac{(73.5 + 0.222h_s)}{u}(Q_h)^{1/4} \qquad (40)$$

for plume rise in a neutral or unstable atmosphere. In case a stable lapse rate exists, $d\theta/dz > 0$, the plume rise as described by eq. 40 is multiplied by

$$G_N{}^{1/4} = \left[\frac{19.8(d\theta/dz)}{u^2}\right]^{1/4} \qquad (41)$$

with u measured at the top of the stack.

The $\frac{1}{4}$-power exhibited in the Lucas-Moore-Spurr formulation agrees with that of Priestley and Ball (54), cited by Strauss in *Industrial Gas Cleaning* (55).

When calculating plume rise for other than electric-power plant stacks, for example, cooling towers, the momentum term is important and the formula containing the momentum term explicitly is required, that is, eq. 38. When calculating plume rise for short time releases with no momentum in the emission, usually at ground level, the momentum term of eq. 38 can be deleted. Experimental data on the rise of such releases has been obtained by Frizzola, Singer, and Smith (47) and the recommended correlation in that article does not differ greatly from that given by the buoyancy term of eq. 38.

C. Topography and Building Effects, Multiple Stacks

Dispersion of air contaminants is strongly dependent on the local meteorology of the atmosphere into which the pollutants are emitted. The mathematical formulation for the design of pollutant dispersal is associated with open ground terrain free of obstructions. Obstructions of either the natural or man-made type alter the atmospheric circulation and with it the dispersion of pollutants. Particularly well cited are the effects of mountain-valley terrain and of buildings (48,49) Lakes and shorelines also establish local wind patterns that alter pollution dispersal (50).

Hewson in 1945 (48) described the Columbia River Valley circulation in the vicinity of the Trail Smelters to explain the high ground-level SO_2 concentrations. This work was the first to draw attention to the fumigation problem and to the diurnal circulation in a deep valley. Wind tunnel simulation studies, analyzing aerodynamic flows over uneven natural or man-made surfaces, have been used to help design stacks of electric power plants (43,51). Since many of the population centers are located on lake or sea shorelines, the lake–land or sea–land circulation associated with both seasonal and diurnal changes must be considered in evaluating pollution problems inland and offshore. Detail design information has yet to be developed but qualitative guidelines which assist in applying the simple available models have been discussed by Hewson and Olsson (50).

Buildings in the lee of a stack or buildings with stacks mounted on them may have aerodynamic downwash of the stack effluent and its pollutants. Wind-tunnel studies of this by Sherlock and Stalker (49), and others have led to some design considerations. Moses, Strom, and Carson (52) cite that a ratio of emission velocity to wind velocity, V_s/u, greater than 1.5 will minimize downwash. They also cite the well-known "$2\frac{1}{2}$ Times Rule," used by the British Electricity Industry for many years. The height of the stack should be $2\frac{1}{2}$ times the height of the building, but stacks as low as $1\frac{1}{2}$ times have been successfully used; so this must not be considered inviolate. The effects of adjacent buildings and stacks, and of nearby heat sources will disturb the pattern (38). A stack design is usually a compromise of many factors, of which ground-level concentration is one.

Multiple stacks reduce the effluent load on a single stack, and in many cases may also reduce the buoyancy and momentum of the rising plume. Multiple stacks in parallel can be used to improve part load performance of a plant by cutting the number of stacks in operation, so that those stacks operating are at their design point for full load. Plume buoyancy calculations for equally loaded stacks is handled by the ASME procedure (8) by using the heat emission rate Q_h divided by the number of stacks, in eqs. 38 or 40. Splitting the load unequally between stacks of uneven height requires consideration of eqs. 38 or 40 only to form a preliminary estimate.

References

1. W. Jost, *Diffusion in Solids, Liquids, and Gases*, Academic Press, New York, 1952.
2. L. Howarth, *Modern Developments in Fluid Dynamics High Speed Flow*, Vol. 2, Oxford, London, 1953, pp. 847–851.
3. H. S. Carslaw and J. C. Jaeger, *Conduction of Heat in Solids*, Oxford, London, 1959.
4. J. G. Knudsen and D. L. Katz, *Fluid Dynamics and Heat Transfer*, McGraw-Hill, New York, 1958.
5. O. F. T. Roberts, *Proc. Roy. Soc.*, *(London)* **A104**, 640–654 (1923).
6. E. L. Deacon, *Quart. J. Roy. Meteorol. Soc.*, **75**, 89–103, (1949).
7. O. G. Sutton, *Micrometerology*, McGraw-Hill, New York, 1953, pp. 85, 283, 291.
8. *ASME Standard No. APS-1, Recommended Guide for the Control of Dust Emissions— Combustion for Indirect Heat Exchangers*, ASME, New York, (1966).
9. C. H. Bosanquet and J. L. Pearson, *Trans. Faraday Soc.*, **32**, 1249–1264 (1936).
10. F. Pasquill, *Meteorol. Mag.*, **90**, 33–49 (1961).
11. F. Pasquill, *Atmospheric Diffusion*, Van Nostrand, New York, 1962, pp. 179–214.
12. Ref. 7, pp. 287, 288.
13. J. V. Uspensky, *Introduction to Mathematical Probability*, McGraw-Hill, New York, 1937.
14. O. G. Sutton, *Proc. Roy. Soc. (London)*, **A135**, 143–165 (1932).
15. O. G. Sutton, *Quart. J. Roy. Meteorol. Soc.*, **73**, 257–281 (1947).

16. O. G. Sutton, *Quart. J. Roy, Meteorol. Soc.*, **73**, 426–436 (1947).

17. T. S. Walters, *Intern. J. Air Water Pollution*, **6**, 349–352 (1962).

18. Ref. 7, p. 291.

19. P. M. Morse and H. Feshbach, *Methods of Theoretical Physics*, McGraw-Hill, New York, 1963, pp. 812–816, 862.

20. E. W. Bierly and E. W. Hewson, *J. Appl. Meteorol.*, **1**, 383–390 (1962).

21. F. Pooler, "Potential Dispersion of Plumes from Large Power Plants," USPHS Publication #999-AP-16, 1965.

22. M. E. Miller and G. C. Holzworth, *J. Air Pollution Control Assoc.*, **17**, 46–50 (1967).

23. G. H. Strom, "Atmospheric Dispersion of Stack Effluents," in *Air Pollution*, Vol. I, A. C. Stern, Ed., Academic Press, New York, 1962.

24. C. H. Bosanquet, W. F. Carey, and E. M. Halton, *Proc. Inst. Mech. Engrs.*, **162**, 355–368 (1950).

25. *Meteorology and Atomic Energy*, AECU 3066, U.S. Atomic Energy Commission, Washington, D.C., 1955.

26. J. H. Perry, *Chemical Engineers Handbook*, 3rd ed., McGraw-Hill, New York, 1950, p. 1021.

27. I. A. Singer, J. A. Frizzola, and M. E. Smith, *J. Air Pollution Control Assoc.*, **16**, 594–597 (1966).

28. J. R. Buss, *Power*, **112**, 72–73 (Jan, 1968).

29. W. A. Hall, *J. Air Pollution Control Assoc.*, **12**, 379–383 (1962).

30. A. Blum, *Engineer*, **186**, 128–130 (1948).

31. I. A. Singer and M. E. Smith, *Intern. J. Air Water Pollution*, **10**, 125–135 (1966).

32. F. A. Gifford, Jr., *Nucl. Safety*, **2**, 47–51 (1961).

33. D. B. Turner, *J. Appl. Meteorol.*, **3**, 83–91 (1964).

34. F. E. Gartrell, F. W. Thomas, S. B. Carpenter, F. Pooler, D. T. Turner, and J. M. Leavitt, "Full-Scale Study of Dispersion of Stack Gases, A Summary Report," USPHS-TVA Rept., August 1964.

35. D. H. Slade, "Dispersion Estimates from Pollutant Releases of a Few Seconds to Hours in Duration," U.S. Dept. Commerce, ESSA Tech Note 2-ARL-1, 1965.

36. T. T. Frankenberg, *Electrical World*, **167**, (12), 104–108 (March 20, 1967).

37. F. W. Thomas, S. B. Carpenter, and F. E. Gartrell, *J. Air Pollution Control Assoc.*, **13**, 198–204 (1963).

38. R. S. Scorer and C. F. Barrett, *Intern. J. Air Water Pollution*, **6**, 49–63, (1962).

39. J. Z. Holland, "Meteorological Survey of the Oak Ridge Area," USAEC Rept. ORO-99, 1953.

40. "ESSA Forecasting for Pollution Potential," U.S. Dept. Commerce, Robert A. Taft Sanitary Engineering Center, 1966.

41. A. Martin and F. R. Barber, *J. Inst. Fuel*, **39**, 294–307.

42. G. N. Stone and A. J. Clarke, *Am. Power Conf.*, **29**, (1967).

43. M. E. Smith, "Reduction of Ambient Air Concentrations of Pollutants by Dispersion from High Stacks," Nat's Conf. on Air Pollution USPHS, Paper B-8, Panel B on Heat and Power Generation, Washington, D.C., (Dec. 13, 1966).

44. "Symposium on Plume Behavior," *Intern. J. Air Water Pollution*, **10**, 393–409 (1966).

45. S. B. Carpenter, J. A. Frizzola, M. E. Smith, J. M. Leavitt, and F. W. Thomas, "Report on Full-Scale Study of Plume Rise at Large Electric Generating Station," Air Pollution Control Assoc., Paper 67–82, 1967.

46. D. H. Lucas, D. J. Moore, and G. Spurr, *Intern. J. Air Water Pollution*, **7**, 473–500 (1963).

47. J. A. Frizzola, I. A. Singer, and M. E. Smith, *J. Air Pollution Control Assoc.*, **14**, 455–459 (1964).

48. E. W. Hewson, *Quart. J. Roy. Meteorol. Soc.*, **71**, 274–277 (1945).
49. R. H. Sherlock and E. A. Stalker, "A Study of Flow Phenomena in the Wake of Smokestacks," Eng. Res. Bull. No. 29, University of Michigan, Ann Arbor, 1941.
50. E. W. Hewson and L. E. Olson, *J. Air Pollution Control Assoc.*, **17**, 757–760 (1967).
51. P. Sporn and T. T. Frankenberg, "Pioneering Experience with High Stacks on the OVEC and AEP Systems," in *The Tall Stack*, National Coal Policy Conference Inc., Washington, D.C., 1967.
52. H. Moses, G. H. Strom, and J. E. Carson, *Nucl. Safety*, **6**, 1–19, (1964).
53. M. E. Smith et al., *Recommended Guide for the Prediction of the Dispersion of Airborne Effluents*, A.S.M.E., New York, May, 1968.
54. C. H. B. Priestley and F. K. Ball," Continuous Convection from an Isolated Heat Source," in *Quart. J. Roy. Meteorol. Soc.*, **81**, 144–157 (1955).
55. W. Strauss, *Industrial Gas Cleaning*, Pergamon, London, 1966.

Symbols

A A numeric, eq. 38

A_p Cross-sectional area of particle

C Arbitrary constant, eq. 2

C_D Drag coefficient, eqs. 25, dimensionless

C_p Specific heat at constant pressure of gas, Btu/g °F

C_y Constant for spreading coefficient in y-direction, eq. 18

C_z Constant for spreading coefficient in z-direction, eq. 18

d diameter of stack at top, m

e Naperian log base $= 2.72$

g Gravitational constant, m/sec²

G_N Factor described by eq. 41

G Ground settling rate, g/m² sec

$\bar{G}$ Average ground settling rate per unit area, over 2/3's range $=$ g/m² sec

h Height of emission source above ground, m

h_s height of stack, m

H Height of reflecting inversion layer above ground, m

Δh Change in emission height above h_s, eqs. 38 and 40, m

k Diffusion coefficient, m²/sec

k_x, k_y, k_z Anisotropic diffusion coefficients per eq. 9, m²/sec

k_0 Diffusion coefficient at z_0, m²/sec

k_{0y}, k_{0z} Anisotropic diffusion coefficients at z_0, m²/sec

L Ground settling rate per unit downwind distance per eqs. 28 and 29, g/m sec

m Boundary-layer exponent per eq. 13, dimensionless

m_p Particle mass, g

n Spreading coefficient exponent for downwind variation eq. 18, dimensionless

p Pressure, g/m sec²

Q Pollutant release rate: continuous point source, g/sec; instantaneous release, g; continuous line source, g/m sec

Q_h Heat rate, continuous point source, Btu/sec

Q_p Pollutant release rate of particles, g/sec

R Gas constant, m²/sec² °R

Re   Reynolds number, eq. 24, dimensionless

r Radial coordinate, $r = [x^2 + y^2 + z^2]^{1/2}$, m

T Temperature, °R or °F

t Time, sec

u Wind velocity (usually at emission height), m/sec

u_0 Wind velocity at z_0, m/sec

V_s Emission velocity from stack, m/sec

w_p Particle settling velocity, m/sec

x Downwind coordinate, m

y Crosswind coordinate, m

z Vertical coordinate, m

z_0 Fixed vertical coordinate, m

α Diffusion coefficient exponent, eq. 19

β Dimensionless settling velocity, eq. 29

γ Adiabatic process exponent, dimensionless

Γ Negative of dry adiabatic lapse rate, $\sim$1.8°F/100 m

∇ Gradient operator, m

$d\theta/dz$ Potential temperature gradient, eq. 39, °F/m

ν Kinematic viscosity, m²/sec

ξ Dimensionless emission height, eq. 28

ρ Density, g/m³

ρ_p Density of particle, g/m³

σ_y Spreading coefficient in y-direction, m

σ_z Spreading coefficient in z-direction, m

σ_a Spreading coefficient of azimuthal angle, degrees

σ_{a0} Spreading coefficient of azimuthal angle at z_0, degrees

σ_e Spreading coefficient of elevation angle, degrees

σ_{e0} Spreading coefficient of elevation angle at z_0, degrees

χ Pollutant concentration, g/m³

χ_i Pollutant concentration, ith pair of sources, eq. 23, g/m³

$\chi_{\max}$ Maximum ground-level concentration, g/m³

The Formation and Control of Oxides of Nitrogen in Air Pollution

J. Bagg

*Department of Industrial Science, University of
Melbourne, Victoria, Australia*

I. INTRODUCTION

The earliest reported measurements of oxides of nitrogen in the atmosphere appear to be those made by Francis and Parsons in London during 1923 (1). These workers reported average concentrations of 1–2 pphm (as NO_2), whereas a maximum concentration of 15–16 pphm was reached on foggy winter days. A few years later these figures were confirmed by Reynolds (2), who also measured the oxidant concentration (0.8 pphm) at sites in London

(2). Oxidant is defined as the class of components capable of producing iodine when passed through potassium iodide solution. In the atmosphere, oxidant is chiefly ozone. In this early work no attempt was made to correlate the concentrations of oxidant and oxides of nitrogen. Paneth in 1938 (3,4) attempted to find a correlation between these two concentrations measured at different times in London but was not successful. None of these early workers placed any particular stress upon the undesirable effects of oxides of nitrogen in the atmosphere, although Reynolds (2) did suggest that their presence, in conjunction with sulfur dioxide and water vapor, could lead to the formation of sulfuric acid droplets.

At high concentrations (greater than 100 ppm), nitrogen dioxide has been well known to be very toxic and has been the known cause of many deaths during the last 50 years when these concentrations have been reached in enclosed spaces (5). Such high concentrations are usually found in chemical plants, but a potentially dangerous source of oxides of nitrogen was identified by Elliot and Berger (6) in 1942 in a study of the ventilation of coal mines. There diesel engines are commonly used to operate equipment in mines, and their exhaust gases were found to contain concentrations up to 600 ppm of the oxides of nitrogen.

By 1949 serious attention was being given to the oxides of nitrogen as one of the most important contaminants of the atmosphere; in particular, pollution from this contaminant was severe in the Los Angeles area, which has a high concentration of internal combustion engines and other energy-producing equipment. One early symptom was a new type of crop damage to citrus fruits found in the Los Angeles area (7) which was caused by what was later called photochemical smog (8). Apart from crop damage, other characteristics of this new type of smog were eye irritation, objectionable odor, decrease in visibility, and the ability to cause cracking in rubber. This smog differed from the more familiar smog found in London, other major cities, and heavy industrial areas, which was known to be a mixture of smoke and sulfur dioxide. London-type smog generally reaches a peak in the early morning, at temperatures of 30–40°F, with high humidity, and accompanied by radiation or surface inversion; photochemical smog reaches a peak at midday, at temperatures of 75–90°F with low humidity, and accompanied by overhead or subsidence inversions. The immediate effect of London smog on human beings is bronchial irritation, while photochemical smog results in eye irritation.

The severity of smog attacks in the Los Angeles area led to a number of studies concerning the nature and formation of this smog, with the hope that eventually its formation might be prevented. These studies showed unequivocally that the emission of oxides of nitrogen into the atmosphere was one essential precondition for this type of air pollution. A major contributor

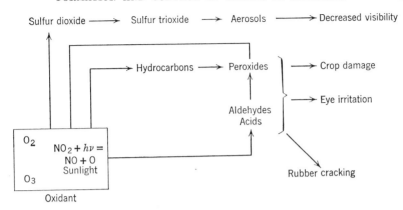

Figure 1.

to these studies was Professor A. J. Haagen-Smit, who, during 1949–1953, clarified the basic mechanism of photochemical air pollution (9–12). The photochemical decomposition of nitrogen dioxide by sunlight initiates a complex series of reactions involving hydrocarbons or other organic material already present in the atmosphere. The end products of these reactions are various and include ozone, formaldehyde, acrolein, and organic peroxides, all of which may produce eye irritation or crop damage. The major features of these reactions and their relation to smog effects are well illustrated in Figure 1, which is due to Haagen-Smit (12).

To show the magnitude of the concentrations of oxides of nitrogen and oxidant (usually considered as ozone) which can be reached in the atmosphere of major urban areas, the values of these for different time-averages, taken in seven cities in the United States, are listed in Table I (13).

TABLE I

Maximum Concentrations of Oxides of Nitrogen and Oxidant in Various Cities (ppm)

Averaging time	Chicago	Los Angeles	New Orleans	New York	Philadelphia	San Francisco	Washington
Oxides of nitrogen							
1 hr	1.06	1.39	0.43	0.65	1.27	0.76	1.41
8 hr	0.62	0.78	0.20	—	0.45	0.52	0.69
1 day	0.50	0.55	0.12	0.22	0.28	0.32	0.38
1 year	0.15	0.12	0.04	—	0.08	0.14	0.07
Oxidant							
1 hr	0.13	0.49	0.17	0.12	0.20	0.17	0.18
8 hr	0.09	0.31	0.11	0.08	0.13	0.10	0.11
1 day	0.07	0.16	0.08	—	0.09	0.08	0.07
1 year	0.02	0.04	0.02	—	0.02	0.02	0.02

Several cities have an hourly average which exceeds 1 ppm; of these cities Los Angeles is exceptional in the very high hourly average of oxidant, 0.49 ppm. Extensive measurements have not been made in other countries but hourly averages of 0.11–0.25 ppm, and long-term averages of 0.02–0.05 ppm have been reported in London (14) and in several large Japanese cities (15,16).

In terms of the maximum concentration attained, the oxides of nitrogen are a major contaminant of the urban atmosphere in the United States. This is demonstrated in Table II (17) which compares the concentrations of the common gaseous pollutants.

In common with other major air pollutants, the oxides of nitrogen arise chiefly from the production of energy in the community, more specifically by the fixation of atmospheric nitrogen in high-temperature regions of furnaces and internal-combustion engines. The emission of oxides of nitrogen and other major pollutants from various sources in the industrial and urban community is shown in Table III (13). In the United States transportation is the major source of both oxides of nitrogen and hydrocarbons which are the two components necessary for photochemical smog.

The approach to the method and level of control of oxides of nitrogen differs from that of other pollutants because of the dual role played by oxides of nitrogen in air pollution. First, as a pollutant in its own right, any reduction in the quantity emitted to the atmosphere will be expected to bring about an improvement in air quality. Second, as the initiator of complex photochemical reactions, changes in the concentration of oxides of nitrogen without concomitant changes in the concentration of hydrocarbons could possibly lead to a deterioration of air quality, despite a reduction in the total quantity of pollutants (18). This second point is still controversial and is discussed later in this review.

TABLE II

The Average Concentration of Various Pollutants in the Atmosphere of Cities in the United States

Pollutant	Range of average conc., ppm	Range of maximum conc., ppm	Number of cities used in average
Oxides of nitrogen	0.02–0.90	0.3–3.5	8
Carbon monoxide	2.10–10.0	3.0–300	8
Sulfur dioxide	0.001–0.7	0.02–3.2	50
Hydrogen sulfide	0.002–0.1	0.005–0.08	4
Ammonia	0.02–0.2	0.05–3.0	8
Hydrocarbons	0.10–2.0	2.0–4.7	(Ref. 59)
Ozone	0.009–0.3	0.03–1.10	8
Hydrofluoric acid	0.001–0.02	0.005–0.08	7

TABLE III

Sources of Pollutants Emitted into the United States Air in 1966

Source	Millions of tons/year					
	Oxides of nitrogen	Carbon monoxide	Sulfur dioxide	Particulates	Hydro-carbons	Miscel-laneous
Transport	3.1	59.6	0.5	1.8	9.7	0.1
Industry	1.6	1.8	8.7	6.0	3.7	1.6
Electricity generation	2.4	0.5	10.2	2.4	0.1	0.1
Space heating	0.8	1.8	3.4	1.2	0.5	0.1
Refuse disposal	0.1	1.3	0.2	0.6	1.0	0.1
Total	8.0	65.0	23.0	12.0	15.0	2.0

The State of California has proposed the following values for nitrogen dioxide and oxidant in its table of ambient air standards (19).

	"Adverse" level	"Serious" level
Nitrogen dioxide	0.25 ppm for 1 hr	3 ppm for 1 hr (Bronchoconstriction in humans)
Oxidant	0.15 ppm 1 hr	Not yet determined

The adverse level for oxides of nitrogen is set by considerations of visibility rather than biological action. Nitrogen dioxide is highly colored and in the atmosphere will reduce the brightness and contrast of distant objects. In addition, it will cause the horizon sky and white objects to appear colored, ranging from pale yellow to reddish brown. Choosing a visibility of 20 miles as a reasonable objective for atmospheric clarity, the presence of 0.25 ppm would not cause unacceptable effects (20). Larsen, in a recent paper, sets a more stringent standard for oxides of nitrogen: 0.1 ppm averaged over 1 hr (13). In order to achieve this level in the atmosphere of U.S. cities by 1985, emissions from stationary sources will have to be less than 0.03 lb NO_2 for a heat release of a million Btu, while the exhaust gases from mobile sources will have to be less than 50 ppm nitrogen dioxide.

It is clear from Table I that the "adverse" level for both oxides of nitrogen and oxidant has been substantially exceeded in many cities of the United States when compared with the State of California standards. It follows that oxides of nitrogen should be controlled even in those cities which currently do not have the problem of photochemical smog, in order to reduce the concentration below the "adverse" level.

Because of the special nature of the pollution caused by the oxides of nitrogen, it will be necessary in this review to briefly discuss the absorption of solar radiation and the nature of the photochemical reactions, in addition to the formation and control of the oxides of nitrogen.

II. THE KINETICS AND EQUILIBRIA OF THE OXIDES OF NITROGEN

There are at least six known stable oxides of nitrogen (and one unstable form, the trioxide) and these are listed here:

Nitrous oxide, N_2O Nitrogen trioxide, NO_3
Nitric oxide, NO   Nitrogen sesquioxide, N_2O_3
Nitrogen dioxide, NO_2 Nitrogen tetroxide, N_2O_4
                                       Nitrogen pentoxide, N_2O_5

Nitrous oxide is a colorless gas which occurs naturally in the lower atmosphere at a concentration of about 0.5 ppm and is said to be formed either by bacterial action or in the upper atmosphere (21). At ambient temperatures it is inert and plays no part in air pollution. *Nitric oxide* is formed almost exclusively during the reaction of oxygen and nitrogen at the high temperatures occurring during combustion of fuels. Nitric oxide, which is colorless, reacts at ambient temperatures with oxygen in the atmosphere to form the reddish-brown nitrogen dioxide. These two oxides are the only ones of importance as air pollutants. The remaining oxides exist in equilibrium with nitric oxide and nitrogen dioxide, but their equilibrium concentrations are extremely small, and can be neglected in air pollution studies (22).

The equilibrium composition of nitrogen–oxygen mixtures calculated from thermodynamic data shows that significant quantities of nitric oxide are not formed except at high temperatures (23). Figure 2 shows the equilibrium concentration of the various species present in air at 1 atm and and temperatures from 1200 to 7000°K (24). In addition to nitric oxide, oxygen atoms and nitrogen dioxide are present in measurable quantities at high temperatures. There is an optimum temperature for the formation of nitric oxide (approximately 3000°) and at this temperature the nitric oxide concentration is 5.2%. At lower pressures, the maximum concentration value is reduced; for example, at 0.01 atm, the maximum concentration is 2.2%.

Equilibrium calculations show that significant amounts of nitric oxide can be formed at combustion temperatures (1300–2500°C); the actual concentrations expected under practical conditions will therefore depend upon the

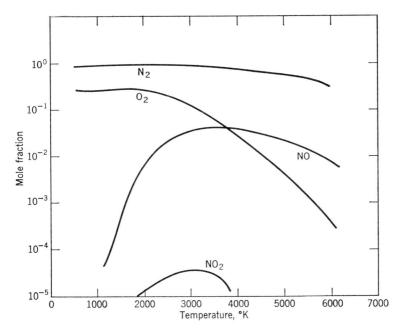

Fig. 2. Composition of air at various temperatures.

rates of formation and decomposition of nitric oxide. Of the many homo-geneous gas-phase reactions that can take place between nitrogen and oxygen, six of the most important are shown in Table IV. The specific rate constants listed here are those recommended by Wray, who has reviewed much of the work in this field (25).

The rate constants of reactions 1–4, 11, and 12 depend to some extent upon the nature of the third body M, and an average value is given in Table IV.

These six equations were used to calculate the final composition of air which had been raised to an elevated temperature, allowed to attain equilib-rium, and then cooled to ambient temperature at various rates (24). If the initial high temperature was 3000°K or lower, then a calculated quench rate of 10^5 °K/sec or greater was necessary for the final composition at ambient temperature to be almost equal to the composition at the temperature of formation. The rate of decomposition of nitric oxide at ambient temperatures is so slow that this final nitric oxide concentration will not change over a measurable period of time. Rapid quenching can occur during combustion in flames, so that the level of nitric oxide in exhaust gases vented to the atmosphere is more likely to be governed by conditions at flame tempera-tures, rather than by lower intermediate cooling or ambient temperatures.

TABLE IV

Reactions and Rate Constants for the Oxygen–Nitrogen System

Reaction	k_f (forward rate)	k_r (reverse rate)
1,2 $O_2 + M = 2O + M$	$1.13 \times 10^{18} T^{-\frac{1}{2}}$ $\times \exp(-118{,}000/RT)$ cc/mole/sec	0.94×10^{15} cc/mole²/sec
3,4 $N_2 + M = 2N + M$	$1.93 \times 10^{17} T^{-\frac{1}{2}}$ $\times \exp(-224{,}900/RT)$ cc/mole/sec	$1.09 \times 10^{17} T^{-\frac{1}{2}}$ cc²/mole²/sec
5,6 $O + N_2 = NO + N$	7×10^{13} $\times \exp(-75{,}500/RT)$ cc/mole/sec	1.55×10^{13} cc/mole/sec
7,8 $N + O_2 = NO + O$	1.33×10^{10} $\times \exp(-7{,}080/RT)$ cc/mole/sec	6.3×10^{14} cc/mole/sec
9,10 $N_2 + O_2 = 2NO$	$9.1 \times 10^{24} T^{-\frac{5}{2}}$ $\times \exp(-128{,}500/RT)$ cc/mole/sec	$4.8 \times 10^{23} T^{-\frac{5}{2}}$ $\times \exp(-85{,}520/RT)$ cc/mole/sec
11,12 $NO + M = N + O + M$	$5.3 \times 10^{15} T^{-\frac{1}{2}}$ $\times \exp(-150{,}000/RT)$ cc/mole/sec	$1.33 \times 10^{16} T^{-\frac{1}{2}}$ cc²/mole²/sec

The details of reactions occurring in flames during the formation of nitric oxide are not clearly understood. In some cases, much higher concentrations of oxides of nitrogen are produced in flames than would be expected from either equilibrium values or nonflame kinetics. In making calculations of nitric oxide formation in flames burning low molecular weight hydrocarbons the most relevant data are those of Zeldovitch (26). Other work has supported his values for the activation energy (27,28) and for kinetic data (29). Singer (30) used the Zeldovitch data to derive the following equation which is in a convenient form to calculate the rate of nitric oxide formation in "lean" (excess air) hydrocarbon–air flames:

$$\frac{dX_t}{dt} = 6.1 \times 10^{14} [N_2][O_2]^{\frac{1}{2}} \frac{1 - X_t^2}{X_e^2} \exp\left(\frac{-128{,}000}{RT}\right) \text{ ppm NO/sec} \quad (1)$$

where t is the residence time (in seconds) at temperature $T°K$; $[N_2]$ and $[O_2]$ are the volume percentages of nitrogen and oxygen and are assumed to be the equilibrium combustion values; and X_t and X_e are the nitric oxide concentrations in ppm at $T°$ at time t and equilibrium, respectively.

For stoichiometric and "rich" flames, the oxygen concentration can be obtained from equilibrium composition tables (31,32) and substitution of this value into the above equation gives (at 1 atm) the following equation:

$$\frac{dX_t}{dt} = 5 \times 10^{16}[N_2] \frac{1 - X_t^2}{X_e^2} \exp\left(\frac{-149,000}{RT}\right) \text{ ppm NO/sec} \qquad (2)$$

If the flow velocity of the burning gases and the temperature profile of the flame are known, then the nitric oxide concentration at various points may be calculated. In Table V, experimental results of several investigators who examined flames burning town or natural gas are compared with calculated values by Singer (30) who used eqs. 1 and 2. The agreement between experimental and calculated values is satisfactory for "lean" and stoichiometric flames, but poor for "rich" flames at sampling distances of 24 in. or less. Singer measured nitric oxide concentrations at different points above a flat propane–air flame and found that the concentration above "lean" and stoichiometric flames was twice as great as the calculated values and above "rich" flames, this increased to six to seven times (30).

The reasons for the poor agreement are not known. Possible reasons are the different fuels (Zeldovitch used a fuel containing about 45% methane), the different flame sizes, burner configurations, and sampling methods. At

TABLE V

Predicted and Experimental Nitric Oxide Concentrations in Flames

Ref.	Fuel/air conc. fraction of stoichiometric	Sampling dist. from flame h in.	Temp. °K Primary flame T_1	Temp. °K Exp. at h	Nitric oxide conc. ppm. Exp.	Nitric oxide conc. ppm. Calc.	Nitric oxide conc. ppm. Equil. at T_1
33[a]	0.9	1	2200	2200	27 ± 3	18	3,870
	1.0	1	2280	2280	27 ± 3	19	2,670
	1.1	1	2230	2230	27 ± 3	14	826
34[b]	0.89	24	2010	1030	33 ± 3	33	2,400
	1.0	24	2120	1030	53 ± 3	53	1,138
	1.1	24	2280	1030	52 ± 5	13	289
35[c]	0.82	100	1940	755	18 ± 1	18	2,660
	1.0	100	2080	755	49 ± 16	49	1,120
	1.1	100	2060	755	16 ± 7	12	148

[a] Town gas–air flame. Combustion products cooled 1 in. from port by water spray.

[b] Natural gas–air flame. Combustor was a steam-cooled copper cylinder 1-in. diameter, 24-in. long. Combustion products rapidly quenched at the exit.

[c] Natural gas–air flame. Combustor 4-in. diameter, 13 ft. long. Combustion products sampled 100 in. from flame holder.

the present time, therefore, the Zeldovitch equations can only be regarded as useful for order-of-magnitude calculations.

Most workers have assumed that nitric oxide is formed initially in the flame, only later oxidizing to nitrogen dioxide when the exhaust gases enter the atmosphere. A recent study of methane–air flames suggests, however, that in some cases this may be an incorrect assumption (36). The burner used in this investigation had a primary feed with 67% of the stoichiometric air and a secondary feed with the remainder. The results are summarized as follows:

1. Nitrogen dioxide was formed in the preheat zone of the flame. This zone extends from the base of the burner to the bottom of the flame.

2. Some of this preformed nitrogen dioxide decomposed to nitric oxide in or just above the flame.

3. Additional nitric oxide was formed in and just above the flame by some mechanism other than (*2*).

4. The total net formation of oxides of nitrogen was complete at heights greater than 1.1 cm above the flame. At an estimated flame temperature of 2300°K, the total concentration of oxides of nitrogen found at heights of 1.1 cm or greater was 12 ppm.

The following set of reactions was proposed as a reasonable scheme to fit the above observations:

$$N_2 + O \rightleftharpoons NO + N \qquad 2O + M \rightleftharpoons O_2 + M$$
$$O_2 + N \rightleftharpoons NO + O \qquad 2N + M \rightleftharpoons N_2 + M$$
$$NO + O \rightarrow NO_2$$
$$2NO_2 \rightarrow 2NO + O_2$$

Using values for rate constants given by Schofield (37), this reaction scheme and a mathematical model of the temperature profile of the flame (38,39) were used to calculate the complete time–temperature–concentration history of the feed mixture. The model also assumed equilibrium between atomic and molecular forms of oxygen and nitrogen. The results of this calculation predicted that only about 1 ppm nitric oxide and almost no nitrogen dioxide should be produced, which is far less than the values actually encountered. The major suspect assumption in this model, which may account for its failure, is the assumption of equilibrium between the atomic and molecular forms of oxygen and nitrogen. While it is known that atomic oxygen and nitrogen are present in greater than equilibrium concentrations in flames (40), the exact concentrations at various points are not known.

The estimated concentrations of oxides of nitrogen in a methane flame were derived by applying the steady-state hypothesis to the following reactions (41):

$$CH_4 + \begin{cases} O \\ H \\ OH \end{cases} \rightarrow CH_3 \begin{cases} OH \\ H_2 \\ H_2O \\ OH \end{cases} \qquad O_2 + H \rightarrow OH + O$$

$$H_2 + \begin{cases} O \\ H \end{cases} \rightarrow H + H_2O \qquad CH_3 + O_2 \rightarrow products$$

The estimated atomic oxygen concentrations calculated were 10–50 times greater than the equilibrium concentrations used in the first set of calculations and at these higher atomic concentrations greater concentrations of oxides of nitrogen should be produced.

From the experiments and calculations described above the conclusion is drawn that a flame does not act solely as a source of heat in the reaction of oxygen and nitrogen to form oxides of nitrogen, but also plays a chemical role in providing reactive free radicals and atomic species. This conclusion is supported by almost all observations made of emissions from furnaces, internal combustion engines, and other combustion equipment. In other words, calculations made using only the concentrations of oxygen and nitrogen in the feed and the temperature of combustion but with no allowance for free radicals and atoms underestimate the actual emission of oxides of nitrogen to the atmosphere.

The structure and chemistry of flames are complex. They vary from one type of fuel to another, from one physical state of the fuel to another, for example, from gaseous to fine droplets, and from one burner design to another. It is not surprising then, that the formation of oxides of nitrogen which will also depend upon these variables has been found to vary widely and has been difficult to predict.

Although nitric oxide may be the major oxide present in the burnt gases immediately after combustion, once these gases are vented into the atmosphere, where excess oxygen is present, oxidation to nitrogen dioxide will take place. The actual mechanism of this oxidation has been controversial. Raschig (42) in 1905, first reported that the reaction was overall third-order; second-order in nitric oxide and first-order in oxygen. The order has been confirmed by subsequent investigations at pressures below 50 mm Hg, but several different mechanisms have been proposed for the reaction (43–46). The reaction has a negative temperature coefficient and the best value for the activation energy appears to be -1.5 kcal/mole (47).

The partial pressures of nitric oxide used in this earlier work were much greater than those found in polluted atmospheres. More recent studies, directed toward air pollution, have been made of the reaction at concentrations as low as a few parts per million (ppm). Glasson and Tuesday, using long-path infrared spectrophotometry, measured the oxidation at concentrations from 2 to 50 ppm and found an initial third-order rate comparable to earlier values (48). A recent careful study of the oxidation made by Morrison, Rinker, and Corcoran seems to be the best source of data for application to air pollution problems (49). Nitric oxide in concentrations of 2–75 ppm reacted with oxygen 0.3–25% (by volume) at 27°C and the products were analyzed by gas chromatography (50). The observed kinetics could be explained by the following reaction scheme:

$$2NO_2 \rightleftharpoons N_2O_4 \tag{1}$$

$$NO + O_2 \rightleftharpoons NO_3 \tag{2}$$

$$NO + NO_2 \rightleftharpoons N_2O_3 \tag{3}$$

$$N_2O_3 + O_2 \rightleftharpoons N_2O_5 \tag{4}$$

$$N_2O_5 \rightarrow NO_2 + NO_3 \tag{5}$$

$$NO + NO_3 \rightarrow 2NO_2 \tag{6}$$

Reactions 1–4 are assumed to be in rapid equilibrium while 5 and 6 are comparatively slow and proceed at measurable rates. The reverse of reaction 5 can be neglected because the rate constant k_5 is approximately two orders of magnitude less than k_6 (51). Reactions 1 and 3 have been extensively studied and are known to be in almost instantaneous equilibrium (52,53). Reaction 2 has been postulated by other investigators (48,54) although no rate constants have been measured. Nitrogen trioxide has been observed spectroscopically (55) and has been found as an intermediate in the decomposition of nitrogen pentoxide (56,57). The total reaction scheme yields the following equation for the overall rate:

$$\frac{d[NO]}{dt} = 2k_6K_2[NO]^2[O_2] + k_5K_3K_4[NO][NO_2][O_2] \tag{A}$$

The kinetic data by Morrison, Rinker, and Corcoran fit this equation very well with the following values for the constants:

$$2k_6K_2 = (1.313 \pm 0.016) \times 10^4 \text{ liter}^2/\text{g mole}^2/\text{sec}$$
$$k_5K_3K_4 = (1.276 \pm 0.028) \times 10^4 \text{ liter}^2/\text{g mole}^2/\text{sec}$$

This equation can be used in a empirical manner to calculate the rate of oxidation of nitric oxide in the atmosphere independently of the correctness or otherwise of the reaction scheme. The initial rates of oxidation in the atmosphere in Table VI are calculated for various initial nitric oxide concentrations, using eq. A.

The maximum concentration of nitric oxide in the atmosphere is approximately 3 ppm and so even at this exceptionally high level only 1%, according

TABLE VI

Rate of Oxidation of Nitric Oxide in the
Atmosphere

Initial nitric oxide conc., ppm	Initial rate of oxidation ppm/min
1	0.0003
10	0.03
100	3.0

to these calculations, would be oxidized within 100 hr. In Los Angeles, however, it has been found that nitric oxide may be converted almost completely to nitrogen dioxide within 3–4 hr, implying that other reactions such as photochemical decomposition are required to account for this rapid oxidation.

The concentration of nitric oxide in the burnt gases of combustion may be as high as 3000 ppm, but because the oxygen concentration is low, oxidation is slight until the gases are vented to the atmosphere. For example, in burnt gases containing 2000 ppm nitric oxide and 0.5–5.0% (by volume) oxygen, the rate of oxidation is only 0.5–5.0 ppm/sec. Normally, exhaust gases are sampled only a fraction of a second after leaving the hot zone and under these conditions almost no nitrogen dioxide would be detected. If, on the other hand, the exhaust gases were stored for appreciable periods and then tested, significant quantities of nitrogen dioxide would be detected. The rate of oxidation of nitric oxide in exhaust gases after they are vented into the atmosphere will depend upon their rate of dilution. If dilution is slow and high local concentration is achieved, oxidation will be rapid; if dilution is rapid, as it might be from the exhaust of a moving car or the high stack of a furnace, then oxidation may be very slow.

During combustion of fossil fuels, carbon monoxide is formed concurrently with oxides of nitrogen; carbon monoxide is the chief air contaminant and so any modification of combustion systems which are proposed to control oxides of nitrogen must be examined carefully for its concomitant effect on carbon monoxide emission. Similar thermodynamic and kinetic considerations to those used in the discussion of oxides of nitrogen formation reveal that the carbon monoxide concentration in the burnt gases is lowest when the air–fuel ratio is high, the flame temperature is high, and the cooling rate of the burnt gases is low (30). These conditions are just those which favor high concentrations of nitric oxide. The concentration of nitric oxide in burnt gases is lowest when air–fuel ratio is low, flame temperature is low, and cooling rate fast, and these conditions are also those which favor high concentrations of carbon monoxide. These predictions only show trends because the inherent chemical action of the flame is not considered, but they do indicate the difficulties that will be associated with methods of control based upon simple modifications of combustion conditions.

It can be concluded from the above that the information necessary to predict the extent of the formation of oxides of nitrogen in flames on a sound theoretical basis is not yet available. More work on the chemistry of flames in relation to this formation is required and until such work is carried out, measurement of the emissions from sources during operation remains the only reliable way of assessing the extent of air pollution by oxides of nitrogen.

III. EMISSION OF OXIDES OF NITROGEN FROM VARIOUS SOURCES IN THE COMMUNITY

The major sources are internal combustion engines, gas turbines, furnaces, and incinerators.

A. Automobiles

High temperatures (at least 1500°F) are reached in the cylinder of an internal combustion engine during the burning of fuel with the consequent formation of oxides of nitrogen. At least 99% of oxides of nitrogen are emitted in the exhaust gases and only 1% or less from the car crankcase (58) so that in all the results to be discussed oxides of nitrogen have been measured only in the exhaust gases.

1. Field Studies

The quantity emitted depends very markedly upon the mode of operating the vehicle, that is, whether accelerating, cruising at steady speed, or decelerating. The emission during different modes, averaged over a number of U.S. automobiles, is given in Table VII (59). The greatest emission occurs during acceleration, due chiefly to the high volume of exhaust gases leaving the engine; next largest are the emissions during steady cruising at high speed.

Because of the variation of emission with operating conditions, a weighting factor for the time spent in different modes during driving over a typical route must be used in order to obtain a meaningful average emission. Two cycles have been used to simulate typical driving conditions, an 8-mode and an 11-mode cycle (58), each with their appropriate weighting factors. Satisfactory agreement has been obtained between averages from city driving in Los Angeles and from the same car run through an 8-mode cycle on a chassis dynamometer (60).

The emissions (8-mode average) from vehicles classed according to weight/ engine displacement ratio (gross weight in lb/engine displacement (in in.³))

TABLE VII

Emission of Oxides of Nitrogen Under Different Operating Conditions

Mode of operation	Air–fuel ratio	Exhaust gas flow, ft³/min	Oxides of nitrogen	
			Conc. ppm, in exhaust	Weight emission, lb/hr
Idle	11.9	6.8	30	0.001
Cruise 30 mph	13.3	24.6	1057	0.19
40	13.6	35.6	1465	0.37
50	13.9	48.5	1450	0.50
Acceleration, wide open throttle				
0–60 mph	12.7	125.0	—	—
0–50	—	105.0	506	0.38
20–45	12.7	90.9	940	0.64
part throttle				
15–30 mph	13.3	42.4	2110	0.64
Deceleration				
Free- at initial 50 mph	11.9	6.8	60	0.003

are shown in Table VIII (61). The emission from vehicles equipped with automatic transmission is independent of weight/displacement ratio and significantly higher than from similar vehicles equipped with manual transmission. The emission from automobiles equipped with manual transmission varied markedly with weight/displacement ratio, and was lowest at the smallest ratios, 8.0–9.9.

The emissions from smaller U.S. and non-American automobiles and light

TABLE VIII

Weighted Average Oxides of Nitrogen Emission for U.S.
Automobiles (ppm NO_2)

Weight/displ. ratio	Automatic transmission			Manual transmission		
	No. of vehicles	Mean	Std dev.	No. of vehicles	Mean	Std. dev.
8.0– 9.9	17	1072	452	3	393	124
10.0–11.9	239	1067	606	29	610	314
12.00–13.9	210	1093	560	97	939	486
14.0–15.9	59	1039	536	71	911	529
16.0–17.9	18	1014	493	15	667	326

trucks measured during two modes of operation, 40 mph cruise and 15–30 mph acceleration, are given in Table IX (62). In this comparison, U.S. compacts equipped with automatic transmission emitted the highest concentrations, close to the average values for U.S. standard automobiles (cf. Table VII). Non-American compact automobiles emitted the lowest concentrations.

Although the concentrations listed in Tables VII–IX show definite trends, further information is necessary before definite conclusions can be drawn about the type of automobile causing the least pollution. The quantity emitted rather than the concentration in the exhaust is the more important factor; for example, a large-capacity engine will produce exhaust gases at a greater rate than a small-capacity engine and therefore, even if the oxides of nitrogen concentrations in the exhaust gases of both engines are equal, the larger engine is a more serious source of pollution. Another feature of the results which makes their interpretation difficult is their large standard deviation. There are large differences between automobiles of the same general type which are presumably due to such factors as different make, different states of maintenance, and different driving patterns. Until these differences between the same type of automobile are clearly understood, only rough comparisons between different types of automobile can be made.

One modification of U.S. automobiles which will become mandatory in the future is a device to reduce carbon monoxide and unburnt hydrocarbon emissions. This device allows the engine to operate with a "lean" air–fuel mixture for more of its operating time and, from the considerations discussed in the previous section, would be expected to increase the oxides of nitrogen emission. This prediction has been verified by a recent study of 300 automobiles from the three major U.S. manufacturers, General Motors, Ford, and Chrysler, all equipped with this type of device (63). Automobiles were tested in five cities, four at low altitude and one at high altitude; all automobiles showed a decrease in hydrocarbons (average decrease -35%) and in carbon monoxide (average decrease -67%) compared to automobiles without control devices. Ford and Chrysler automobiles, in both high- and

TABLE IX

Oxides of Nitrogen Emission (ppm) for Light Trucks and Compact Cars

Trans.	Auto.		U.S. compact cars, manual		Non-U.S. compact cars, manual		U.S. light trucks, manual	
Mode, mph	15–30	40	15–30	40	15–30	40	15–30	40
No. of vehicles	5	5	10	10	50	50	10	10
Mean	2133	1608	1452	1375	1213	753	1456	1019
Std. dev.	1063	708	525	500	581	543	492	460

low-altitude cities, showed increased oxides of nitrogen emissions (average increase $+44\%$), whereas Cheverolet had a similar increase at high altitudes, but a slight decrease (-8%) at low altitudes.

A rough estimate of the total oxides of nitrogen emitted by automobile exhausts into the atmosphere of a city may be made by multiplying the total motor fuel consumption by an average emission factor lb NO_x/gal. This factor has been calculated for the Los Angeles area in the following way: the average air–fuel ratio for a random sample of automobiles was 12.9 (64), and at this particular ratio the average concentration of oxides of nitrogen in the exhaust gases was 975 ppm (65). Taking a density of 6.2 lb/U.S. gal for gasoline, an emission factor of 0.13 lb NO_x/gal was calculated. This figure is probably applicable to other U.S. cities, but on the basis of the results for non-U.S. cars, should be divided by a factor of up to 2, to be applicable to cities outside the United States because of the smaller four-cylinder cars which are widely used in these other cities.

2. Laboratory Studies

The remainder of this section will be taken up with consideration of the specific factors which control oxides of nitrogen emission from the internal combustion engine, rather than the general aspects discussed previously. The major factor governing the production of oxides of nitrogen in internal combustion engines is the air–fuel ratio of the mixture entering the combustion chamber (66–68). Small concentrations are observed in the exhaust gases when the air–fuel ratio is low; as the air–fuel ratio increases so does the concentration, finally reaching a maximum; further increase in air–fuel ratio above this optimum value causes the concentration to fall. This behavior is clearly shown in Figure 3. Other effects illustrated in this figure are the increase in concentration with advance in ignition timing or decrease in manifold vacuum (66). Some additional factors which increase the emission are increase in compression ratio (66), increase in coolant temperature, and buildup of deposit in the combustion chamber (67).

In all the work described so far in this section oxides of nitrogen have merely been measured in the exhaust gases. The theoretical considerations discussed earlier (Sec. II) make it likely that the oxides of nitrogen concentrations at ambient temperature are very close to those at the temperature of formation. Direct measurements have been made in the combustion chamber of a single-cylinder engine which confirm this theoretical conclusion (69). An engine was equipped with sampling ports through which the composition of the burning gases could be determined at different distances from the ignition point and at different times during combustion. Only in the flame front was nitric oxide formed to any extent and to a degree determined by the local adiabatic temperature. Some slight decomposition of

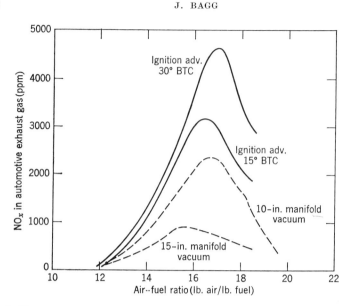

Fig. 3. Emission of NO_x as a function of air–fuel ratio, ignition advance, and manifold vacuum.

nitric oxide may have occurred in the postflame gases but this rapidly stopped as the temperature decreased during the expansion stroke.   Another method of nitric oxide determination was used by Newhall and Starkman (70) to study its formation in the combustion chamber.   The cylinder of an internal combustion engine was fitted with a magnesium oxide window which transmitted infrared radiation emitted by the hot nitric oxide molecules.   This experiment confirmed that nitric oxide was formed in the combustion zone and that little or no decomposition took place outside the combustion zone because of rapid cooling during expansion.   The concentration of nitric oxide was governed chiefly by the peak flame temperature and the available oxygen in the combustion chamber at that temperature.

The observed variation of oxides of nitrogen emission with the air–fuel ratio can now be explained as follows: at low ratios both the available oxygen and flame temperatures are low; as the ratio increases so do the available oxygen and flame temperature; finally at very high ratios the flame temperature may decrease due to dilution of the burning gases by some excess air.

B. Diesel Engines

Diesel engines are used widely in trucks, buses, and in fixed units which operate generators or other machinery   Diesel engines are operated at higher air–fuel ratios than gasoline spark-ignition engines and this difference leads to different concentrations of contaminants in their respective exhaust gases. Table X shows the concentration and quantity of oxides of nitrogen emitted by a typical diesel engine in a vehicle during various modes (59)   The concentrations emitted during acceleration or cruise are $\frac{1}{2}-\frac{1}{5}$ those emitted by a gasoline engine under the same conditions; however, because the exhaust gas flow is much greater from a diesel engine than a gasoline engine, the quantity emitted is greater than from a gasoline engine.   Under typical driving conditions the average weight emission from diesel-powered vehicles will be equal to or slightly greater than gasoline-powered vehicles (71).

Detailed studies of the formation of oxides of nitrogen during combustion in diesel engines have been made by McConnell (72) and Middleditch (72a).

McConnell (72) has made a careful study of three different types of diesel engines: a four-stroke, direct-injection engine, a two-stroke, direct-injection engine, and a four-stroke, indirect-injection engine.   The results and conclusions of this work are described in some detail because of their relevance to the fundamental mechanism of formation not only in diesel engines, but also in gasoline engines.

Figure 4 shows the concentration of oxides of nitrogen in the exhaust of the three types of engine as a function of the position of commencement of combustion and engine speed.   The maximum concentration is observed when combustion starts well before the piston reaches TDC (top dead-center). When fuel starts to burn before the piston reaches TDC, the temperature rise of the charge due to the heat of combustion is augmented by the compression of the rising piston.   The effect of engine speed on emission is different for each engine, but can be explained in the following way: there are two effects that engine speed can have on the temperature attained in the combustion chamber.   First, as the engine speed is increased at fixed injection time, combustion will start later in the cycle when the compression

TABLE X

Exhaust Emissions from a Diesel Engine

Mode of operation	Exhaust gas flow ft³/min	Oxides of nitrogen	
		ppm	lb/hr
Idle	114	59	0.05
Cruise 30 mph	345	237	0.59
Acceleration, part throttle, 15 mph	479	849	2.90
Deceleration constant at 25 mph	302	30	0.06

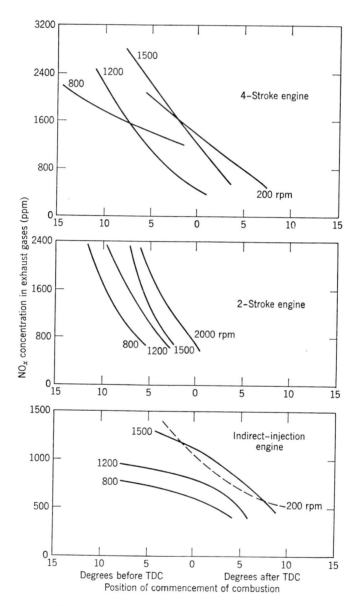

Fig. 4. Emission of NO_x from three types of diesel engine.

effect will not be so marked. The maximum temperature reached would be expected to be reduced by these circumstances. Second, as the engine speed increases, there will be less time to dissipate the heat evolved, and this effect will tend to increase the maximum temperature. From Figure 4, it appears that the first effect is important for the two-stroke and the indirect-injection engine, but that both factors may be important for the four-stroke, direct-injection engine, accounting for the complex variation observed.

The overall emissions from the indirect-injection engine were lower than those from direct-injection engines. This result was further confirmed when the variation of emission with air-fuel ratio was measured and these results are shown in Figure 5. The behavior of the direct-injection engines is very similar to that observed with gasoline engines and by other workers testing diesel engines (73). On the other hand, the emissions from the indirect

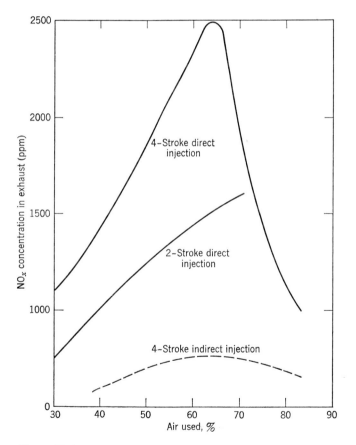

Fig. 5. Emission of NO_x from three types of diesel engine at 1500 rpm.

engine are much lower and insensitive to air–fuel ratio. This independence can be partially attributed to the late injection time; even at low power output, when flow rates of fuel are low, about two-thirds of the fuel injected per cycle is added after TDC, and this proportion will increase at high power output. Under these conditions fuel is being injected at a time when the gas is being increasingly cooled by expansion. The net effect is to make the peak temperature almost independent of the quantity of fuel injected.

The length of the injection period was found to influence the emission; for a given quantity of fuel injected the lowest emission occurred when the fuel was injected as slowly as possible. During a slow injection there should be less oxides of nitrogen formation due to lower peak temperatures and the possibility of dissociation of oxides of nitrogen during prolonged high temperatures in the burning gases.

A tentative theory was proposed for formation in the cylinder of a diesel engine. Combustion starts with several independent flames moving with an orderly "swirl" motion at points distributed throughout the charge being compressed. Oxygen is consumed from the vicinity of the flames. Those parts of the charge remote from the burning fuel droplets retain their original oxygen concentration but are heated by increasing compression and by radiation from the flame. In these remote regions oxides of nitrogen are formed because these regions have relatively high oxygen concentrations in contact with nitrogen at temperatures in excess of $1200°C$. As combustion proceeds those parts of the charge retaining their original oxygen concentration represents a rapidly diminishing fraction of the total charge. Thus, despite rising flame temperatures toward the completion of combustion the rate of oxides of nitrogen formation may be decreasing. It follows, on this hypothesis, that most of the oxides of nitrogen finally emitted in the exhaust gases were formed relatively early during combustion. Another conclusion is that the oxygen concentration calculated from flame equilibria is not the determining value for oxides of nitrogen formation because a significant proportion of nitric oxide is formed just outside the flame in a region of higher oxygen concentration. This hypothesis might also be expected to apply to gasoline engines and does not contradict the explanation put forward earlier for those engines but does add some details of the process.

McConnell (72) points out the difficulties in defining the appropriate temperature for calculations of nitric oxide formation in diesel or gasoline engines. Because of the cyclic nature of operation, the average temperature, determined by the total heat of combustion and the weight of the charge to be heated, is not the correct temperature for this purpose. The average temperature, however, is of some importance because it governs the overall level of the cyclic temperatures. Another temperature that could be considered is the temperature of the flame front.   Large variations in oxides of nitrogen were obtained when the injection time was altered at constant fuel and air delivery.   Under these conditions the flame temperature should be substantially constant.

The last temperature which can be defined easily is that associated with the pressure and volume of the compressed gas.   No oxides of nitrogen were detected when air was merely compressed and expanded in the cylinder in the absence of fuel.   However, when combustion took place to give a calculated peak temperature equal to that due to compression alone, 120 ppm oxides of nitrogen were detected; even with combustion temperatures lower than those due to compression 88 ppm were detected.   This result is a striking illustration of the chemical role of a flame in oxides of nitrogen formation.   It is clear that no single temperature can characterize the formation in engine cylinders but that complete time–temperature–concentration profiles within the combustion chamber are necessary.

Examination of the carbon monoxide concentration in diesel exhaust has shown the same inverse relationship to the nitric oxide concentration as that found in gasoline engines (73a).   When conditions favor carbon monoxide production nitric oxide formation is not favored, and vice versa.

C. Gas Turbines

Gas turbines are used in many commercial aircraft and have been used experimentally in automobile propulsion.   Oxides of nitrogen are emitted in the exhaust gases of turbines; the concentration in the exhaust from an aircraft engine (Pratt and Whitney JT8, rated at 14,000 lb thrust) is shown in Figure 6 as a function of air–fuel ratio (74,75).   The oxides of nitrogen concentration has been adjusted to a stoichiometric ratio (1.0 on the scale used in Fig. 6); for example, an adjusted value shown in Figure 6 as 300 ppm at a fuel–air ratio of 0.2 corresponds to a measured value of 60 ppm in the exhaust.   This adjustment was made to permit direct comparison on a weight basis with conventional gasoline engines which operate at close to a stoichiometric ratio; that is, the adjustment allows comparison at equal exhaust gas flow.   Emissions from aircraft engines are much lower than from automobile engines of the same aggregate capacity.   The difference is even more marked between an experimental automobile (Rover) powered by a gas turbine and a conventional automobile, the emissions being 50 and 1000–2000 ppm, respectively, for these two types.

The factor which may account for the marked difference is the maximum temperatures attained by the burning gases.   These range from 600 to 900°F in the gas turbine, whereas they attain temperatures greater than 1500°F in the conventional engine.

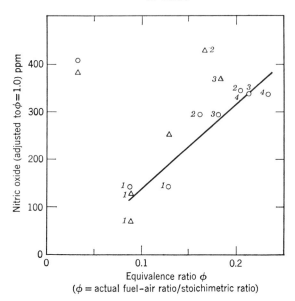

Fig. 6. Nitric oxide in exhaust of aircraft turbine engines. Engine operating conditions: *1*, idle; *2*, cruise; *3*, maximum continuous power; *4*, maximum power; (○) Ref. 74; (△) Ref. 75.

D. Oil-Fired Furnaces

Oil-fired plants are very widely used to generate power and heat. In such equipment fuel oil is either vaporized by heating within the burner unit or mechanically treated to form very small droplets which burn readily. The vaporized or atomized oil is mixed with air and burnt to give a flame whose temperature and structure is dependent upon the type of burner, oil, amount of air, and droplet size.

Oxides of nitrogen concentrations in the waste gases of two designs of furnace have been reported. In the tangential design, the flame and combustion products rotate in a spiral upward and around the walls of the firebox. In the horizontally fired furnace, the flame is at right angles to the walls of the firebox; this type of firing tends to concentrate the hot gases at the center of the firebox.

Sensenbaugh and Jonakin (76,77) have compiled many values of the emission rates from large, tangentially fired units and their results are shown in the form of a histogram in Figure 7. The most frequently reported values were 220–200 ppm in the stack gases, corresponding to an emission of 6–7 lb NO_x/1000 lb oil-fired. A similar survey for horizontally fired units, shown in the same way in Figure 8, gave a most frequently reported value of 460–480 ppm, corresponding to 13–14 lb NO_x/1000 lb oil-fired (78,79). The horizontally fired units, therefore, emit approximately twice as much as the tangentially fired units.

Most furnaces in the United States are horizontally fired and the following discussion refers solely to this design. The variation of emission with firing rate, that is, rate of fuel consumption, has been found to fit the following equation (80):

$$NO_x \text{ emitted lb/hr} = \left[\frac{\text{firing rate (lb oil/hr)}}{248}\right]^{1.18}$$

The oil is assumed to contain 86% carbon. Other things being equal, an increase in firing rate increases the flame temperature and, therefore, the rate of formation. Large boilers usually have to meet a fluctuating power demand and run at approximately 85% load. In keeping with the effect of firing rate an increase of 0.6–1.1% emission per 1% increase in load above 70% and a decrease of 0.6–0.9% per 1% decrease in load below 70% (78,81) has been observed.

In power plants the amount of excess air used in the furnaces may vary from 8 to 30% in a given plant. The excess in many plants is 16–20%, producing 14% carbon dioxide in the stack gases. An approximately linear

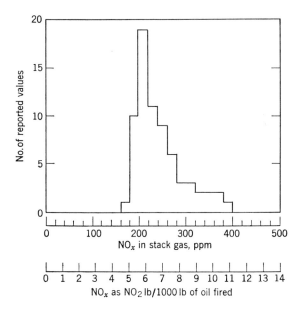

Fig. 7. NO_x emission from large, tangentially fired units.

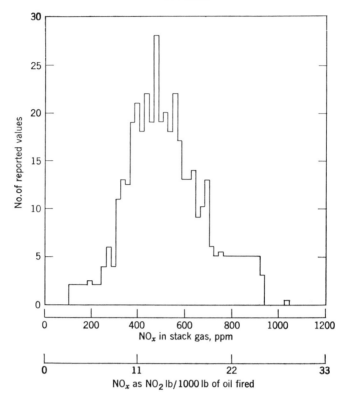

Fig. 8. NO_x emission from large, horizontally fired units.

relationship is found between the carbon dioxide and oxides of nitrogen in the stack gases (76). As the carbon dioxide concentration decreases (excess air increasing, the oxides of nitrogen concentration increases; for example, at 70% load, when the carbon dioxide concentration decreases from 15 to 12% the oxides of nitrogen increased from 500 to 680 ppm.

Oxides of nitrogen formation increase if deposits on boiler tubes are not removed frequently, and a possible explanation is that heat transfer is improved by cleaning, so reducing flame temperature (76,78).

The discussion so far has been restricted to large installations which produce well in excess of 1000 hp, but there are many small units in service for domestic and industrial heating and power generation. In these smaller units the flame temperature is usually lower and so consequently is the oxides of nitrogen emission. A wide range of values up to 18 lb NO_x/1000 lb oil-fired have been reported but the most reliable average seems to be 8–9 lb NO_x/1000 lb oil-fired (82).

E. Coal-Fired Furnaces

Despite the extensive use of oil in large industrial plants, coal still remains the major fuel, particularly for power generation. The reported range of oxides of nitrogen in the stack gases of large coal-fired plants is 100–1460 ppm, and when these concentrations are normalized to an equivalent stack gas containing 12% carbon dioxide (from bituminous coal,) they represent an emission range of 0.17–2.6 lb $NO_x/10^6$ Btu or 1.7–26 lb $NO_x/1000$ lb coal-fired (using a heating value of 10,000 Btu/lb coal) (83–85).

Different designs of furnace emit different quantities of oxides of nitrogen as shown in Table XI (85).

Small domestic stoves used for cooking and heating emit only 0.0014–0.047 lb $NO_x/10^6$ Btu (86). When average values are required for calculating emissions from different units the following values have been recommended (87):

Large units ($> 10^6$ Btu/hr)   0.8 lb $NO_x/10^6$ Btu

Small units ($< 10^6$ Btu/hr)   0.2 lb $NO_x/10^6$ Btu

One detailed study has been made of the formation of oxides of nitrogen in a coal-fired furnace (88). This particular furnace was fired vertically downward into a primary combustion chamber and could burn 1–4 lb/hr of pulverized coal (89). Combustion air was metered and then divided into three streams, the primary transporting the coal, the secondary and tertiary

TABLE XI

Emissions of Oxides of Nitrogen from Furnaces Fired with
Pulverized Coal

Load	Burner configuration or type	Oxides of nitrogen lb/10^6 Btu	
		Before fly-ash collector	After fly-ash collector
Full	Vertical	0.38	0.55
	Corner	0.95	0.71
	Front wall	0.68	0.95
	Spreader stoker	0.65	0.76
	Cyclone	2.5	2.2
	Horizontally opposed	0.65	0.59
Partial	Vertical	0.28	0.31
	Corner	0.73	0.57
	Front wall	0.82	0.74
	Spreader stoker	0.73	0.68
	Cyclone	1.9	1.8
	Horizontally opposed	0.66	0.56

J. BAGG

TABLE XII

Concentration of Oxides of Nitrogen at Different Points in a Coal-Fired Furnace, NO_x in ppm; $T°F$

	Port. 1			Port. 2			Port. 3			Stack		
Excess air	NO_x	$T°$	$O_2\%$	NO_x	$T°$	$O_2\%$	NO_x	$T°$	$O_2\%$	NO_x	$T°$	$O_2\%$
Feed rate, 2 lb/hr, excess air%												
0	630	2480	1.2	320	2190	0.4	255	1850		105	800	0.2
5	600	2375	0.7	440	2275	0.4	300	1900	1.9	210	975	1.1
22	690	2335	3.5	780	2440	2.0	590	1950	4.0	550	975	4.0
Excess air 22% feed rate, lb/hr												
1	660	2280	3.2	415	1860	4.0	355	1580	5.0	320	635	4.2
2	690	2335	3.5	780	2440	2.0	590	1950	4.0	550	975	4.0
4	815	2800	0.5	830	2670	2.5	685	2215	3.8	745	1255	4.3
Feed rate, 2 lb/hr, Excess air%												
0 in primary	610	2510	0.9	405	2270	1.4	22%	air injection		265	870	4.0
5	670	2510	1.8	420	2185	1.4	17%	air injection		210	930	3.9
5	610	2400	1.4	17%	air injection		395	2040	5.3	320	840	3.9

controlling the flame pattern and maintaining ignition stability. Sampling was carried out at various points between the base of the flame and the base of the stack which vented burnt gases into the atmosphere. A high-volatiles coal (heating value 13,670 Btu/lb) was used to avoid clogging of the sample probes by molten ash. The effects of air–fuel ratio, feed rate, and secondary air injection are shown in Table XII. The position of the sample points referred to in this table are as follows: (1) the base of the flame, (2) the tip of the flame, (3) and (4) the intermediate points between the flame and the base of the stack. An increase in the air–coal ratio increases the overall temperature of the flame and the available oxygen concentration and the oxides of nitrogen emission increases correspondingly. An increase in the feed rate increases the flame temperature and also the emission.

Normally, pulverized coal furnaces are operated with air in 20% excess above stoichiometric to ensure complete combustion and because of heat-transfer considerations. The total combustion air (120% stoichiometric) was divided among the three inlets in the following way: 22% primary, 5% secondary, and 73% tertiary. In the first two sets of experiments in Table XII, the excess air was distributed in the same way as the total combustion air, but in the third set of experiments most of the *excess* was in the secondary feed. Under these conditions the stack concentration was only 210 ppm compared to 550 ppm when the excess was distributed uniformly. Two-stage operation had little effect on the combustion efficiency which remained above 99%.

Recirculation of stack gases is often used in boiler operation to maintain thermal efficiency under reduced load. Recirculation of stack gases did not produce any significant reduction in emissions.

The concentrations of oxides of nitrogen produced by coal–air flames were very much higher than in gas–air flames. For example, at a flame temperature of 1700°C the concentration was 91 ppm in the burnt gases from a propane–air flame (30); at a flame temperature of 1300°C the concentration was 780 ppm in the stack gases from a coal–air flame. If the coal–air flame was treated solely as a hot zone the predicted concentration would be less than 1 ppm.

The coal used in these experiments contained 1.3% nitrogen and this nitrogen content may account for the high oxides of nitrogen emission. In other work oxides of nitrogen have been observed in the gases given off during the first stage of heating coal to form coke (90). The amounts evolved depended upon the type of coal, its granular structure, and the temperature rise; emission appeared complete when the temperature reached 400°C. At such a low temperature no measurable quantities of oxides of nitrogen could have been formed by thermal reaction of oxygen and nitrogen. Decomposition of nitrogen-containing compounds in the coal seems to be the only explanation for the results.

F. Gas-Fired Appliances

Gas-fired equipment is used widely for domestic heating and cooking and in many industrial installations. The emission of oxides of nitrogen from gas-fired equipment is shown in Table XIII (91–93). Comparison with

TABLE XIII

Oxides of Nitrogen from Natural Gas-Fired Appliances

Type of equipment	Oxides of nitrogen	
	ppm (as NO_2)	lb $NO_2/10^6$ Btu
Bunsen burner	21	0.07
Range-top burner	22	0.03
Range-oven	15	0.05
Water heater (20 gal)	25	0.05
Water heater (100 gal)	45	0.09
Floor furnace	30	0.07
Forced-air furnace	50	0.09
Steam boiler (10^7 Btu/hr)	90	0.16
Industrial burners	216	—
Boilers and process heaters	—	0.21

previous tables shows that, in general, gas-fired equipment produces smaller quantities of oxides of nitrogen than oil or coal-fired furnaces. In a direct comparison of similar industrial heaters and process equipment oil-firing produced twice the quantity of oxides of nitrogen that gas-firing produced (93,94).

Although the emission from domestic gas appliances is low compared to other sources, these appliances are used in enclosed spaces where the ventilation rates may also be low. Consequently, the local concentrations inside domestic dwellings could conceivably reach undesirably high levels.

G. Incinerators

Incineration is the traditional method for refuse disposal. The waste is converted chiefly into carbon dioxide and water, leaving a small residue of ash.

Incinerators can be divided into two classes, large industrial or municipal incinerators, and small domestic incinerators. Emissions from these two types of incinerator are given in Table XIV (78,95,96). It appears that the emission depends upon the type of material being burnt, and that the introduction of a scrubber reduces the emission. The temperature attained by the particular refuse during combustion and the nitrogen content of the refuse will be major factors in determining the emission.

H. Chemical Plant and Miscellaneous Processes

The manufacture of nitric acid is the major source of oxides of nitrogen emitted to the atmosphere by chemical plant (98). Nitric acid is

TABLE XIV

Oxides of Nitrogen Emission from Various Types of Incinerator

Type	Oxides of nitrogen ppm (avg. during burning)	lb/ton refuse burnt
Municipal	24–58	—
Glendale, Calif., with scrubber	24–58	—
Glendale, Calif., without scrubber	58–92	—
Alhambra, Calif., with spray	64	2.1
Los Angeles	—	2.5
Flue-fed incinerator		
16-story apt. bldg.	—	0.1
Domestic		
Gas-fired AGA (97) mixed rubbish	22–24	2.6–3.8
paper	6–13	1.0–2.2
Backyard 6 ft³ mixed rubbish	1	10.6
paper	1	0.1

manufactured by passing mixtures of ammonia and air over platinum catalysts to form nitric oxide. The nitric oxide is then further oxidized to nitrogen dioxide and absorbed in water to form nitric acid. In a typical plant in the United States, a preheated mixture of 90% air–10% ammonia (by volume) is passed through a catalytic reactor at 1700°F and 110 psig. The resulting mixture of nitric oxide, water vapor, oxygen, and nitrogen is passed through heat exchangers to an absorption tower. The gases finally discharged to the atmosphere are mainly nitrogen with small amounts of oxygen and nitric oxide. These waste or tail-gases contain approximately 0.3% oxides of nitrogen which corresponds to a weight emission of 50 lb NO_x/ton of nitric acid (100%).

Other chemical processes which can produce local high concentrations of oxides of nitrogen are sulfuric acid manufactured by the lead chamber process (99), nitration of organic compounds, and dissolution of metals in nitric acid. These sources, however, are only a very small fraction of the total oxides of nitrogen emitted by all sources and are not difficult to control.

One common combustion process which has not been mentioned in this review is that occurring during cigarette smoking. Medical research workers have examined cigarette and cigar smoke and found oxides of nitrogen in concentrations from 144 to 655 ppm (100,101). On inhalation, 94% of the oxides of nitrogen were absorbed in the lungs; on the basis of 20 cigarettes a day this is equivalent to an absorption of 3.0 mg per day, which is in excess of the recommended limit of 1.2 mg. It seems likely that the oxides of nitrogen originate from the decomposition of nitrogeneous compounds in the tobacco rather than by fixation of atmospheric nitrogen.

IV. BIOLOGICAL EFFECTS OF THE OXIDES OF NITROGEN

Nitrogen dioxide is a phytotoxic substance; that is, it can cause damage to vegetation. From the limited amount of information available, it appears that sensitive plants only show symptoms of acute damage when concentrations exceed 2.0–2.5 ppm (102). Milder effects, such as significant growth suppression, chlorosis, and perhaps premature abscissions of the leaves, have been found at concentrations below 1 ppm (102).

Nitric oxide has not, so far, been regarded as a highly toxic gas; the suggested MAC (maximum allowable concentration) is 25 ppm (5). Nitric oxide can react with hemoglobin in the blood stream to form metheglobin which is not suitable for oxygen transport. Paralysis and convulsions have been reported after exposing animals to the gas, but no cases of poisoning in man have been reported. At the concentrations found in polluted atmospheres, nitric oxide does not seem to be dangerous for moderate time exposures.

Nitrogen dioxide is among the most toxic of the oxides of nitrogen. The effects proved on man and the lower animals are confined almost entirely to the respiratory tract. With increasing dosage the following sequence is observed: odor perception, nasal irritation, discomfort in breathing, acute respiratory distress, pulmonary oedema and, finally, death. Detailed reviews of these effects are found in the following references (59,103,104). For transient exposures to nitrogen dioxide, there is no evidence that any damage occurs at concentrations below 3.5 ppm even with susceptible species and individuals and, accordingly 3 ppm is suggested as the MAC for periods up to 1 hr (105). For long-term exposures without periods of recovery the average concentration should not exceed 0.5–1.0 ppm (105).

Further evidence on the very long-term effects, say more than 20 years, would be desirable, but, using the available evidence, it seems that dangerous levels are not reached in the atmospheres of most cities.

Although the oxides of nitrogen may not reach dangerous levels, there is abundant evidence that, in the Los Angeles area, the products of photochemical reactions initiated by nitrogen dioxide do reach levels which are harmful to human, animal, and plant life. Photochemical pollution damage to crops was first recognized by Middleton in 1944 (7) and he reviewed subsequent work in 1961 (106). Extensive work at the University of California has shown that ozone and peroxyacetyl nitrate, at concentrations similar to their concentrations in smog, can cause damage to vegetation identical with that caused by smog. The chief symptom observed in human beings after exposure to smog is eye irritation. The following irritants have all been detected in smog: ozone, acrolein, formaldehyde, and peroxyacetyl nitrate, but there is still some doubt as to which compounds are actually responsible for the observed irritation. In addition to being an irritant, ozone is a very toxic gas and concentrations greater than 1 ppm are a definite health hazard (107). A general review of the effects of photochemical pollutants on man and animals is given in the following reference (108).

V. OXIDES OF NITROGEN IN THE FORMATION OF PHOTO-CHEMICAL SMOG

A. Smog Formation in Los Angeles

Photochemical air pollution is caused by the atmospheric reaction of organic substances and nitrogen dioxide under the influence of sunlight. On a typical day of pollution in Los Angeles the unreacted pollutants usually reach a maximum by 8.00 A.M. because at this time motor vehicle emission reaches a maximum and also meteorological conditions restrict dilution. As time passes the intensity of sunlight increases and reaction begins. Hydrocarbons are consumed and nitric oxide disappears with the attendant

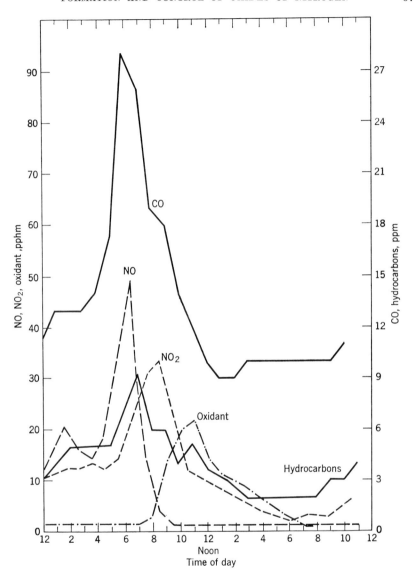

Fig. 9. Hourly average concentration of pollutants in Los Angeles atmosphere (September 3, 1964).

and parallel appearance of nitrogen dioxide. Nitrogen dioxide increases, passes through a maximum concentration, and then decreases with concurrent development of ozone. Nitric oxide and ozone can only exist together at very low concentrations (109) so that the production of ozone testifies to the nearly complete conversion of nitric oxide to nitrogen dioxide. This complex sequence of events is shown in Figure 9 (110).

From this brief description of smog formation, it can be concluded that the intensity of solar radiation, absorption rate of radiation by nitrogen dioxide presence of hydrocarbons in the atmosphere, and the prevailing meteorological conditions are the major factors controlling its formation.

B. Solar Radiation

Solar radiation is treated in detail in a book by Leighton (111) and the presentation in this section is based upon that treatment. The lower atmosphere receives visible and ultraviolet radiation not only directly transmitted from the sun, but also from sunlight which has been scattered from the sky or reflected from the earth's surface. The quantity of radiation received in the lower atmosphere depends upon the following factors: solar spectral irradiance outside the lower atmosphere, solar zenith angle, nature and extent of scattering, diffusion and absorption of radiation by the total atmosphere, and the fraction of light reflected from the surface beneath the region of interest.

The rate of the primary photochemical reaction is controlled by the amount of radiation absorbed by the reacting species and this in turn depends not only upon the total quantity of radiation, but also upon the effective path length, concentration of the species in the polluted atmosphere, and the absorption coefficients at different wavelengths of that species.

The spectrum of the sun in the visible and ultraviolet region is essentially that of a black-body radiator whose temperature is approximately $6000°K$. The solar spectral irradiance has been investigated for many years (112,113) and in terms of energy units, per cent wavelength interval, the maximum of solar intensity is at approximately 4500 Å or, in terms of number of photons, at approximately 6000 Å. The Bouger–Lambert law can be used to calculate the amount of solar radiation at a given wavelength transmitted directly through the atmosphere:

$$_\lambda I_t = {_\lambda I_0} \exp\left(-\sigma_\lambda m\right)$$

where $_\lambda I_t$, $_\lambda I_0$ are the transmitted and incident intensities of solar radiation wavelength λ; σ_λ is the attenuation coefficient at that wavelength; m is the ratio of the length of the path of the direct solar radiation to the vertical path and is a function of the solar zenith angle, and, therefore, of the latitude, time of day, and year.

There are several ways in which the solar radiation passing from the outer limits of the atmosphere to the earth's surface can be attenuated and the total attenuation coefficient, σ_λ, can be expressed as the sum of three individual coefficients: $_m\sigma_\lambda$ is the coefficient of molecular scattering; $_p\sigma_\lambda$ is the coefficient of particulate scattering; and $_0\sigma_\lambda$ is the coefficient of absorption by ozone in the upper atmosphere. A layer of ozone is formed in the upper atmosphere (10–20 miles) by the photochemical dissociation of oxygen (114) and this ozone acts as a filter, strongly modifying the spectral distribution of the radiation received by the lower atmosphere (<1 mile). The most important effect is to reduce to insignificant quantities any radiation of wavelength shorter than 3000 Å (115) and consequently only reactions initiated by radiation of longer wavelength can play any part in photochemical air pollution.

The treatment of molecular scattering is standard and estimates of $_p\sigma_\lambda$ are available so that, in principle, σ_λ can be calculated (116,117).

By using eq. 1 with the appropriate values the rate of absorption of radiation by any given species at a concentration c and in a polluted atmosphere of height h may be calculated. The rate of absorption has the dimensions photons/cm³/sec and is the maximum rate possible for the primary reaction. For a clear sky, i.e., no particulate scattering, the following expression for the rate of absorption is derived (111):

$$R = \sum 2.303\alpha_\lambda c J_\lambda$$

where J_λ is the actinic irradiance, α_λ is absorption coefficient, and c is concentration. The values of J_λ are applicable to any species and provided the concentration c and the absorption coefficient of the species are known, then R may be readily calculated. To illustrate the variation of J_λ with wavelength and zenith angle, Table XV shows J_λ at wavelengths from 2900–8000 Å and zenith angle from 0 to 80° (111). The table shows, first, the sharp cut-off at 2900 Å, and secondly, that J_λ does not decrease markedly until the zenith angle is less than 60°.

The values of J_λ given in Table XV must be modified if they are to be used to calculate absorption when scattering by particles in a polluted atmosphere occurs. During smog attacks in Los Angeles, reductions in radiation intensity from 10 to 80% can take place due to the presence of aerosols in the atmosphere (118). When the effects of particle scattering are calculated it seems that J_λ, even in extreme conditions, may be increased by a maximum of 30% above the value under clear skies and usually the increase will be much less (111).

The total irradiance in the lower atmosphere will be the sum of the radiation incident from the upper atmosphere, plus radiation reflected from the

TABLE XV

Estimated Actinic Irradiance in the Lower Atmosphere
for Clear Sky, $J_\lambda \times 10^{-15}$ photons (cm^{-2} sec^{-1} 100 Å^{-1})

Wavelength in Å	Zenith angle				
	0°	20°	40°	60°	80°
2900	0.0014	0.0009	0.0002	—	—
3500	2.06	1.98	1.72	1.18	0.36
4000	3.10	3.01	2.70	2.00	0.62
4500	5.00	4.88	4.48	3.51	1.19
5000	4.95	4.84	4.51	3.68	1.40
6000	5.32	5.24	4.95	4.21	1.86
7000	5.05	5.00	4.82	4.29	2.38
8000	4.55	4.51	4.37	3.98	2.47

surface below the lower atmosphere. The amount of radiation reflected back
into the atmosphere depends upon the type of terrain, but will probably be
(approximately) 0.1–0.2 of the radiation incident on a city (119).

Clouds cause a very marked reduction in the intensity of solar radiation
entering the lower atmosphere. The intensity is reduced to 0.24 of the
incident intensity by stratus and to 0.85 by cirrus clouds (120). Photochemical smog is usually observed on clear days so that the reduction of
intensity by clouds presumably decreases the rate of photochemical reaction
below some critical level.

C. Primary Photochemical Reaction

Nitric oxide does not absorb radiation in the ultraviolet and visible range
of the solar spectrum and, therefore, does not participate in photochemical
reactions.

Nitrogen dioxide, on the other hand, absorbs radiation throughout the
whole range of the solar spectrum (121,122). The primary reaction is the
photochemical decomposition of nitrogen dioxide into nitric oxide and
oxygen atoms,

$$NO_2 + h\nu = NO + O$$

This reaction has been very carefully studied by a variety of methods and
all the evidence shows that nitrogen dioxide is dissociated by radiation
whose wavelength is less than 4000 Å (111,123,124). This wavelength is in
excellent agreement with the wavelength calculated from the dissociation
energy, 71–72 kcal/mole, of nitrogen dioxide into nitric oxide ($^2\Pi$) and
oxygen atoms (^{3}P) (124).

TABLE XVI

Rate of Decomposition of Nitrogen Dioxide by Solar Radiation (Clear Sky, Quantum Yield Assumed Unity)

Zenith angle	Initial rate (2900–3850 Å) pphm/hr
$0°$	260
$20°$	250
$40°$	220
$60°$	150
$80°$	45

By using the values of the actinic irradiance, J_λ, and the absorption coefficients of nitrogen dioxide in the region below 4000 Å, the rate of decomposition can be calculated. The initial rates of reaction at an initial concentration of 10 pphm are shown in Table XVI (111). These rates of reaction are by far the fastest of all known primary photochemical reactions of any relevance to air pollution (111). The small change in rate when the zenith angle varies from 0 to 40° implies that the rate of reaction is almost constant from say, 8.00 A.M. to 2.00 P.M. in a city on latitude 35°N, for example, Los Angeles. At the summer solstice the rate of reaction would vary by only 5% in a city on latitude 20°N to one on latitude 50°N during the hours from 8.00 A.M. to 4.00 P.M. (111); therefore, if smog depended solely on this rate, its formation should be possible on clear, summer days in many cities in North America and Europe.

D. Secondary Reactions

Although the primary reaction is well understood, the secondary reactions which produce the harmful products in smog are extremely complex and still the subject of extensive investigation. The excellent reviews by Leighton (111) and Altshuller (125) are recommended for the detailed chemistry of these reactions. In the absence of other species the following reactions occur:

$$NO_2 + h\nu = NO + O$$

$$O + O_2 + M = O_3 + M$$

$$O_3 + NO = NO_2 + O_2$$

The presence of organic compounds greatly alters the sequence of reactions, and some of the reactions which have been proposed are given below:

$$O + H_2O = OH + OH$$
$$OH + RH \text{ (hydrocarbon)} = H_2O + R \text{ (free radical)}$$
$$O_3 + RH = R''CHO + R''COO$$
$$R''COO + NO = NO_2 + R''CO$$
$$R''CO + NO_2 + O_2 = R''CO_3NO_2$$
$$RH + O_2 = RHOO$$
$$RHOO + O_2 = RHO + O_3$$

Most of the information about photochemical reactions has been obtained from laboratory experiments in which mixtures of nitric oxide and hydrocarbons or automobile exhaust gases have been irradiated by ultraviolet light. Large chambers, up to 100-ft^3 capacity, have been used to contain the reactants and the effects produced in these chambers were similar to those observed in the atmosphere of Los Angeles. Despite the similarity in effects, there are many difficulties in extrapolating the results of chamber experiments to explain the behavior of the city atmosphere.

The atmosphere surrounding a city is a dynamic system receiving emissions from many sources, irradiated by sunlight of varying intensity, and fluctuating in temperature and humidity. Wind and inversion conditions may change from day to day, even from hour to hour. Chamber experiments have usually been performed under static conditions with predetermined quantities of reactants irradiated at a constant intensity and without any reaction products being removed from the system. Even in dynamic experiments the reactants are transported at a constant rate through the chamber under constant irradiation and these conditions are only a crude approximation of a city atmosphere. No systematic study appears to have been made of the influence of temperature and humidity upon the rates and products of reaction. It is difficult to purify the chamber air to such a level that all the observed effects can be attributed solely to the reactants which have been deliberately added. The concentrations of any added reactants are normally in the ppm or pphm range, and under these conditions the blank correction of the chamber atmosphere is sometimes as great as the effect being measured. The presence in the atmosphere of small particles which can act as nuclei appears to play an important part in the formation of aerosols by a photochemical reaction (126). These aerosols result from reactions involving nitric oxide, sulfur dioxide, and hydrocarbons, and can cause marked reductions in visibility. The nature and number of these particles in the atmosphere and their effect on the reactions not concerned with aerosol production is not known at present, but clearly chamber and city atmosphere might differ considerably with respect to these particles.

The considerations of the paragraph above must be further borne in mind when laboratory data are assessed for relevance to photochemical air pollution. From the large mass of experimental data, the following conclusions have been drawn and are accepted by the majority of workers in the field:

1. Both hydrocarbons and oxides of nitrogen are necessary for the smog reaction. Branched chain saturated hydrocarbons and unsaturated hydrocarbons react more rapidly and produce more aldehydes than do saturated linear hydrocarbons (111).

2. The magnitude of smog effects produced in the laboratory is not directly proportional to the concentrations of either hydrocarbon or oxides of nitrogen alone. For a specific concentration of either reactant there is a limited range over which the other concentration may vary and still produce maximum effect. This behavior is illustrated by Figure 10*a–c*, which show the yields of peroxyacetyl nitrate, formaldehyde, and ozone produced by the irradiation of propylene–nitric oxide mixtures in air (127–129). In each case, a maximum is observed, most marked for the production of ozone and peroxyacetyl nitrate, less for formaldehyde. Figure 10*a* and *c* show that at a constant nitric oxide concentration a reduction in the hydrocarbon concentration will reduce the contaminant concentration. Depending upon which side of the maximum the initial nitric oxide concentration lies, reduction of nitric oxide concentration may produce an increase or decrease in contaminant concentration.

Measurements made in the atmosphere of Los Angeles support the laboratory findings that a critical hydrocarbon/nitric oxide ratio exists at which maximum smog effects are found. This ratio of hydrocarbons (ppm carbon)/oxides of nitrogen ppm varied throughout the year from 6.7 to 25.0 (130), but during periods of most severe eye irritation, averaged over 5 days during 1962, the ratio was almost constant at 11.1 ± 2 (131). On some days when no eye irritation was observed the oxides of nitrogen level was, in fact, higher than on days of severe eye irritation.

In a few cases a maximum has not been observed, but an additive effect of hydrocarbon and nitric oxide, for example in the photochemical reaction of nitric oxide and 3-methyl pentane (132) and in irradiated exhaust gases containing oxides of nitrogen and hydrocarbons (133), has been observed.

In laboratory tests the existence of an optimum ratio for the production of irritants seems to be well established; in both laboratory (131,134) and field tests, the existence of an optimum ratio for maximum eye irritation is also established. The correlation between the irritants produced in laboratory

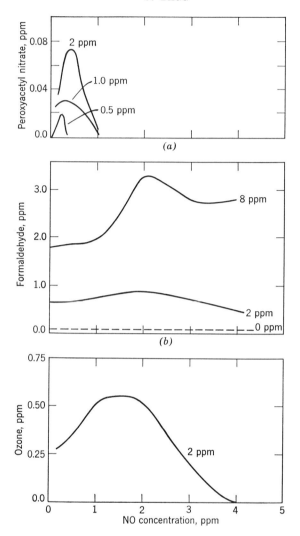

Fig. 10. Products from irradiation of propylene–nitric oxide–air mixtures. Propylene concentration shown on each curve.

tests and the observed eye irritation, however, is poor and still requires more investigation (111).

3. During smog attacks there is generally a reduction in visibility due to the presence of very fine particles or aerosol in the atmosphere. The particulate matter contains appreciable quantities of sulfate (20%) and organic matter (33%) (111). There is strong evidence to suggest that this

aerosol is produced by a reaction between nitric oxide, sulfur dioxide, and hydrocarbons. Schuck, Ford, and Stephens observed that the addition of sulfur dioxide to automobile exhaust–air mixtures caused a marked increase in the rate of aerosol formation during irradiation (135). When hydrocarbon/nitric oxide/air mixtures were irradiated, little or no light scattering took place, but upon the addition of sulfur dioxide to the mixture very strong light scattering due to aerosol formation was observed (136–138).

Careful work by Goetz et al. has gone some way to elucidating the complex mechanism of aerosol formation (126,139). It appears that nucleating particles must be present for appreciable reaction to occur and that reaction continues preferentially on the surfaces of the growing nuclei. The nature of the nuclei is unknown but there are always many small particles in the urban atmosphere. Humidity is a very important factor in aerosol formation; addition of sulfur dioxide at a concentration equal to that of nitric oxide in a system at low humidity increased the aerosol concentration by five times, but at high humidity the increase was only 50%.

The discussion so far has shown the importance of sunlight, hydrocarbons, and nitric oxide in the formation of smog, but has not considered meteorological factors. In other cities the oxides of nitrogen concentration can reach that of Los Angeles, for example, in Chicago, Philadelphia, and Washington. During the summer the solar irradiation is very similar to that in Los Angeles and yet in these other cities little or no photochemical smog effects have been reported. The probable reason is that the Los Angeles area is subject to frequent temperature inversions at times when the wind velocity is low (140). The city is surrounded by elevated ground which further hampers the dispersion of pollutants, allowing high concentrations to be maintained for appreciable periods of time. The other cities mentioned do not have the same meteorological and geographical conditions; for example the average wind velocity in Chicago is 10.1 mph, compared to 6.2 mph in Los Angeles (141), and so concentrations may not reach critically high levels.

Because of the complexity of the subject, predictions of the effects of controlling the emission of oxides of nitrogen and hydrocarbons on the manifestations of smog are difficult. The Department of Health in the State of California came to these conclusions on the basis of the available experimental evidence (142):

1. The control of the emission of hydrocarbons, even without concurrent oxides of nitrogen control, will reduce the effects of smog and the improvement is expected to be proportional to the degree of control achieved.

2. If the hydrocarbons are controlled, then the further improvement by controlling oxides of nitrogen is not expected to be proportional to that control. Little improvement is to be expected from a reduction in the oxides of nitrogen concentration until some critical concentration is reached. Reduction below this critical value will produce marked reduction in smog effects.

3. If a marked reduction in the amounts of hydrocarbons emitted to the atmosphere was achieved, then oxides of nitrogen concentration would have to be reduced to 0.1 ppm or less to obtain further reduction in smog effects. The less effective the control of hydrocarbons, the higher will be this critical concentration for oxides of nitrogen, below which further improvement is achieved.

4. Moderate control of hydrocarbon emissions without control of oxides of nitrogen will lead to an *increased* nitrogen dioxide dosage (concentration × time). The time–concentration behavior of nitrogen dioxide in the atmosphere is controlled by nitric oxide emission, meteorological, and photochemical factors. Changes in the photochemical reactions caused by decreasing the hydrocarbon concentration will alter the balance to increase the nitrogen dioxide dosage for a given emission of nitric oxide.

VI. MEASUREMENT OF OXIDES OF NITROGEN IN THE ATMOSPHERE

Standard colorimetric methods for determining nitrogen dioxide employ the reagents phenol-disulfonic acid (143) or *m*-xylenol (144). These methods are not satisfactory for the very low concentrations found in polluted air. The major disadvantage of the standard reagents is their slow rate of absorption of nitrogen dioxide from the large volumes of air which must be sampled and so a reliable determination takes an inordinately long time. A satisfactory reagent has been developed by Saltzman for nitrogen dioxide and is now widely used in studies of air pollution (145–147). *N*-(1-Naphthyl)-ethylene diamine hydrochloride is the color-forming reagent. If air is bubbled through a solution of the Saltzman reagent for 10 min at a rate of 0.4 liter/min, a sensitivity of a few parts per billion is possible. Nitric oxide does not give a color and for its determination must be first oxidized by passage over some strong oxidizing agent such as chromium trioxide.

Automatic equipment with which a Saltzman determination can be made every 10 min has been developed (148) and is available commercially (149, 150).

At the much higher concentrations found in the exhaust gases of automobiles, 100–2000 ppm, the ultraviolet absorption of nitrogen dioxide may be used for a quantitative measurement and this method is a very suitable method for continuous monitoring (151,152).

Other methods which have been used, and will probably be developed further, are gas chromatography (50,153), long-path infrared spectrophotometry (154), and mass spectrometry (155).

VII. METHODS OF CONTROL

A. General

The development of control methods for the oxides of nitrogen is less advanced than for other major atmospheric contaminants. Most of the oxides of nitrogen emitted to the atmosphere are produced during the combustion of fossil fuels. However, most of the existing abatement technology has been developed to deal with the emission from nitric acid manufacture which is only a small fraction of the total emission.

There are two approaches to control, either to modify the source in such a way that the quantity of oxides of nitrogen emitted to the atmosphere is reduced or, where it is difficult or inconvenient to modify the source, to extract the oxides of nitrogen by some means from the exhaust gases before allowing them to enter the atmosphere. Before describing the various methods that have been suggested or used in detail, we list the methods under two headings. This list is based on one suggested by Henschel (156).

Modification of original source

1. Mobile sources
   a. Changes in engine design, for example, timing or type of fuel injection
   b. Exhaust gas recirculation
   c. Replacement of conventional gasoline or diesel engines by gas turbines
   d. Replacement of internal combustion engines by electric motors
2. Stationary sources
   a. Change in furnace design, for example, tangential firing instead of horizontal
   b. Two-stage combustion
   c. Change in type of fuel, for example, from oil to gas
   d. Inclusion of additives in the fuel which would reduce oxides of nitrogen formation

Removal of oxides of nitrogen from exhaust or waste gases

1. Mobile or stationary sources
   a. Sorption by liquids, for example, ferrous sulfate or calcium hydroxide solutions
   b. Sorption by solids, for example, silica gel or molecular sieves

   c. Catalytic reduction to oxygen and nitrogen with or without the
       addition of a reducing gas such as carbon monoxide
   d. Thermal reduction by burning in a fuel-rich flame
   e. Vapor-phase reaction with other compounds to produce materials
       more readily removed from the gas stream, for example, formation
       of a gummy substance by reaction with hydrocarbons.

B. Modifications of Mobile Sources

The design of internal combustion engines has been developed over many years to the present level of high performance and reliability. The design variables which are known to decrease oxides of nitrogen emission, for example, air–fuel ratio or timing, cannot easily be altered without loss of performance or increase in carbon monoxide emission. With both gasoline and diesel engines, the smallest capacity engine possible consistent with required performance is recommended. In diesel engines indirect injection with the maximum possible injection time of the fuel is one modification which will reduce oxides of nitrogen emission without deleterious effect on the performance.

Recycling a fraction of the exhaust gases has been shown to be an effective method of reducing the oxides of nitrogen emission from both gasoline and diesel engines (72,157–160). The Atlantic-Richfield Co. (161) have developed this method and carried out practical tests on a number of U.S. automobiles. Exhaust gas was taken internally from the heat riser area, passed through a cooling system located within the manifold, and finally returned to the manifold for distribution after passing through a control valve outside the manifold. The temperature of the recycled exhaust gases was above ambient to prevent condensation and improve fuel vaporization. The performance of the engine was acceptable for recycled gas temperatures from 100 to 475°F, but not at lower temperatures. The quantity of gas recycled varied from 15 to 22% of the volume of the air–fuel mixture; at lower percentages, less reduction of the oxides of nitrogen emission took place and at higher percentages the engine performance was poor, with the onset of "roughness," surging, and loss of power.

The valve controlling the admission of exhaust gases was designed to remain closed during idle, deceleration, and full throttle. During these first two modes of operation emission is very low and does not require control. During full throttle acceleration, although emissions were substantial, no possible loss of power, at a time when maximum performance might be needed, could be tolerated.

During steady-state cruising, automobiles with exhaust recirculation emitted at least 80% less oxides of nitrogen than similar automobiles without this modification. The percentage of other components in the exhaust gas was also changed: carbon monoxide decreased by 50% and hydrogen increased from 0.1% to 0.4%. Recycling caused a small decrease in manifold vacuum, small increase in air–fuel ratio, and a reduction of over 200°F in the spark plug temperature, compared to operation without recycling. Fuel consumption at constant speed was almost unchanged, to within 2%, by recycling. No abnormal deposits appeared to be formed in the combustion chamber, on the spark plugs, or around the exhaust valves after driving many thousands of miles. Recycle was, in fact, advantageous because ring wear was reduced by 50–90% on the pistons of engines using recycle (162).

The results of theoretical calculations (160) and of experiments in which combustion cycles were photographed and combustion times estimated (163) suggest that recycling leads to lower combustion temperatures, lower oxygen concentrations, and longer burning times, which allow appreciable decomposition of nitric oxide to take place.

Exhaust recycling appears to be a promising method for the reduction of oxides of nitrogen emission from automobiles because only a slight modification to the engine is required; maintenance of such a system should be minimal and the carbon monoxide emission is also reduced.

The replacement of conventional internal combustion engines in vehicles by gas turbines would lead to marked reductions in oxides of emission and in other major pollutants (164). However, there are technical and economic problems to be overcome before this becomes a practical solution. The use of electric propulsion in vehicles would eliminate oxides of nitrogen pollution from this source. The problems of electric propulsion, such as the range between charging batteries, the power–weight ratio of the vehicle, and the cost of batteries, make this a long-term solution, when better batteries will be available rather than one for the immediate future. The point should be made that, if the electricity to operate these vehicles is generated by combustion of fossil fuels, then oxides of nitrogen will still be produced, although not at street level and will still have to be controlled.

C. Modification of Stationary Sources

Earlier in this review it was shown that tangential firing is the method preferred over horizontal firing of oil-burning furnaces. Tangentially fired furnaces produced, on the average, only half the quantity produced by horizontally fired furnaces (76–79).

Significant reductions in emission can be achieved by changing the method of mixing air and fuel and burning the fuel in two stages. Normally the air (115–130% of the stoichiometric amount required for complete combustion) is added as primary and secondary air with the oil before entering the burner.   In two-stage combustion only 90–95% of the total air is introduced at the burner with the fuel and the remainder enters as tertiary air through auxiliary ports in the walls of the firebox.   In horizontally fired units, two-stage combustion can reduce the oxides of nitrogen emission by 50% compared to single-stage operation (78,165,166), and in tangentially fired units the corresponding reduction was 22% (166).   In an experimental coal-fired furnace a reduction of 50% was achieved by using two-stage combustion (88).   An additional advantage of this system is that the radiation from the partially burnt fuel is increased, increasing the heat transfer.

The processes occurring during two-stage combustion compared to single-stage require further investigation, but one factor leading to reduce emission may be a decrease in the concentration of available oxygen in the hottest part of the flame.

For partial control of emission from furnaces two-stage combustion is an attractive method because it requires small changes in existing equipment and does not require the large capital outlay for catalytic systems or removal and disposal of waste products extracted from burnt gases.

If a choice of fuel is possible, the order of preference for emission with low oxides of nitrogen is gas, oil, and least desirable, coal.   This choice of fuel is not usually possible because economic and design factors govern the type of fuel used.   Air is used exclusively as the oxidant in the combustion mixture in all types of installations, but the oxides of nitrogen could be eliminated by using pure oxygen instead of air.   If there were ever an absolute need to control oxides of nitrogen, then the cost of using pure oxygen would have to be compared to the costs of removing oxides of nitrogen by other methods.

D. Removal of Oxides of Nitrogen from Effluent Gases

Attempts to remove oxides of nitrogen by sorption in liquids have been directed chiefly toward purifying tail-gas from nitric acid plants.   Using water as a sorbent, even with the most efficient scrubbing towers, the effluent gases still contain 0.3% (by volume) oxides of nitrogen.   Ammoniacal solutions of various types have been proposed as sorbents; waste caustic ammonia liquor (167), ammonium bicarbonate solution (168,169), and ammonium sulfite or bisulfite solutions (170).   It is claimed that the oxides of nitrogen concentration can be reduced to 100 ppm in tail-gas by scrubbing with ammonium bisulfite and here ammonium sulfate can be recovered as a useful byproduct (170).   An aqueous suspension of lime has also been suggested as a sorbent and, in this case, calcium nitrate and nitrite can be recovered for fertilizers (171).

In one patent relating to reducing of oxides of nitrogen from the chamber process for the manufacture of sulfuric acid it was claimed that passing the waste gases from the chamber through concentrated sulfuric acid (90%) reduced the oxides of nitrogen concentration to a very low value (172).

The U.S. Bureau of Mines carried out a series of experiments where the exhaust gases of diesel engines were scrubbed with various liquids and found that ferrous sulfate solution was the most effective in reducing the oxides of nitrogen concentration (173). The concentration was reduced by 15–50% after a single-stage scrubbing using bubble-cap towers or fritted bubblers (174). Other workers obtained reductions of 20% by using potassium hydroxide or permanganate solutions (175).

The disadvantages of sorption by liquids as a control method are the low efficiency, at best 50%, the cost of the solutions, and the problems of disposing of the large volumes of waste solutions that would be produced. If a useful product such as a fertilizer could be recovered from the waste solution, then the method might be economically feasible, but in general this method seems unsuitable for the control of oxides of nitrogen emitted by large installations or vehicles.

The use of solid sorbents was first reported in 1925 when the sorption of nitrogen dioxide by silica gel at low temperatures and the subsequent desorption at 100°C was studied (176). Sometime later other workers used a silica gel bed to extract nitric oxide from flue gases and this gas was recovered as nitrogen dioxide. More recently, zeolites have been examined as possible sorbents for oxides of nitrogen (178). A zeolite was packed in a vertical column and tested under the following conditions: Temperature 109°F, space velocity 1250 ft^3/hr bed, inlet concentration of NO_x 1800 ppm, outlet concentration 200 ppm.

Water was removed simultaneously with nitrogen dioxide and under the above conditions the capacity of the column was 2.16 lb NO_x/100 lb zeolite and 2.72 lb water/100 lb zeolite. The oxides of nitrogen could be regenerated by passing hot air or hot air plus steam through the column. In the first instance, the recovered product was a mixture of oxides of nitrogen and nitric acid and in the second, almost entirely nitric acid. In both instances, recovery was 80–90% of the total sorbed.

Active carbon has been used to remove oxides of nitrogen from inert gases. The efficiency of the carbon in removing oxides of nitrogen was initially high but fell after further contact to a steady value of 55% reduction in concentration of oxides of nitrogen, at a temperature of 700°C and space velocity 1800 hr^{-1} (179). Because almost all exhaust and stack gases contain small concentrations of oxygen, active carbon would not be a suitable sorbent in practice.

The disadvantages of solid sorbents for large-scale control are very similar to those of liquid sorbents. It is true that the solid sorbent can be used many times by sorption–desorption cycle but the problems of disposing of the desorbed nitrogen dioxide or nitric acid still remain.

The catalytic decomposition of oxides of nitrogen is a much more attractive method of controlling oxides of nitrogen emissions because the products of the decomposition, nitrogen and oxygen, can be vented directly into the atmosphere, avoiding disposal problems. Green and Hinshelwood (180) found the decomposition of nitric oxide catalyzed by platinum at 1000–1500°C was unimolecular and retarded by the presence of oxygen. Platinum–rhodium alloys were effective catalysts for this decomposition in the same temperature region (181). Base metal oxides will also catalyze the decomposition (182); e.g., copper oxide deposited upon silica gel was found to be a suitable catalyst for the decomposition of oxides of nitrogen at concentrations comparable to those in exhaust gases (183). In a stream of nitrogen a 69% reduction in the initial nitric oxide concentration of 892 ppm was observed by passing over this catalyst at 510°C and space velocity of 1320 hr^{-1}. No marked decrease in activity of this catalyst was detected after 300 hr.

At the present time the only large-scale application of catalytic removal of oxides of nitrogen is to the tail-gas of nitric acid plants. The tail-gas is mixed with a reducing gas such as carbon monoxide or methane before being passed over the catalyst. The addition of the fuel gas allows the reaction to proceed at lower temperatures and with the evolution of recoverable heat, compared to the higher temperatures and input of heat required by the catalyzed reaction in the absence of fuel gas. Typical

TABLE XVII

Operating Conditions for Treating Tail-Gas

Process conditions	To produce heat & reduce to NO	To produce heat & reduce to N_2			To selectively reduce NO & NO_2
Usable fuels	Natural gas, hydrogen, carbon monoxide, coke oven gas, vaporized kerosene				Ammonia
Assumed fuel	Natural gas				Ammonia
O_2 in tail-gas %	3–5	2–3		3–5	2–5
Number of stages	1	1		2	1
			1st stage	2nd stage	
Preferred catalyst	Pd	Pd	Pd	Pd	Pt
Space vel. SCFH/CF	20,000–40,000	60,000	20,000–40,000	60,000	30,000
Min. inlet temp., °C	440–500	460	440–500	460	180
Amt. fuel vol. % of tail-gas stream	0.5–1.4	1.1–1.7	0.8–1.3	0.9–1.4	0.3
Temp. rise °C approx.		130 to remove all O_2 from tail-gas containing 1% O_2.			20–40

reactions in the presence of reducing gases are shown below:

$$2NO + 2CO = N_2 + 2CO_2$$
$$2NO_2 + 4CO = N_2 + 4CO_2$$
$$4NO + CH_4 = 2N_2 + CO_2 + 2H_2O$$
$$2NO_2 + CH_4 = N_2 + CO_2 + 2H_2O$$

Although the idea of catalytic treatment of tail-gas is not new (184), not until 1961 did a study by Andersen et al. show that the method was practical for large-scale removal of oxides of nitrogen (185). The platinum metals were found to be most effective catalysts for the large-scale decomposition. Typical operating conditions suggested by Andersen et al. are summarized in Table XVII. Over 90% of the oxides of nitrogen in the tail-gas are decomposed by this process. The particular process conditions chosen depend upon the usefulness of recovering heat, the amount of capital available for the cost of heat exchangers and related equipment, and the availability and cost of fuels. Depending upon the type and quantity of fuel added, treatment can be one of the following:

1. Decolorization of tail-gas; the rate of reaction of oxides of nitrogen is such that nitrogen dioxide is first reduced to colorless nitric oxide. The next stage of reduction to nitrogen does not take place to any appreciable extent until all the oxygen present in the tail-gas has reacted with the fuel gas. Therefore, sufficient fuel gas must be added to react with the oxygen as well as the nitrogen dioxide if complete reduction is required. In some cases, formation of colorless nitric oxide and dispersion from a tall stack without forming an objectionable brown plume is regarded as satisfactory. To achieve this end, the amount of fuel added is less than the stoichiometric sum of the nitrogen dioxide and oxygen. Less heat is evolved than in complete reduction, which can be an advantage where heat disposal is a problem.

2. Complete reduction of oxides of nitrogen; fuel equal to or greater than the stoichiometric sum of oxygen plus nitrogen dioxide is added. The heat evolved is usually recovered and made to do useful work operating compressors and generators. This reduction is limited to an upper temperature of 850°C; if reduction has to be carried out in the presence of high oxygen concentrations, the reduction must be performed in two stages to prevent the temperature rise exceeding the stated limit.

3. Selective removal of oxides of nitrogen; ammonia will react selectively with oxides of nitrogen but not with oxygen:

$$8NH_3 + 6NO_2 = 7N_2 + 12H_2O$$
$$4NH_3 + 6NO = 5N_2 + 6H_2O$$

The advantage of this method is that the amount of heat evolved is much less than with, say, carbon monoxide and so is useful in situations which do not justify the installation of heat recovery equipment. The purified tail-gas vented to the atmosphere after this selective reaction contains 80 ppm oxides of nitrogen and 10 ppm ammonia.

For a given fuel, there is a minimum ignition temperature required to initiate the reaction. Once reaction has started the heat of reaction will maintain the temperature. Ignition temperature is lowest for hydrogen and carbon monoxide, 150°C, higher for kerosene, 300°C, and highest for methane, 400°C.

Supported palladium is the most suitable catalyst for use with all common fuel gases except ammonia (186). Originally the support for the catalyst was ceramic pellets or spheres on which the catalyst was deposited. Recently a new type of support has been developed which is much superior to these other supports (186). The new support is the unitary honeycomb ceramic cartridge which has a ceramic structure identical in geometry to the honeycomb occurring in nature. The honeycomb is made of alumina or mullite with an individual cell dimension of $\frac{1}{8}-\frac{3}{4}$ in. Comparison is made in Table XVIII (186) of the honeycomb and older particulate supports. The major advantages of the honeycomb compared to the particulate catalysts are easier packing, higher effective surface area, and lower pressure drop across the catalyst bed, allowing higher space velocities. Palladium may be deposited

TABLE XVIII

Comparison of Some Properties of Honeycomb and Particulate Supports

	Honeycomb		Particulate	
	7 cpi[a]	$4\frac{1}{2}$ cpi	$\frac{1}{8}$ in.	$\frac{1}{4}$ in.
Surface area, m²/liter				
Superficial	1.98	1.17	1.11	0.48
Catalytic	4×10^5	4×10^5	1.28×10^5	3.44×10^5
Voids in packed bed, %	65–70	82	38.5	41.5
Pressure drop/ft depth at linear vel. 20 ft/sec. Pressure 100 psi; temp. 500°C	0.6 psi	0.3 psi	16.5 psi	10.1 psi
Ratio of vessel dia. for pressure drop 1 psi at space vel. 100,000	1.25	1.0	4.0	3.25
Commercially usable space vel. SCFH/CF	90,000–140,000	90,000–140,000	20,000–60,000	20,000–60,000
Bed orientation limits	None: can be mounted in any orientation		In general, catalyst must be contained in a vertical column	

[a] cpi, channels per linear inch.

efficiently on the inner walls of the honeycomb to produce a highly effective catalyst for the control of oxides of nitrogen. This type of catalyst has been developed by the Du Pont and Engelhard companies (186,187) to the point where by 1966 it was in use on at least 18 plants in the United States manu facturing a total of 1,650,000 tons of nitric acid (186).

The cost of platinum metal catalysts makes an uneconomical solution to the removal of oxides of nitrogen from automobile exhausts. In this instance, recoverable heat cannot be used to offset the initial capital cost. Automobile exhaust gases contain approximately 4–6% carbon monoxide, 1% oxygen, and 0.1% oxides of nitrogen so that the catalytic reduction should take place readily without additional reducing gases. Taylor searched for other catalysts and found that copper-containing catalysts appeared to be the most promising (188). Copper–chromite catalysts were tested by passing mixtures of carbon monoxide/nitric oxide in nitrogen over catalysts at various temperatures and analyzing the products (189). At a space velocity of 10,000 hr^{-1} and catalyst inlet temperature of 200°C, 91% removal was achieved and at 1500 hr^{-1} and 107°C, 92% removal was achieved. The following scheme was proposed to account for the experimental results:

$$CO + \text{oxidized catalyst} = CO_2 + \text{reduced catalyst}$$
$$NO + \text{reduced catalyst} = \tfrac{1}{2}N_2 + \text{oxidized catalyst}$$

The catalyst should attain a steady-state oxidation level which varies with the composition of the gases passing over the surface. When the ratio of the partial pressure of carbon monoxide/nitric oxide was changed from 10 to 0.25, no significant change in the reduction rate of nitric oxide was observed which could be attributed to changes in oxidation state of the catalyst. Changes in the oxidation level of the catalyst should not, therefore, be a limitation to the practical use of this catalyst (190). The oxidation–reduction cycle did cause mechanical attrition of the unsupported catalyst but better mechanical properties could be obtained by supporting the catalyst on alumina.

The kinetics of the reaction catalyzed by a barium-promoted copper–chromite catalyst were analyzed by Ayen and Yu-Sim Ng who concluded that the rate-determining step was reaction between adsorbed carbon monoxide and nitric oxide molecules on neighboring sites (190). At temperatures below 160°C appreciable quantities of nitrous oxide were formed (192).

The oxygen concentration in the exhaust gases is important in controlling the extent of reduction of nitric oxide. Copper–chromite is also an excellent catalyst for the oxidation of carbon monoxide, and unless the carbon monoxide present exceeds the stoichiometric equivalent of the oxygen, little or no reduction of oxides of nitrogen will occur. The oxygen concentration in exhaust gas is approximately 0.5%, so that at least 1.5% carbon monoxide must be present to remove oxides of nitrogen. Untreated exhaust gases contain 4–6% carbon monoxide and even if the carbon monoxide was reduced to the level recommended by the State of California, 1.5%, there should still be enough present to reduce oxides of nitrogen.

When water vapor was present in the carrier gas, ammonia was formed as well as nitrogen by catalytic reduction over copper–chromite (191). More information about the extent of this reaction is desirable because mobile exhaust does contain water vapor and the production of significant quantities of ammonia is not acceptable.

Some tests have been carried out using copper-containing catalysts under conditions approaching automobile operation. Exhaust gases from a Ford V-8 engine mounted on a fixed bed, using standard leaded gasoline as fuel, were passed through a reactor containing the catalyst under test and then analyzed (191). The most effective catalyst was a mixture of copper oxide, cobalt oxide, and aluminum oxide compacted into $\frac{1}{8}$-in. pellets and preheated at 500°C in carbon monoxide for 8 hr. The concentration of oxides of nitrogen was reduced by 90% using this catalyst throughout a test lasting 350 hr, the equivalent of 12,000 miles driven at 35 mph. At the end of this long test, the attrition of the catalyst was only mild. A copper-chromite catalyst was as effective as the first catalyst in a test lasting 270 hr but showed more attrition at the end of this period. In both catalyst tests, space velocities of 10,000 hr^{-1} and temperatures of 480–510°C were employed.

Design calculations based upon the copper–cobalt oxide catalyst showed that a reactor containing 0.1 ft^3 of catalyst (12 lb) placed 3-in. from the manifold would reduce the nitric oxide concentration by 86%, without thermal overload. The heat from the exhaust gases would initiate the reaction. The location of the reactor may be important because if catalytic reactors for carbon monoxide require additional air to be mixed with the exhaust gases before entering the reactor, and if this additional air were added before the exhaust gases had passed through the nitric oxide reactor, the reduction of nitric oxide would be prevented. A suggested location for the carbon monoxide reactor is close to the muffler, allowing easy placement of an air inlet between the two reactors.

Although the tail-gases from nitric acid manufacture and automobile exhaust do not contain significant quantities of sulfur compounds, the waste gases from oil and coal-fired furnaces which burn fuels containing sulfur do contain appreciable quantities of sulfur dioxide as well as the oxides of nitrogen. Some preliminary work has been carried out on catalysts for the simultaneous reduction of nitric oxide and sulfur dioxide by carbon monoxide (193). Reductions of 90% in the concentrations of both nitric oxide and sulfur dioxide were achieved by passing these gases, mixed with carbon monoxide, over copper supported by alumina at a temperature of

1000°F.   There are three separate equilibria that must be considered in choosing the correct composition of gases to be treated:

$$2CO + SO_2 \rightleftharpoons 2CO_2 + S$$

$$CO + S \rightleftharpoons COS$$

$$CO + NO \rightleftharpoons CO_2 + \tfrac{1}{2}N_2$$

The production of carbonyl sulfide is undesirable and should be minimized by controlling the ratio carbon monoxide/sulfur dioxide.   The optimum concentration of carbon monoxide for minimum carbonyl sulfide is stoichiometric for sulfur dioxide plus nitric oxide.   To achieve this condition a furnace should be operated close to the stoichiometric air–fuel ratio or even slightly fuel "rich."

The Air Correction Division (Universal Oil Products Co.), has developed a "rich" firing, direct-flame burner to reduce the essentially pure oxides of nitrogen emitted during the nitration of organic compounds or acid etching of metals (194).   The oxides of nitrogen are taken to the burner and mixed with natural gas and some supplementary combustion air before entering the flame.   During combustion approximately 95% of the oxides of nitrogen are reduced to nitrogen.   If the final oxides of nitrogen concentration are to be reduced below 100 ppm, then the gases leaving the flame must be further treated by passing over a platinum catalyst before entering the atmosphere.

A 10–20% decrease in oxides of nitrogen concentration was observed when waste gases from combustion were passed through an electrostatic purifier (195).   The addition of unsaturated hydrocarbons greatly increased the effectiveness of this removal, for example, in the presence of cyclopentadiene the extent of removal was four times greater than in the absence of this hydrocarbon.   Presumably the nitric oxide forms some compound with the hydrocarbon which is deposited on the walls and so removed from the gas stream.

E. Summary

The choice of the method or methods of control will depend upon several factors which include the degree of control required, the cost, and the particular source of emission.

The degree of control required depends upon the concentration of oxides of nitrogen in the atmosphere which is considered tolerable for long-term exposures.   The effects of oxides of nitrogen pollution in the atmosphere are well defined in the Los Angeles area, where the problem of photochemical pollution has been under investigation for a number of years.   Despite this intensive investigation the degree of control of oxides of nitrogen necessary to eliminate photochemical smog is still a matter of some controversy. The difficulty in making this decision arises from the complexity introduced by the reactions of hydrocarbons in the atmosphere. At the present time, in fact, the most effective method of controlling photochemical air pollution seems to be to control primarily the emission of hydrocarbons into the atmosphere rather than oxides of nitrogen.

Even if hydrocarbon emission is controlled and the major effects of photochemical air pollution reduced, there still remains the problem of the oxides of nitrogen which will still be present in the atmosphere of Los Angeles. This problem also exists in those other cities which do not have a photochemical smog problem but do have substantial oxides of nitrogen concentrations in their atmospheres. Before an acceptable level of oxides of nitrogen concentration can be decided upon, information is required about the effects on human, animal, and plant life of exposure to oxides of nitrogen at low concentrations over long periods, say 10 years; the effects on building, construction materials, and fabrics; and possible synergistic effects with other atmospheric pollutants.

In the absence of this desirable information, the level suggested by Larsen (13) of 0.1 ppm (averaged over 1 hr) seems to be a good choice for the oxides of nitrogen concentration. If this level is to be achieved, then in many cities the total emission of oxides of nitrogen will have to be reduced by over 90%.

If the sources of oxides of nitrogen emission are compared with the sources of two other major pollutants, carbon monoxide and sulfur dioxide, a marked difference in distribution emerges. In the United States, over 90% of carbon monoxide is emitted from mobile sources (automobiles) and over 95% of sulfur dioxide from stationary sources. For each of these pollutants control of one type of source will therefore essentially eliminate their presence in the atmosphere. On the other hand, the oxides of nitrogen emission are divided into 40% from mobile and 60% from stationary sources so that control methods must be devised for both types of source to solve this pollution problem.

If the goal of over 90% reduction is to be achieved, then at the present state of knowledge catalytic reduction of oxides of nitrogen seems to be the most effective method, both for standard and mobile sources. Because the control methods are still in development (with the exception of the treatment of the tail-gas from nitric acid plants) it is not yet possible to make satisfactory estimates of cost of control.

References

I. A. G. Francis and A. T. Parsons, *Analyst*, **50**, 262 (1925).
2. W. C. Reynolds, *J. Soc. Chem. Ind.*, **49**, 168 (1930).
3. F. A. Paneth, *Nature*, **142**, 112 (1938).

4. J. L. Edgar and F. A. Paneth, *J. Chem. Soc.*, **1941**, 519.

5. H. B. Elkins, *The Chemistry of Industrial Toxicology*, Wiley, New York, 1958.

6. M. A. Elliot and L. B. Berger, *Ind. Eng. Chem.*, **34**, 1065 (1942).

7. J. T. Middleton, J. B. Kendrick, and H. W. Schwalm, *Plant Disease Rept.*, **34**, 245 (1950).

8. L. H. Rogers, *J. Chem. Ed.*, **35**, 310 (1958).

9. A. J. Haagen-Smit, "The Analysis of Air Contaminants," in *Report to the Los Angeles Country Air Pollution Control District*, 1949.

10. A. J. Haagen-Smit, *Eng. Sci. (Calif. Inst. Tech.)*, **14**, 1 (1950).

11. A. J. Haagen-Smit, E. F. Darley, M. Zaitlin, M. Hull, and W. Noble, *Plant Physiol.*, **27**, 18 (1952).

12. A. J. Haagen-Smit, *Ind. Eng. Chem.*, **44**, 1342 (1952).

13. R. I. Larsen, *J. Air Pollution Control Assoc.*, **17**, 823 (1967).

14. C. F. Barrett and L. E. Reed, *Intern. J. Air Water Pollution*, **9**, 357 (1965).

15. S. Hayushi, *Kūki Seijo (Tokyo)*, **4**, 8 (1964).

16. O. Tadu, *J. Sci. Labour (Tokyo)*, **42**, 263 (1966).

17. A. C. Stern, *Air Pollution*, Vol. I, Academic Press, New York, 1962.

18. Department of Public Health, Bureau of Air Sanitation, State of California, *The Oxides of Nitrogen in Air Pollution*, 1966, p. 80.

19. Department of Public Health, Bureau of Air Sanitation, State of California, *California Standards for Ambient Air Quality*, March, 1967.

20. See ref. 18, p. 62.

21. A. P. Altshuller, *J. Air Pollution Control Assoc.*, **6**, 97 (1956).

22. See ref. 18, p. 2.

23. *JANAF Thermochemical Data*, Dow Chemical Co., Midland, Mich., 1963.

24. P. R. Ammann and R. S. Timmins, *A.I.Ch. E. (Am. Inst. Chem. Engrs.) J.*, **12**, 956 (1966).

25. K. L. Wray, *Res. Repr. 104*, Avco Everett Research Lab., Mass., 1961.

26. Y. B. Zeldovitch, *Acta Phys. (U.S.S.R.)*, **21**, 577 (1946).

27. P. V. Marrone, *Anti-Missile Res. Advisory Council Proc.*, **4**, 353 (1961).

28. H. S. Glick, J. J. Klein, and W. Squire, *J. Chem. Phys.*, **27**, 850 (1957).

29. C. P. Fenimore and G. W. Jones, *J. Phys. Chem.*, **61**, 654 (1957).

30. J. M. Singer, E. B. Cook, E. M. Harris, V. R. Rowe, and J. Grumer, Bureau of Mines Rept. 6958, 1967.

31. G. I. Koslov, *Seventh Symp. (Intern.) on Combustion*, Butterworth Scientific Publications, London, 1959, p. 152.

32. R. W. Smith Jr. and E. B. Cook, Bureau of Mines Rept. 6672, 1965.

33. P. A. Cubbage and K. C. Ling, *Composition of Combustion Products*, The Gas Council (England), R.C., GC 100, 1963.

34. R. C. Seagrave, H. H. Reamer, and B. H. Sage, *Combust. Flame*, **9**, 7 (1965).

35. G. N. Richter, H. H. Reamer, and B. H. Sage, *J. Chem. Eng. Data*, **8**, 215 (1963).

36. R. B. Rosenberg and D. H. Larson, American Gas Assoc. Inc., Basic Research Symp., Institute of Gas Technology, Chicago, 1967.

37. K. Schofield, *Plant. Space Soc.*, **15**, 643 (1967).

38. S. A. Weil, *Inst. Gas Technol. Res. Bull.*, **35** (1964).

39. S. A. Weil, W. R. Staats, and R. B. Rosenberg, A. C. S. Division of Fuel Chemistry, Preprints No. 10, 84 (1966).

40. A. G. Gaydon and H. G. Wolfhard, *Flames*, Chapman and Hall Ltd., 1960, p. 111.

41. R. B. Rosenberg, S. A. Weil, and D. H. Larson, private communication, 1968.

42. F. Raschig, *Z. Angew. Chem.*, **18**, 1281 (1905).

43. G. Lunge and E. Berle, *Z. Angew. Chem.*, **19**, 861 (1906).

44. M. Bodenstein, *Helv. Chim. Acta*, **18**, 743 (1935).

45. E. M. Stoddard, *J. Chem. Soc.*, **1939**, 5.

46. F. B. Brown and R. H. Crist, *J. Chem. Phys.*, **9**, 804 (1941).

47. A. F. Trotman-Dickenson, *Gas Kinetics*, Butterworth Scientific Publications, London, 1955, p. 264.

48. W. A. Glasson and C. S. Tuesday, *J. Am. Chem. Soc.*, **85**, 2901 (1963).

49. M. E. Morrison, R. G. Rinker, and W. H. Corcoran, *Ind. Eng. Chem.*, *Fundamentals*, **5**, 175 (1966).

50. M. E. Morrison, R. G. Rinker, and W. H. Corcoran, *Anal. Chem.*, **36**, 2256 (1964).

51. I. C. Hitsatsune, G. Crawford, Jr., and R. A. Ogg, Jr., *J. Am. Chem. Soc.*, **79**, 4648 (1957).

52. P. Gray and A. D. Yoffe, *Chem. Rev.*, **55**, 1069 (1955).

53. F. H. Verhoek and F. Daniels, *J. Am. Chem. Soc.*, **53**, 1250 (1931).

54. J. C. Treacy and F. Daniels, *J. Am. Chem. Soc.*, **77**, 2033 (1955).

55. E. J. Jones and O. R.-Wulf, *J. Chem. Phys.*, **5**, 873 (1937).

56. R. L. Mills and H. S. Johnston, *J. Am. Chem. Soc.*, **73**, 938 (1951).

57. G. Schott and N. Davidson, *J. Am. Chem. Soc.*, **80**, 1841 (1958).

58. See ref. 18, p. 7.

59. U.S. Dept. of Health, Education and Welfare, Report of the Surgeon General to the U.S. Congress, *Motor Vehicles, Air Pollution and Health*, 1962.

60. C. S. Tuesday, Rept. to the Steering Committee, Los Angeles Test Station Project, 1963.

61. C. E. Zimmer, Rept. to the Steering Committee, Los Angeles Test Station Project, 1963.

62. See ref. 18, p. 13.

63. W. F. McMichael, R. E. Kruse, and D. M. Hill, *J. Air Pollution Control Assoc.*, **18**, 246 (1968).

64. See ref. 18, p. 15.

65. A. H. Rose, R. Smith, W. F. McMichael, and R. E. Kruse, *J. Air Pollution Control Assoc.*, **15**, 362 (1965).

66. L. F. Gilbert, D. A. Hirschler, and R. C. Getoor, paper presented at 132nd Meeting, American Chemical Society, New York, 1957.

67. G. J. Nebel and M. W. Jackson, paper presented at 132nd Meeting, American Chemical Society, New York, 1957.

68. T. A. Huls and H. S. Nickol, S.A.E., Publ. 670482 (1967).

69. M. Alperstein and R. L. Bradow, S.A.E., Publ. 660781 (1966).

70. H. K. Newhall and E. S. Starkman, S.A.E., Publ. 670122 (1967).

71. J. R. Kinosian, J. A. Maga, and J. R. Goldsmith, Rept. to the Calif. Legislature, Dept. of Public Health, Calif., "The Diesel Engine and its Role in Air Pollution," December 1962.

72. G. McConnell, *Proc. Inst. Mech. Engr.*, **178**, 1001 (1964).

72a. H. Middleditch, *Proc. Inst. Mech. Engr.*, **179**, 1097 (1964).

73. M. A. Elliot, S.A.E., Quart. Trans., **3**, 490 (1949).

73a. L. E. Reed, *Proc. Inst. Mech. Engr.*, **178**, 1010 (1964).

74. D. S. Smith, R. F. Sawyer, and E. S. Starkman, *J. Air Pollution Control Assoc.*, **18**, 30 (1968).

75. R. E. George and R. M. Burlin, *Air Pollution from Commercial Jet Aircraft in Los Angeles County*, Air Pollution Control District, Los Angeles County, 1960.

76. J. D. Sensenbaugh and J. Jonakin, A.S.M.E., Paper 60-WA-334, 1960.
77. J. D. Sensenbaugh, paper presented at 5th Ann. Mt. (New England Section) Air Pollution Control Assoc., Bloomfield, Conn., 1961.
78. J. L. Mills, K. D. Leudtke, P. F. Woolrich, and L. B. Perry, *Emissions of Oxides of Nitrogen from Stationary Sources in Los Angeles*, Rept. No. 3, Los Angeles County Air Pollution District (1961).
79. R. L. Chass, R. G. Lunche, N. R. Schaffer, and P. S. Tow, *J. Air Pollution Control Assoc.*, **10**, 351 (1960).
80. P. F. Woolrich, *Am. Ind. Hyg. Assoc. J.*, **22**, 481 (1961).
81. W. S. Smith, U.S. Dept. of Health, Education and Welfare, Publ. No. 999-AP-2, 1962.
82. "Emissions of Oxides of Nitrogen from Stationary Sources," Rept. No. 2, Los Angeles County Air Pollution District, 1960.
83. H. C. Austin, *J. Air Pollution Control Assoc.*, **10**, 292 (1960)
84. H. C. Austin and W. L. Chadwick, *Mech. Engr.*, **82**, 63 (1960).
85. S. T. Cuffe and R. W. Gerstle, paper presented at Ann. Mtg. of Am. Ind. Hyg. Assoc., Houston, Texas, 1965.
86. P. H. Crumby and A. E. Fletcher, *Inst. Fuel J.*, **30**, 608 (1957).
87. W. S. Smith and C. W. Gruber, U.S. Dept. of Health, Education and Welfare, Publ. No. 999-AP-24, 1966.
88. D. Bienstock, R. L. Amsler, and E. R. Bauer, Jr., paper presented at Symp. on Fossil Fuels and Air Pollution, American Chemical Society Meeting, Pittsburgh, Pa., 1966.
89. R. C. Kurtzrock, D. Bienstock, and J. H. Field, *Inst. Fuel J.*, **36**, 55 (1963).
90. W. Thürauf, Brennstoff-Chem., **9**, 270 (1966).
91. F. E. Vandaveer and C. G. Segeler, *Ind. Eng. Chem.*, **37**, 816 (1945).
92. Stanford Research Inst., *The Smog Problem in Los Angeles*, 1954, p. 130.
93. *Emission in the Atmosphere from Petroleum Refineries*, Rept. No. 7, Los Angeles County Air Pollution District, 1958.
94. Final Rept. No. 9, Los Angeles County Air Pollution District, 1958.
95. R. B. Engdahl, *Air Pollution*, Vol. 2, Academic Press, New York, 1962.
96. E. R. Kaiser, J. Halitsky, M. B. Jacobs, and L. C. McCabe, *J. Air Pollution Control Assoc.*, **10**, 183 (1960).
97. F. E. Vandaveer, *J. Air Pollution Control Assoc.*, **6**, 84 (1956).
98. U.S. Dept. of Health, Education and Welfare, Publ. No. 999-AP-27, 1966.
99. U.S. Dept. of Health, Education and Welfare, Publ. No. 999-AP-13, 1965.
100. A. J. Haagen-Smit, *Arch. Ind. Health*, **20**, 399 (1959).
101. C. Bokhoven and H. J. Niessen, *Nature*, **192**, 458 (1961).
102. O. C. Taylor, see ref. 18, p. 65.
103. J. R. Goldsmith, *Health Effects of Nitrogen Dioxide and Nitric Oxide*, Calif. State Dept. of Public Health (1962).
104. H. E. Stokinger, *A.M.A. Arch. Ind. Health*, **15**, 181 (1957).
105. See ref. 18, p. 106.
106. J. T. Middleton, *Ann. Rev. Plant Physiol.*, **12**, 431 (1961).
107. L. H. Roger, *J. Air Pollution Control Assoc.*, **8**, 124 (1958).
108. L. S. Jaffe, *Arch. Environ. Health*, **15**, 78 (1967).
109. H. S. Johnston and H. J. Crosby, *J. Chem. Phys.*, **22**, 689 (1954).
110. See ref. 18, p. 73.
111. P. A. Leighton, *Photochemistry of Air Pollution*, Academic Press, New York, 1961.
112. E. O. Hulburt, *J. Opt. Soc. Am.*, **37**, 405 (1947).

113. F. S. Johnson, J. D. Purcell, R. Tousey, and N. Wilson, *Rocket Exploration of the Upper Atmosphere*, Pergamon, New York, 1954.

114. S. Fritz, *Solar Radiant Energy and its Modification by the Earth and its Atmosphere*, Compendium of Meteorology, Am. Met. Soc., Boston, Mass., 1951.

115. H. U. Dütsch, *Chemical Reactions in the Lower and Upper Atmosphere*, Interscience, New York, 1961, p. 167.

116. R. Pendorf, *J. Opt. Soc. Am.*, **47**, 176 (1957).

117. H. C. Van de Hulst, *Light Scattering by Small Particles*, Wiley, New York, 1957

118. R. Stair, *Proc. Natl. Air Pollution Symp.*, *3rd*, Pasadena, Calif., **1955**.

119. R. Geiger, *The Climate Near the Ground*, Harvard University Press, Cambridge, Mass., 1950.

120. B. Haurwitz, *J. Meteorol.*, **5**, 110 (1948).

121. J. K. Dixon, *J. Chem. Phys.*, **8**, 157 (1940).

122. T. C. Hall Jr. and F. E. Blacet, *J. Chem. Phys.*, **20**, 1745 (1952).

123. T. C. Hall Jr., Doctoral thesis, University of California, Los Angeles, 1953.

124. S. Sato and R. J. Cvenanovic, *Can. J. Chem.*, **36**, 279, 970 (1958).

125. A. P. Altshuller and J. J. Bufalini, *Photochem. Photobiol.*, **34**, 245 (1965).

126. A. Goetz and R. Pueschel, *J. Air Pollution Control Assoc.*, **15**, 90 (1965).

127. See ref. 18, p. 70.

128. W. A. Glasson and C. S. Tuesday, paper presented at 148th National Meeting, American Chemical Society, Chicago, 1964.

129. See ref. 18, p. 80.

130. See ref. 18, p. 47.

131. M. W. Korth, A. H. Rose, Jr., and R. C. Staham, *J. Air Pollution Control Assoc.*, **14**, 168 (1964).

132. A. J. Haagen-Smit and M. M. Fox, S.A.E. Annual Meeting, Detroit, Michigan, January 1955.

133. W. J. Hamming and J. E. Dickinson, paper presented at 57th Annual Meeting, Air Pollution Control Assoc., Houston, Texas, 1964.

134. P. W. Leach, L. J. Leng, T. A. Bellar, J. E. Sigsby, and A. P. Altshuller, *J. Air Pollution Control Assoc.*, **14**, 176 (1964).

135. E. A. Schuck, H. W. Ford, and E. R. Stephens, Rept. No. 26, Air Pollution Foundation, San Marino, Calif., 1958.

136. E. A. Schuck and G. J. Doyle, Rept. No. 29, Air Pollution Foundation, San Marino, Calif., 1959.

137. N. A. Renzetti and G. J. Doyle, *J. Air Pollution Control Assoc.*, **8**, 293 (1959).

138. N. J. Renzetti and G. J. Doyle, *Intern. J. Air. Water Pollution*, **2**, 237 (1960).

139. A. Goetz and R. Pueschel, *Atmos. Environ.*, **1**, 287 (1967).

140. E. Robinson, Symp. Air Pollution Control, A.S.T.M., Spec. Tech. Publ. No. 281 1951.

141. A. Lewis, *Clean the Air*, McGraw-Hill, New York, 1965.

142. See ref. 18, p. 88.

143. R. L. Beatty, L. B. Berger, and H. H. Schrenk, U.S. Bureau of Mines Rept. of Invest. No. 3687, 1943.

144. A. C. Holler and R. V. Huch, *Anal. Chem.*, **21**, 1385 (1949).

145. B. E. Saltzman, *Anal. Chem.*, **26**, 1949 (1954).

146. B. E. Saltzman, *Anal Chem.*, **32**, 135 (1960).

147. B. E. Saltzman and A. L. Mendenhall Jr., *Anal. Chem.*, **36**, 1300 (1964).

148. M. D. Thomas, J. A. Macleod, R. C. Robbins, R. C. Goettelman, and R. W. Eldridge, *Anal. Chem.*, **28**, 1810 (1956).

149. Atlas Electric Devices Co., Bulletin No. 1240.
150. Precision Scientific Development Co., Bulletin No. 716A.
151. S. W. Nicksic and J. Harkins, *Anal. Chem.*, **34**, 8 (1962).
152. Beckman Instruments Inc., Bulletin 4070.
153. M. E. Morrison and W. H. Corcoran, *Anal. Chem.*, **39**, 255 (1967).
154. E. R. Stephens, P. L. Hanst, R. C. Doerr, and W. E. Scott, *Ind. Eng. Chem.*, **48**, 1498 (1956).
155. R. M. Campau and J. C. Neerman, S.A.E. Publ. 660116, January 1966.
156. D. B. Henschel, private communication, 1958.
157. H. D. Daigh and W. F. Deeter, 27th Mid-Year Meeting, Am. Petroleum Inst., San Francisco, Calif., May 1962.
158. R. D. Kopa and H. Kimura, paper presented at 53rd Annual Meeting, Air Pollution Control Assoc., Cincinnati, Ohio, May 1960.
159. Y. S. Yee, W. Linville, and L. A. Chambers, "The Effect of Exhaust Gas Recirculation on Oxides of Nitrogen," *Prelim. Report*, Los Angeles County Air Pollution Control District, March 1960.
160. H. K. Newhall, S.A.E., Publ. No. 670495, 1967.
161. ARCO Chemical Co., *The Atlantic Richfield Nitric Oxide Reduction System*, Anaheim, Calif., September 1966.
162. R. D. Kopa, R. G. Jewell, and R. V. Spangler, Southern Calif. Section, Soc. Automotive Engrs., Los Angeles, Calif., March 1962.
163. J. T. Wentworth and W. A. Daniel, *J. Air Pollution Control Assoc.*, **5**, 91 (1955).
164. A. O. London, Air Pollution Conf., University of Calif., Los Angeles, Calif., December 1961.
165. G. C. Jefferis and J. D. Sensenbaugh, A.S.M.E. Publication., October 1959.
166. D. H. Barnhart and E. K. Diehl, *J. Air Pollution Control Assoc.*, **10**, 397 (1960).
167. H. R. L. Speight, *Can. J. Chem. Eng.*, **36**, 3 (1958).
168. H. Rudorfer and K. Kremser, Austrian patent 184,992, 1956.
169. R. Schonbeck, Austrian patent 188,723, 1957.
170. Societa Industriale Cataverse Societa per Azioni, French patent (cl.C.01c), October 1966.
171. Italian patent 523,953, 1955.
172. East German patent (cl.C.01b), October 1966.
173. Proceedings 1st Congr. on Air Pollution, New York, N.Y., 1955; *Problems and Control of Air Pollution*, Reinhold Publ. New York, 1955, p. 74.
174. M. S. Peters, *Chem. Eng.*, **62**, 197 (1955).
175. P. G. Taigel, Safety in Mines Estab. Rept., No. 48, September 1952.
176. J. A. Almquist, V. L. Gaddy, and J. M. Braham, *Ind. Eng. Chem.*, **17**, 599 (1925).
177. E. G. Foster and F. Daniels, *Ind. Eng. Chem.*, **43**, 986 (1951).
178. B. B. Sundaresan, C. I. Harding, F. P. May, and E. R. Hendrickson, *Environ. Sci. Tech.*, **1**, 151 (1967).
179. G. Bedjai, H. K. Orbach, and F. C. Riesenfeld, *Ind. Eng. Chem.*, **50**, 1165 (1958).
180. T. E. Green and C. N. Hinshelwood, *J. Chem. Soc.*, **1926**, 1709.
181. P. W. Bachman and G. B. Taylor, *J. Phys. Chem.*, **33**, 447 (1929).
182. J. M. Fraser and F. Daniels, *J. Phys. Chem.*, **20**, 22 (1952).
183. S. Sourirajin and J. L. Blumenthal, *Actes Congr. Intern. Catalyse*, 2, **1960**, 2521.
184. G. Fauser, U.S. patent 1,487,647, March 1924.
185. H. C. Anderson, J. G. Green, and D. R. Steel, *Ind. Eng. Chem.*, **53**, 199 (1961).
186. H. C. Anderson, P. L. Romeo, and W. J. Green, Engelhard Ind. Inc., Tech. Bull. 7,100, 1966.

187. E. I. du Pont de Nemours and Co., Ind. and Biochemicals Dept., "Torvex Ceramic Honeycomb," Wilmington, Del., April 1967.
188. F. R. Taylor, Rept. No. 28, Air Pollution Foundation, San Marino, Calif., September 1959.
189. F. R. Roth and R. C. Doerr, *Ind. Eng. Chem.*, **53**, 293 (1961).
190. R. J. Ayen and Y. S. Ng, *Intern. J. Air Water Pollution*, **10**, 1 (1966).
191. R. A. Baker and R. C. Doerr, *Ind. Eng. Chem. Process Design Develop.*, **4**, 178 (1965).
192. R. A. Baker and R. C. Doerr, *J. Air Pollution Control Assoc.*, **14**, 409 (1964).
193. P. R. Ryason and J. Harkins, *J. Air Pollution Control Assoc.*, **17**, 796 (1967).
194. L. D. Decker, Paper No. 67-148, UOP Air Correction Division, Universal Oil Products Co., Greenwich, Conn., 1967.
195. W. Ehnert, *Brennstoff-Chem.*, **9**, 273 (1966).

The Control of Sulfur Emissions from Combustion Processes

Werner Strauss

Department of Industrial Science, University of Melbourne, Victoria, Australia

I. INTRODUCTION

Sulfur dioxide is the chief gaseous impurity emitted by combustion processes other than those in internal combustion engines. The largest contributors to the total emission are electric power stations fired by fossil fuels, although individual higher concentrations* can be obtained from some

---

* Note on gas concentration units: 1% by volume is equal to 10,000 parts per million (ppm). In some cases, grains/100 ft³ are used. If the volumes have been reduced to standard conditions, gr/100 s ft³ is used.

smelters of sulfide ores even when some attempt is made to recover the sulfur dioxide. Conventional sulfuric acid plants using the contact process, oil refineries, and other sources also contribute to the sulfur dioxide pollution of the atmosphere.

A. Extent of Sulfur Dioxide Emissions

Detailed estimates of the sulfur dioxide emissions in the United States were made for the years 1963 and 1966 (1) and these are shown in Table I. These data indicate marked increases from a number of sources over the three-year period, particularly the smelting of ores, and a decrease only in the amount from the burning of coal refuse banks. The emission of sulfur dioxide from the operation of oil refineries and of refuse incinerators has not increased during this time; this is probably due to the improved control of refinery emissions in the one case, and the increased use of sanitary land-fill as a means of refuse disposal in the other. Rohrman, Ludwig, and Steigerwald (1), who made the estimates in 1963 and 1966, earlier predicted a 1966 total emission value of 25,245,000 tons of sulfur dioxide. As the table shows, the predicted value was exceeded by more than 3 million tons, or about 13%.

TABLE 1

Sulfur Dioxide Emissions in the United States, 1963–1966[a,b]

	1963		1966		Per cent increase
	'000 tons	%	'000 tons	%	
Burning of coal (total)	14,029	60.0	16,625	58.2	
Power generation	9,580	41.0	11,925	41.6	24.5
Other	4,449	19.1	4,700	16.6	5.6
Oil combustion (total)	4,817	20.6	5,604	19.6	16.4
Power generation	714	3.0	1,218	4.3	70.6
Other	4,103	17.6	4,386	15.3	6.9
Smelting of ores	1,735	7.4	3,500	12.2	101.7
Refinery operations	1,583	6.8	1,583	5.5	0
Coke processing	462	2.0	500	1.8	8.2
Sulfuric acid	451	1.9	550	1.9	22
Coal refuse banks	183	0.8	100	0.4	−83
Refuse incineration	100	0.4	100	0.4	0
Miscellaneous sources	20	0.1	25	0.1	25
Totals	23,380	99.9	28,637	100.1	22.5

[a] It is estimated that 6,000,000 tons of SO_2 are emitted by thermal power stations in the U.K. each year.

[b] From the tables of Rohrman, Ludwig, and Steigerwald (1).

TABLE II

Fuel Consumption by Thermal Power Stations in
England and Wales[a,b]

Year	No. of stations	Fuel consumption: tons $\times 10^6$	
		Coal	Oil
1920	—	4.52	0.09
1930	429	7.96	0.03
1938	323	13.20	0.02
1950	275	29.38	0.05
1955	278	38.87	0.02
1960	234	47.08	5.28
1965	234	64.06	5.78

[a] The fuel consumption for Scotland in 1965 amounted to 4.8 million tons of coal and 0.44 million tons of oil.

[b] From ref. 5.

Estimates for the United Kingdom indicate that in 1952 4.3 million tons of SO_2 were emitted, of which 1.8 million tons were from thermal power stations. A more recent estimate for the United Kingdom for the latter contribution can now be made using data for oil and coal consumption (Table II). This is of the order of 4.5 million tons of SO_2 per year if an average of 3% sulfur content is assumed. Similar values for the Federal Republic of Germany for 1958 show that the sulfur dioxide from the combustion of coal amounted to 2,382,000 tons (2).

TABLE III

Anticipated Power Plant Coal and Oil Consumption
in the United States to 2000[a]

Year	Coal, tons $\times 10^6$	Oil, U.S. barrels $\times 10^6$	Tons $\times 10^6$
1966	267	142	21
1970	375	175	26
1975	550	225	33
1980	750	240	35
1990	875	250	37
2000	800	230	34

[a] From ref. 4.

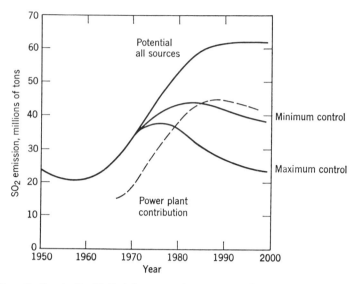

Fig. 1. SO_2 pollution in the United States to the year 2000 (4). Minimum control: 1% control from 1970, 5% from 1975 on new power plants: 80% control from 1990 on new plants. Maximum control: 75% on new plants after 1975, increasing by 1% to a maximum 90% in 1990.

Workers in the United States (3,4) have calculated the potential rate of increase in sulfur dioxide emissions up to the year 2000 (Fig. 1) and have found that these will more than double in the next 20 years, even if reasonable controls are enforced. The primary source of these emissions will still be from the combustion of oil and coal in power plants (Table III). The rate of increase (Fig. 1) will decline after 1970 because of the successive introduction of nuclear power plants as a replacement for obsolete fossil fuel-fired plants. Other calculations estimate the emission levels with the application of various degrees of control with respect to sulfur dioxide, but even with advanced techniques used to the fullest extent emission levels will increase by 25% before 1975, while more realistic estimates are of the order of a 75% increase. Unfortunately, similar predictions for other countries are not available.

B. Air Quality Criteria for Sulfur Oxides

The tolerance level of plants and animals to pollutants such as sulfur dioxide depends on a number of factors: the concentration of the pollutant and the exposure time, the type of plant or animal, and its condition and age. Thus the maximum allowable concentration of sulfur dioxide in which it is considered possible for a healthy man to work for eight hours (MAC value) is 5.0 ppm. However, during the 1952 London smog, which caused 4000 deaths, the maximum recorded concentration was only 1.7 ppm, with average values of 0.7 ppm, indicating that longer exposures, particularly of people with bronchial disorders, may lead to deaths at much lower concentrations. In fact, increased cardiovascular morbidity is shown to occur when the sulfur dioxide concentration in the atmosphere exceeds 0.015 ppm for a 12-month period, which is a value exceeded in many cities in the United States and elsewhere.

Work on the damage to vegetation by sulfur dioxide was initially carried carried out in areas affected by fumes from smelters roasting sulfide ores, where damage was severe. The degree of injury follows a time (t in hours)–concentration (c in ppm) relation with a finite threshold. This is expressed in a simple linear relation, first obtained by O'Gara in 1922, and subsequently generalized by Thomas and Hill (7,8). These workers found that traces of leaf destruction could be observed when the dosage and time exceeded that calculated from

$$(c - 0.24)t = 0.94$$

This shows that small exposures with long times are equally as destructive as large exposures for short times.

It has also been observed that if small quantities of an oxidant such as ozone is present (0.03 ppm) sulfur dioxide in concentrations as low as 0.10 ppm can cause the same damage as concentrations of 0.30 ppm would without the presence of oxidants.

In view of these factors, the U.S. Department of Health, Education and Welfare (7) has set a series of ranges of sulfur dioxide concentrations for acceptable air quality (Table IV). It is interesting to note that the 24-hr average value is similar to that set for the same period in the Soviet Union, that is, 0.056 ppm (10).

It should also be noted that the levels achieved by the best industrial practice, such as that enforced by the United Kingdom Alkali Inspectors,

TABLE IV

Criteria for Acceptable Air Quality Sulfur Dioxide
Levels[a]

Time period	Maximum ppm	1 percentile ppm
24-hr average	0.05–0.08	0.04–0.06
1-hr average	0.12–0.20	0.05–0.11
5-min average	0.10–0.50	0.05–0.14

[a] From ref. 7.

give values for individual plants within these limits. For example, the recommended chimney heights for contact sulfuric acid plants burning sulfur are based on calculations assuming a three-minute mean ground level concentration of sulfur dioxide of 0.20 ppm. However, the maintenance of air quality in an industrial complex or a city environment with many sources may prove extremely difficult in practice.

C. Possible Processes for Reduction of Sulfur Oxides

The growing awareness that comparatively low concentrations of sulfur dioxide may be harmful, and the knowledge that the quantities emitted, particularly from thermal power stations, are rapidly increasing, have made the introduction of processes to remove sulfur oxides of great urgency. An additional stimulus is the growing shortage throughout the world of sulfur for fertilizers and chemicals; the recovery of sulfur from the fossil fuel, the gasified fuel, or combustion gases, would make a significant contribution to the sulfur available (12) for chemical processing. For example, a 2000-MW power station, burning 5.2 million tons of coal containing 1.5% sulfur could produce 75,000 tons of sulfur each year.

Sulfur can be removed from the original fossil fuel, but the sulfur is chemically bound in such a way that removal is very difficult. This would have the advantage that the relative quantity of material to be handled is the minimum possible. It is possible to gasify a fossil fuel, either coal or oil, and to release the sulfur compounds, usually as hydrogen sulfide, which can then be removed before the fuel gas is burnt. A third possibility is to burn the fuel and then recover the sulfur oxides which have been formed. In this case, the relative concentration of sulfur dioxide in the gas is much lower, but the total volume of gas to be treated is much greater. Processes have been suggested and investigated in recent years for the removal of sulfur oxides at all stages of the combustion process, and this paper reviews these in detail and comments on their technical and economic feasibility.

II. REMOVAL OF SULFUR FROM FOSSIL FUELS

Both coal and fuel oil contain sulfur in varying concentrations. This concentration depends primarily on the source of the fuel, but also, in the case of fuel oil, on the refinery processes because the sulfur compounds tend to be concentrated in the heavy fuel oil fraction.

A. Removal of Sulfur from Coal

Coal, as mined, contains generally between 0.2 and 7% sulfur by weight on a dry basis; the high-sulfur coals are defined as those which have more than 3% sulfur. In the United States about 23% of the coal mined in 1963 fell into the high-sulfur category, with exceptionally high sulfur content in the coals from Iowa (4.44%) and Missouri (4.8%), although these latter coals comprise only about 1% of the total mined (13).   The average sulfur content of the bituminous coals mined was about 2%, while that of the coal burned was 2.41%, the low-sulfur coals being used for coking and export. European coals tend to have slightly lower sulfur contents, while Australian coals (except for Greta seam coals in New South Wales) (14) usually have below 1%.   Some brown coals, which are used on a large scale in Germany and also in Victoria (Yallourn and Morwell open cut) are exceptionally low in sulfur, averaging 0.3–0.5% (16), although in other seams the sulfur content can be as high as 5 or even 8%.

The sulfur in the coal is present in both organic and inorganic compounds, the latter generally as metal sulfates and sulfides.   In coals with a high sulfur content, the major constituent is iron disulfide (FeS_2), which is present either as pyrite or, to a lesser extent, marcasite.   Smaller quantities of sphalerite (zinc sulfide, ZnS), as well as traces of other sulfides, are also found in some coals.   Sulfates, which are present in much smaller quantities than the sulfides, are usually from the alkaline earth group—gypsum ($CaSO_4 \cdot 2H_2O$) and barite ($BaSO_4$).

Pyrites and other inorganic sulfur compounds occur discretely as either fine grains (50 μm or less), or in some cases in layers some millimeters or centimeters thick in the coal seams (17).   High-sulfur coals are likely to contain 50–60% of the sulfur as pyrites, and in exceptional cases 80%, while sulfur in the sulfate form is generally much lower.   Low-sulfur coals, on the other hand, are likely to contain appreciable quantities of organic sulfur, which is chemically bound to the coal and cannot be removed without drastic treatment.   Typical analyses are shown in Table V.

TABLE V

Typical Analyses of the Forms of Sulfur in Coals[a]

Source of coal	Sulfur form, %			
	Pyritic	Sulfate	Organic	Total
Illinois	2.4	0.06	2.6	5.1
West Kentucky	2.1	0.1	2.5	4.7
Missouri	4.4	0.1	2.2	6.7
West Virginia	3.2	0.01	0.5	3.7
Greta Seam (Australia)	2.21	0.03	1.52	3.76
Greta tops. shale (Australia)	0.63	0.0	2.20	2.83
Ravensworth (Australia)	0.52	0.05	—	0.47

[a] From ref. 16.

Coal washing is common practice, particularly with machine mined, high-ash coals. Because pyrite and marcasite have a specific gravity of approximately 5, compared to 1.25 of naturally clean coal, washing can reduce the sulfur content of the coal by as much as one-third in certain cases, although in practice the average reduction is unlikely to reduce the sulfur by more than one-tenth (18). However it has been found that if instead of the commonly used coarse crushing used for ordinary coal washing, crushing to below 3 mm was carried out; then the sulfur content may be reduced even further, possibly by as much as 50% in some cases (18).

However even this degree of crushing is unlikely to release the very fine grains of sulfate and sulfide. It should be noted that coal is crushed to below 200 mesh (A.S.T.M. 74 μm) for firing in pulverized fuel boilers and this would release the fine grains of sulfur minerals which could be separated by further suitable means of separation. Several systems have been investigated for this separation.

The simplest, studied by Bituminous Coal Research Inc. (B.C.R.) (105), uses an air classifier with recycle and two stages of classification. The fines, which contained less pyrite after crushing and air separation, were fed to the boiler, while the coarser fractions were subjected to a sink/swim washing, using a fluid with a specific gravity of 1.6. Significant reduction in the sulfur content was achieved by this method.

In a more sophisticated treatment, the pyrite and marcasite can be converted into the magnetic pyrrhotite form by oxidation at high temperatures. In a series of laboratory studies (19) coal was heated to 320–340°C and then treated with an air–steam mixture or air for 2–5 min. This converts the surface layers of the grains to the magnetic form, so that effective magnetic separation of these particles could be carried out using a magnetic field of 10^4 Gauss (19). Pilot plant studies with somewhat higher temperatures and reaction times of 15–30 sec also gave favorable results.

Triboelectric effects, that is, the inducing of electric charges on the different constituents by rubbing, were also tried in an attempt to separate the pyrites in a nonuniform field. Tests using a 12-kV generator proved disappointing (105), and effective separation would require recycling. It was concluded that an unrealistic number of cycles would be needed to precipitate the pyrite selectively from the coal.

The use of nonuniform electrostatic fields has been reported by B.C.R. (105). It was noticed that when samples were placed on plates subjected to high (12 kV) inductive charging the coarse coal fraction was differentially attracted to the charged field, and this caused a "beneficiation" of pyrite at one end of the plate. With one coal (No. 6 Seam Illinois Coal) containing 3.2% pyrite sulfur; 56% of "clean" coal (1.1% pyrite sulfur, 4.4% ash), 37% middling coal (3.0% sulfur, 16.3% ash) and refuse (13.1% sulfur, 42,4%

ash) could be obtained.   However, these results could not be obtained with other coals.

Bacterial oxidation of the iron sulfides to dilute sulfuric acid has been considered as a possible method of removing sulfur.   Experiments (21) which were carried out largely to show which sulfides produced the acids in mines indicated that Ferrobacillus ferro-oxidants would accelerate the oxidation of coarsely crystalline marcasite and different pyrites except coarsely crystalline pyrite, unless this was very finely crushed and exposing maximum surface area.   However, the slow rate of this process is unlikely to make it suitable for treating the large quantities involved in mined coal. Furthermore, the disposal either by neutralization or concentration of the dilute acid is unlikely to make it economically feasible.

Solvent extraction processes have been suggested which will remove the organic sulfur as well as the inorganic inclusions.   Virtually all coals, except anthracite, will dissolve in a solvent oil at pressures of about 1000 psi and at 450°C.   In a process suggested by Spencer Chemical Co. (22) (Fig. 2), coal ground to below 200 mesh is slurried with the solvent oil, a small amount of hydrogen (about $\frac{1}{2}\%$) is added, and the mixture is then treated at high temperature and pressure in order to dissolve the coal.   The hydrogen performs

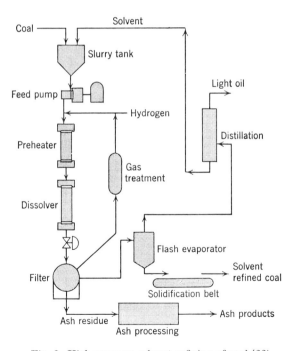

Fig. 2. High-pressure solvent refining of coal (22).

the dual function of preventing the coal from polymerizing as well as interacting with the organic sulfur. Some of this is converted to hydrogen sulfide, which can be extracted. Pressure filtration eliminates both ash and pyritic sulfur, and subsequent flash evaporation is used to recover the solvent. No additional makeup solvent is required in this process to replenish losses as some solvent oils are released by the coal. The hot liquid coal residue is cast into moulds of desired shape, to give brittle "briquettes" which melt at about 150°C. Alternatively, the coal residue can be handled as a liquid at about 300°C. The residual product is higher in heating values than raw coal (about 16,000 Btu/lb/ft^3, 10,000–13,000 Btu/lb), and contains less than 1% sulfur even when it is made from a high-sulfur coal. Economic considerations indicate that while this solvent-refined coal may have advantages for certain applications, it is about $2\frac{1}{2}$ times as expensive as raw coal (based on equivalent heating values), and so solvent extraction is unlikely to be a major process for reducing sulfur emissions from coal-fired power plants.

B. Removal of Sulfur from Oil

Heavy fuel oil, commonly referred to as residual fuel oil, contains quantities of sulfur usually varying between 0.5 and 5% depending on the source and subsequent refinery treatment (2,15). It is designated as API grades 5 and 6 (U.S.), Bunker C (U.K. and for marine use), or Heating Oil S (Germany).

The crudes from some African fields (Nigeria and Libya) are very low in sulfur, as are some of the U.S. crudes, a typical value for Pennsylvania being as low as 0.08%. Generally, however, the sulfur concentration of most crude oil imported into the United States is much higher. As a general rule, Venezuelan and West Indian crudes average 2.5%, Arabian crudes 3.5%, while Mexican crudes, although averaging 4% sulfur, have been up to 5.18%. Crudes from the East Indies are generally low in sulfur (below 1.5%). There is no upper limit on the sulfur content of commercial A.P.I. grades 5 and 6 under the N.B.S. standards (23), although the permissible maximum for Heating Oil S (German Industry Standard DIN 51603) is 4.2%.

At present, New York City legislation requires that fuel oil burned must contain less than 2% sulfur but after May 1971 the sulfur must be reduced to below 1%. Similarly, the bi-state New York/New Jersey area has recommended that after October, 1969, power plants will be restricted to fuel oil containing less than 1% sulfur, while other commercial and industrial users must use oil with less than 0.3% sulfur. Construction of new power plants is also restricted unless fuel of this quality is available.

While short-term needs can possibly be met by use of low-sulfur fuels, desulfurizing of the oil will be the long-range requirement. Effective processes, which have been developed over recent years, essentially involve the treatment of the residual oil with hydrogen at high pressures in the presence of suitable catalysts and are called "hydrodesulfurization" or HDS processes.

The sulfur in fuel oil is present as either straight chain or cyclic sulfur compounds, the general trend being an increase in the proportion of aromatic sulfur compounds in the high boiling fraction, and a predominance of thiols, sulfides, and disulfides in the lower boiling distillates. The gas oils from the Middle East and Texas have over 60% of their sulfur as benzothiophenes and dibenzothiophenes. The crudes contain sulfur chiefly in the form of condensed thiophenes such as anthranaphthene and a range of complex polycyclic sulfur compounds (24,25).

Sulfur can be removed from fuel oil by several different processes of varying effectiveness which remove different proportions of sulfur. Thus, a minor degree of desulfurization may be achieved by a simple distillation process while delayed coking, a solvent deasphalting, and residual HDS processes will give increasing degrees of sulfur removal. For the most intensive treatments, combinations of these, such as solvent deasphalting and residual HDS, may be used. An extensive list of these processes (to 1955) is given by Kalichewsky and Kobe (26).

Mercaptans can be decomposed by passing the vaporized fuel over silicate bauxite, or alumina catalysts, but these will not decompose the aromatic thiols (27). Light distillates can be effectively treated by reacting them with hydrogen over a nickel catalyst for 12–17 min at 375–550°C and 200 psig, which reduces the sulfur from 1.15 to 0.05% (Szayna process) (27).

Conventional HDS processes for fuel oils generally consist of treating the oil with hydrogen over a catalyst at temperatures of 400–550°C and pressures of 500–1000 psig. In some experiments pressures up to 2000 psig have been used (24). The volumes of hydrogen used are considerable, and vary, depending on the crude, between 500 and 1800 ft^3/barrel. Carpenter and Cottingham (24) have surveyed these processes in some detail, and two typical processes will be described here. The Gulf HDS process operates at 430–455°C, pressures of 500–1000 psig, with a hydrogen recycle rate of 2500–10,000 ft^3/barrel over an undisclosed catalyst. The hydrogen consumption is 499 ft^3/barrel for a Kuwait crude, and the average sulfur content was reduced from 2.61 to 0.46%. For a bottoms product (boiling above 355°C) the reduction was from 4.11 to 0.84%, while the 210–355°C boiling fraction was reduced from 1.54 to 0.39%. The H-Oil process, for which a 2500 barrel/day plant has been built at Lake Charles, La., uses a pressure of 1500 psig and hydrogen consumption is 1000–1800 ft^3/barrel with similar temperatures to the Gulf HDS process. A larger H-Oil plant with a capacity of

16,500 barrels/day is being built by Humble Oil and Refining Co. at its Bayway Refinery in New Jersey. Japan is currently building 6 HDS plants with a capacity of 187,500 barrels/day.

In a typical residual fraction (boiling range 344–525°C) the sulfur content was reduced from 5.3 to 1%. The process was more effective with lower boiling fractions and less effective (about 50%) with higher boiling fractions. At present HDS processes are expensive. However, if a low-pressure process, using moderate temperatures, and hydrogen quantities approaching stoichiometric values (calculated to be 158 ft³/barrel) could be achieved, the desulfurization of furnace oils may become of prime importance commercially.

Extensive economic studies of desulfurizing fuel oils have been carried out on Kuwait crudes, California crudes (28), and, more recently, Caribbean crudes (29). The last is of particular importance since the residual fuel oils from this source are the major fuel for thermal power stations on the eastern seaboard of the United States.

These economic studies which were carried out by the Bechtel Corporation for the American Petroleum Institute (29) were based on a "typical" refinery, being the average for seven major operating refineries in the Caribbean area. This "refinery" processes 300,000 barrels/day of a 23.6° A.P.I. crude and produces as its chief product, 57.4% residual fuel oil (A.P.I. grade 6) with 2.6% sulfur. This proportion of furnace oil is of course not typical elsewhere. Domestic refineries in the United States average 11%, whereas some produce as little as 8% residual fuel oil.

The Caribbean residual oil sells for $U.S. 2.00–2.25/barrel. Some desulfurization involving a solvent deasphalting process removes about 30% of the sulfur, leaving about 1.8% and can be carried out at a cost of 30 ¢/barrel or 37 ¢, if this is based on equivalent heating values. Further treatment, involving HDS of the residual oil, reducing the sulfur to 1% is possible at a cost of 58 ¢/barrel (72 ¢ equivalent heating values). Further reduction of the sulfur to 0.5% sulfur costs 80 ¢/barrel (97 ¢ equivalent heating value). So effective desulfurization to such a value adds about 50% to the fuel cost. The calculations were carried out for a number of different processes and combinations of these, and the results have been summarized in graphical form (Fig. 3). This clearly shows the approximately linear relation of the increasing costs involved in decreasing the sulfur content.

As an alternative to improved HDS methods, with more selective catalysts, lower hydrogen consumption, and lower pressures, other novel methods of removal of sulfur may also be possible. One of these is the use of metallic sodium, which could remove the sulfur as sodium sulfide, but the present cost of sodium prevents this from being seriously considered. New methods of desulfurizing are badly needed if this is to be carried out economically.

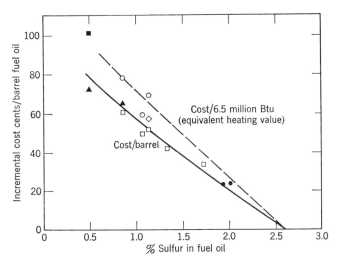

Fig. 3. Desulfurization cost studies (27). (●) Distillation, 2 cases; (□) solvent deasphalting, 5 cases; (○) residual desulfurization, 3 cases; (◇) delayed coking, 1 case; (▲) solvent deasphalting and coking, 2 cases; (■) solvent deasphalting and residual desulfurization, 1 case.

III. "CLEAN GAS" PROCESSES: THE GASIFICATION OF SOLID AND LIQUID FUELS, AND THE REMOVAL OF SULFUR COMPOUNDS FROM FUEL GASES

The removal of organic sulfur in both oil and coal requires the decomposition of the complex organic sulfur molecules by disruption of the fuel structure at the molecular level. High-pressure solvent treatment of coal or hydrodesulfurization of fuel oil produces fuels comparatively low in sulfur, and physically not unlike the starting compounds. In contrast, complete gasification of the fuels would release all the sulfur, converting most of it into hydrogen sulfide.

Gasification can essentially be regarded as the breaking of the carbon rings and subsequent reaction with hydrogen. If coal and oil are compared, far more hydrogen must be reacted with coal to reduce the carbon–hydrogen ratio to the value for natural or town gas (Table VI). It is therefore apparent that, in general, it will be more difficult to gasify coal than fuel oil. Despite this difficulty, the gasification of coal has been carried out for much longer periods and more extensively than oil because of the need for coke as well as the wider distribution of coal.

Fuel gas can be treated either by scrubbing processes at ambient temperatures or by high-temperature adsorption processes, and in both cases the efficiency of removal of sulfur compounds is better than 99% (see Sec.

TABLE VI

Carbon/Hydrogen Ratios for Solid, Liquid,
and Gaseous Fuels

Carbon coke	20:1
Coking coal	15:1
Heavy (residual) fuel oil	8:1–10:1
Light distillate feedstock	5:1–7:1
Natural or town gas	3:1–5:1

III-C-2). This efficiency should be contrasted with the 90% efficiency of removal of sulfur dioxide from flue gases which is generally assumed for the processes for SO_2 removal. Furthermore, the quantity of gas treated is less because no nitrogen dilutent gas is present. The "clean fuel" can be used in gas turbines as well as in conventional steam power systems, which gives this some economic advantages which must, however, be largely cancelled by the capital and operating costs of the gasification plant. A number of the "clean fuel" cycles have been discussed by Squires (30) and these may offer economic advantages under certain conditions.

Another recent example of a "clean fuel" system, which combines coal gasification, treatment, and a gas turbine steam turbine combination, is being studied by the Babcock and Wilcox Company [See Fig. 4 (108)].

A. Gasification of Coal

Although coal has been gasified in order to obtain coke and gas for about 200 years, the complete gasification of coal is of more recent origin. The first

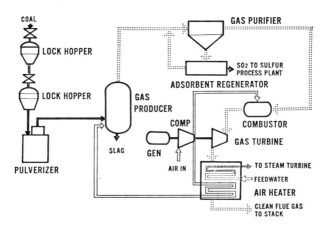

Fig. 4. Coal-gas purification system combined with advanced gas turbine–steam turbine power generation system (108).

of the complete gasification processes used pressures not far from atmospheric and air–steam mixtures or alternating air and steam were used to produce producer and water gas of comparatively low calorific value (125–230 Btu/ft³). The coals used for this process were limited either to weakly agglomerating types, or to fusing and nonagglomerating types (33).

Recent developments in complete gasification, however, have introduced systems where, in some cases, high pressures and oxygen–steam mixtures are used, rather than air–steam mixtures. The fuel can be in a fixed bed, a fluidized bed, or in suspension, and the heating value of the fuel gas produced varies between about 200 and 450 Btu/ft³. These processes are fully reviewed elsewhere (31,32) and only a brief relevant discussion is presented here.

The Lurgi process has been widely applied on a commercial scale with plants in South Africa, Australia, and Scotland as well as in Germany. It is a fixed bed process using either low-ranking noncoking coals or brown coal briquettes, and steam–oxygen mixtures at generator pressures of 30 atm and temperatures of 850–1000°C. For satisfactory operation, the Lurgi process requires coal in a limited size range, and the need for non-coking coals may limit its applicability. Furthermore, there is considerable tar and phenol formation, which may lead to a serious liquid effluent problem if inadequate treatment plants are not provided. Lurgi gas, after purification, has a calorific value of 386–460 Btu/ft³.

Fluidized bed gasification at ambient pressures has been carried out in the Winkler process and in its modifications, the Flesch-Winkler (single generator) and the B.A.S.F.—Flesch-Demag (twin generator) process. Steam/air/oxygen blasts are used and in the two later versions the flow is cyclic to reduce carryover, which is an inherent problem with the fluidized bed type process. For high gasification rates and minimum carryover the feed should contain less than 30% of less than 1 mm coal for coking coals, or 50% in the case of noncoking coals. The calorific value of the gas is about 150 Btu/ft³.

High ash coals, particularly if the ash has a low fusion temperature (which can be assisted by the addition of dolomite as a flux), can be gasified in slagging gasifiers in which the coal as a powder is suspended in the gas stream. In the Koppers-Totzec process the coal is introduced together with the oxygen, and steam is added just before the entry, in two horizontally opposed nozzles. In the Rummel process the coal is introduced separately from the air/oxygen/steam mixture in four (or more) tangential entry gasifiers, such as the Babcock and Wilcox-DuPont unit, which has been tried on a pilot-plant scale. The heating value of the fuel varies between 270 and 300 Btu/ft³. Both the Koppers-Totzec and Rummel plants have comparatively high steam and oxygen consumption rates and serious carryover problems. The efficiencies in conversion of coal into gas are about 72%, based on heat content, but much higher overall efficiencies—about 90%—are obtained if

TABLE VIIa

Sulfur Compounds in Coal Gas

Gas	Average values in impure town gas (Powell) (34)			Metallurgical (35) coke oven gas, gr/100sft³	Synthesis gas from pressure gasifier, gr/100sft³	Koppers-Totzek (31), gr/100sft³	U.S. Bur. Mines (31) slagging gasifier, gr/100sft³
	%	ppm	gr/100sft³				
H_2S	0.3–3	3000–30,000	250–2500	516–675	107–117	130–175(3000)[a]	80–400
CS_2	0.007–0.07	7–700	13.3–133.7		0–2.4		
COS	0.009	9	13.6		12.6–18.0		
RSH (Thiols)	0.003	3	4.6		0.20		
C_4H_4S (Thiophene)	0.010	10	21.1				
HCN	0.1–0.25	1000–2500	67.8–170				
Total org. S				13–39	17–19.5	0.0005–6(21.9)[a]	4–40

[a] Ref. 32.

waste heat and chemical byproducts are also considered. These processes are reviewed in detail elsewhere (31–33).

The sulfur compounds in gas from coal depend on the sulfur in the original coal, but some idea of the concentration ranges can be obtained from the typical data in Table VIIa.

B. Gasification of Oil

Because of the lower carbon/hydrogen ratio in fuel oils, the gasification of oils is much easier than of coal. Furthermore, the gas does not contain ammonia and hydrogen cyanide as well as the sulfides which have to be removed, thus reducing the capital cost of gasification plant. Additional reasons, such as the ready availability of petroleum feedstocks, compared with the shortage of suitable gas-making coals and the limited market for byproduct coke and ammonium sulfate, which played an important part in the economics of traditional gas making, have also contributed to the wide commercial applications of oil gasification processes. Oil gasification has been the subject of a recent review article (32) and a book (33).

The gasification processes essentially involve the thermal or catalytic cracking of the oil molecules to produce gas molecules and carbon. Hydrogen can be added during this process either directly or in the form of steam. Carbon is produced in the process. In some of the processes it is collected as carbon black, whereas in others it is deposited on the catalyst, which is subsequently removed either by using a cyclic process, with a fixed catalyst, burning off carbon with air during a "blow," or by recycling continuously in a fluid bed type plant, removing part of the catalyst, and burning off the carbon in a separate vessel. Commerical plants of different designs of all these types of processes are at present in operation.

The simple thermal cracking processes, the first of which was the "Jones" process, produces town gas (550–750 Btu/ft³) and in some modifications, carbon black, in a cyclic system, and uses chambers with a nonactive cracking. The "Hall" process, which is similar, but with less rigorous cracking conditions, produces gas with much higher calorific values (above 1000 Btu/ft³) (Table VIIb). Continuous processes of this type using a bed of fluidized pebbles have also been used. One, called the Lurgi-Ruhrgas-Sandcracker uses sand, while another, developed by the Mobil Oil Corp. and Surface Combustion Corp. uses $\frac{1}{4}$–$\frac{3}{4}$ in. coke pebbles.

Hydrogasification processes, which give almost complete conversion of the oil to gas, have been developed by the Institute of Gas Technology (I.G.T.), Imperial Chemical Industries (I.C.I.), and the British Gas Council (B.G.C.). Some of these processes require comparatively light hydrocarbon feedstocks, and produce gas with calorific values between 500 and 1000 Btu/ft³. The

TABLE VIIb

Sulfur Compounds in Gas Made from Residual Oils

	Hall type (33) cyclic pyrolysis process, gr/100sft³	ONIA-GEGI (36) process, gr/100sft³	SEGAS (33) process, gr/100sft³
H₂S	600	100–180	50–250
Org. sulfur	40	6.5–10.5	30–40
CS₂	—	—	15–20
S in oil, %	—	1.65–1.70	
Cal. value of gas Btu/ft³	1000–1140	523–527	473

B.G.C. and one of the I.G.T. processes can cope with residual fuel oils. The B.G.C. process uses a recirculation fluidized bed of powdered coke, pressures of 300–750 psig, and temperatures from 700 to 925°C. The I.G.T. process is a cyclic one with fixed refractory beds, pressures below 50 psig, and similar temperatures to the B.C.G. process.

Catalytic steam cracking processes are most likely to be applicable to the gasification of high-sulfur residual fuels, and can be either continuous or cyclic. The continuous processes are able to use light oil or gaseous feeds, low in sulfur for cracking and reforming operations, while the cyclic processes can use a wide range of fuels, including residuals, with no limit to their sulfur content. These latter include the ONIA-GEGI, SEGAS, and MICRO-SIMPLEX processes. They are very widely used for commercial and town gas and large plants have been built with single-unit capacities up to 20 million ft³/day.

Essentially, the hydrocarbon reacts with steam to form carbon monoxide, methane, and hydrogen at temperatures of 700–1000°C (at the bottom of the catalyst) and at atmospheric pressure. The catalysts are specific to the processes, but generally contain nickel. The ONIA-GEGI process uses irregular ½-in. pebbles of magnesite impregnated with nickel, giving an average bulk concentration of 3–4% NiO, although about 10% at the surface, while the SEGAS process uses either nickel-impregnated silica nodules, or cylindrical magnesia pellets with free lime, as the active material. In both ONIA-GEGI and SEGAS plants, air blasts burn off the carbon deposited during the gas-producing runs. Sulfur is also deposited on the catalyst so that the hydrogen sulfide concentration is lower in the product gas. On the other hand, blow air exhausted to the atmosphere will contain some sulfur dioxide as well as other sulfur compounds.

The quality of the gas produced by these processes varies; thus the

)NIA-GEGI process can produce a range of calorific values from 300 to ₁000 Btu/ft³, the SEGAS process, 473 Btu/ft³ (fuel oil feed, lime catalyst), ₁nd the MICRO-SIMPLEX process, 300–540 Btu/ft³ (typically 500).

An alternative to the thermal cracking processes are the partial oxidation processes, in which part of the fuel oil is burnt to supply the heat requirement of the cracking. Thus the Koppers-Totzec process previously mentioned

TABLE VIII

Efficiency of Gasification Processes Based on Total Heat Content of Fuel and Other Energy Requirements of the Processes (33).

	Thermal yield per 100 therms of total materials used		Efficiency of purified gas production, %
	Gas therms	Gas plant net product, therms	
Conventional coal gasification			
Producer gas (cold gas)	78	78	72
Carbonizing process	25	85	60
Blue water gas process	64	64	58
C.W.G. with gas oil enriching to 500 Btu/ft³	70	74	66
C.W.G. with light distillate enriching and reforming to 500 Btu	76	76	69
Integral process	59	66	59
Unconventional coal gasification			
Lurgi	77	90	—
Koppers-Totzec	72	72	—
Hydrocarbon gasification			
Cyclic reforming			
Heavy oil (500 Btu)	70	78	67
Light distillate (500 Btu)	82	82	72
Refinery gas (clean gas)	82	82	—
Continuous tubular reforming			
Methane	83	83	—
Light distillate	80	80	—
Shell & Texaco partial oxidation			
Heavy oil (500 Btu gas)	84	87	—
Hydrogenation			
Non cat. light distillate max. benzene	58	85	—
Light distillate, two-stage catalytic	90	90	—

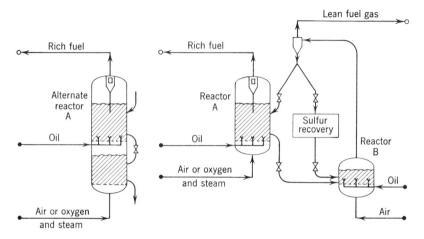

Fig. 5. High-temperature cracking process using fluidized beds of dolomite (37).

for coal has also been successfully used for cracking residual oils. There are also a number of other partial oxidation processes, such as the Shell and Texaco processes in this category. A comparison of the thermal efficiencies of these gas-making processes is given in Table VIII.

A high-temperature cracking process which simultaneously removes the the hydrogen sulfide, suggested by Squires (37), utilizes fluidized beds of calcined dolomite (Fig. 5). The upper bed is an oil-cracking zone while the lower bed is a coke-gasification zone. In the lower bed, which operates at 1000°C, air at 550°C and 40 atm is used for fluidizing and gasification of the coke which is formed in the cracking zone in the upper bed. The dolomite, which contains crystals of CaO, MgO, CaS, and $CaCO_3$, reacts with the hydrogen sulfide formed, converting CaO to CaS. The sulfur is subsequently recovered in the sulfur desorber as hydrogen sulfide, which is finally converted to sulfur in a Claus system plant. The proposed system would provide clean gas for a gas turbine plant, clean liquid fuel for an existing conventional power station, and recover byproduct sulfur. The overall efficiency of such a system, assuming two ideal energy balances, has been estimated by Squires as 95% or better. The process using dolomite for recovery of hydrogen sulfide will be discussed more fully below.

C. Processes for Removing Hydrogen Sulfide and Organic Sulfur Compounds

All the processes for gasification of oil or coal, with the exception of the Squires dolomite process, do not remove the sulfur compounds produced as part of the gasification stage, and it is therefore necessary to introduce a stage for removing hydrogen sulfide and organic sulfur compounds. These organic sulfur compounds are present in much lower quantities than hydrogen sulfide and are largely carbon disulfide, (CS_2), carbonyl sulfide (COS), and thiophene

(Tables VIIa and VIIb).

If the gas produced by oil or coal gasification is to be used directly for firing either boilers or gas turbines it would be desirable, in order to avoid sensible heat losses during cooling and reheating the gas, to clean the gases at temperatures close to the gasification temperature, that is, at about 700°C. Although there are a large number of processes for the removal of hydrogen sulfide at ambient temperatures, only one process, which involves the adsorption of H_2S on ferric oxide in a fluidized bed at 400°C, has been successfully used on a commercial scale at high temperatures. Several other processes have, however, been proposed which may be equally suitable for the adsorption process and offer better methods for the subsequent recovery of sulfur.

1. High-Temperature Processes for the Adsorption of Hydrogen Sulfide from Fuel Gases

Squires (38) has proposed the cyclic use of calcined dolomite for the desulfurization of gases as well as fuel oils. The first stage is the removal of H_2S:

$$[CaO + MgO] + H_2S \rightarrow [CaS + MgO] + H_2O$$

the H_2S is then recovered by reacting the [CaS + MgO] compound with steam and CO_2 at 15 atm:

$$[CaS + MgO] + H_2O + CO_2 \rightarrow [CaCO_3 + MgO] + H_2S$$

The dolomite [$CaCO_3$ + MgO] is calcined to give the adsorbent material:

$$[CaCO_3 + MgO] \rightarrow [CaO + MgO] + CO_2$$

The MgO fraction of the dolomite does not play a part in the chemical reaction but it is important in maintaining the physical stability of the adsorbent.

Squires reports laboratory experiments as well as thermodynamic data which indicate the feasibility of the process. At temperatures of 600–650°C and pressures of 10–15 atm an initial concentration of 1% H_2S (i.e., 10,000 ppm) in a gas mixture containing H_2 and CO was reduced to concentrations

varying between 2 and 140 ppm. Furthermore, the reacted dolomite would
react with a gas containing 82% CO_2, 9% CO, 9% H_2, and steam (giving a
CO_2:H_2O ratio of 1:1.75) at 550–600°C and 15 atm to give an effluent gas
mixture of 20–24% H_2S which is suitable for a feed gas for a Claus sulfur
recovery system.

Thermodynamic equilibrium data for the adsorption reaction (38) indicates
that satisfactory removal levels for H_2S would still be achieved at 850°C.

Temperatures of the order of 400°C at ambient pressures have been successfully used for treating coal gas with 600 gr/100 ft^3 (i.e., 0.72%) H_2S and one-
tenth or less of this of organic sulfur in the Appleby-Frodingham process
(35,39). The gases are treated in a four-stage fluidized bed absorber of iron
oxide (crushed ferric oxide ore) which removes 99.7–99.9% of the H_2S
(final concentration 10 or 20 ppm), 70–80% of organic sulfur (not including
thiophene), and 30–45% of the thiophene.

The partly sulfided and reduced ferric ore from the fluid bed is regenerated
by roasting in air at 800°C, producing sulfur dioxide for sulfuric acid manufacture. Equilibrium calculations indicate that satisfactory amounts of
H_2S may still be removed at 600°C, although the processing conditions would
be more difficult and could make this process less attractive.

A number of commercial catalysts, which may be of the zinc oxide,
promoted iron oxide, chromic alumina, cobalt–molybdenum, or activated
charcoal types, are available commercially for removing very small quantities
of H_2S and organic sulfides at temperatures ranging from ambient to 450°C,
depending on the catalyst. Until recently, they did not appear to be economical for treatment of the large quantities and comparatively high concentrations involved.

In one particular example of such a process, developed by the United
Kingdom Gas Council, organic sulfides, COS and CS_2 are converted to H_2S
over a catalyst consisting of uranium oxides (U_2O_3 and U_2O_8) supported on a
refractory bed at temperatures up to 500°C. The hydrogen sulfide is
adsorbed on extruded pellets of iron oxide from spent bauxite (Luxmasse),
which will retain 40% of its own weight of H_2S at temperatures up to 350°C.
In contrast, zinc oxide catalysts will only retain up to about 15% of their
weight.

The work by Munro and Masdin (40) with activated alumina and molecular
sieve (alkali metal alumino silicates) catalysts, indicates that a continuous
removal by these may be feasible on a commercial scale. Their investigations show that 70–95% of the H_2S (at 0.5% concentration) can be converted
continuously to elemental sulfur by catalytic reaction with sulfur dioxide:

$$2H_2S + SO_2 \rightleftharpoons 2H_2O + 3S$$

The temperature for this reaction, which is essentially the Claus-Chance process, is 250°C, and it proceeds satisfactorily with low CO content and less than 5% water vapor.   The SO_2 is added in stoichiometric quantities to the gas, and can be made by combustion of either sulfur or H_2S.

2. Ambient Temperature Absorption Processes for Hydrogen Sulfide

Numerous absorption processes for the removal of H_2S from fuel gases at ambient temperatures are described in the literature (41,42).   These involve one of two basic principles:

a. The gases are scrubbed with an alkaline reagent such as sodium carbonate, ammonia, or one of the ethanolamines, followed by the regeneration of the reagent with the production of sulfur.

b. The sulfur compounds are adsorbed on a solid adsorbent such as hydrated ferric oxide or activated carbon, with subsequent regeneration.

The latter processes were widely used for the purification of coal gas using the oxide contained in boxes, trays, or towers, finely distributed over wood shavings or mixed with peat or fibrous matter.   The sulfur was subsequently recovered by combustion or solvent extraction.   These processes, however, are not generally used in new plants.   On the other hand, scrubbing processes have been widely applied in the purification of natural gas, gas from hydrocarbon cracking, and coal gas.

The most common process for refinery or natural gas purification is the Girbotol process (Girdler Corp.) (41) using monoethanolamine (MEA) in a 15–20% aqueous solution.   Essentially, the H_2S (and CO_2) are absorbed in a plate column by the formation of a complex with the MEA.   The MEA is then passed to a stripping column where the H_2S acid gases are regenerated by heating with steam.   The H_2S concentration is adequate for conversion to sulfur.   The MEA forms a heat-stable compound with carbonyl sulfide, resulting in some loss of MEA.   The gas treated by this process, however, contains less than 1 ppm of H_2S.

A similar process, developed more recently for the higher concentration of H_2S from the natural gas on the Lacq field, is a diethanolamine process of the Société Nationale des Petroles d'Aquitaine, the SNPA-DEA process (43). The process operates in a similar way to the Girbotol process, but uses higher pressures and a higher amine concentration.   Advantages claimed are a smaller plant, less capital cost, less problems with foaming, and a ready absorption of carbonyl sulfide, which is subsequently released in the flash gases without degradation of the DEA.   The purified gas contains less than 2 ppm of H_2S.

Alkali carbonates have been used in a number of commercial processes for the removal of H_2S.   The simplest of these processes, which uses dilute sodium carbonate, is the Seaboard process (41). Regeneration is obtained by blowing air through the solution, and absorption is such that removal efficiencies of 85–95% are achieved in single-stage plants. A development of the Seaboard process is the Koppers Vacuum Carbonate process, where stripping in a vacuum reduces the steam requirement to one-sixth of the atmospheric pressure process.

The hot potassium carbonate or Benfield process (103) developed by the U.S. Bureau of Mines produces very low H_2S concentrations similar to the MEA process, but with lower steam consumption. Because hot solutions are used, increased carbonate concentrations are possible (up to 40%) and CO_2 concentrations can also be reduced.

An alternative to processes based upon carbonate solutions is the Shell potassium phosphate process in which 40–50% K_3PO_4 solution is used. It is suitable, in the same way as the hot carbonate process, for operation at higher temperatures. These processes all produce H_2S, which then has to be treated in a subsequent plant to convert it to sulfur by the Claus process.

The modified Claus (or Claus-Chance) process is of great importance in converting H_2S in acid gases from removal processes into sulfur, with overall efficiencies of 95% (44,45). The feed gas for the split flow process is typically 20% H_2S, 79% CO_2, and 1% hydrocarbons (dry basis). The split flow process is in two stages: in the first, one-third of the H_2S is oxidized to SO_2 in the combustion chamber of a modified fire tube boiler; because of the highly exothermic reaction:

$$\tfrac{1}{3}H_2S + \tfrac{1}{2}O_2 \rightarrow \tfrac{1}{3}SO_2 + \tfrac{1}{3}H_2O$$

The SO_2 is then reacted with H_2S over a catalyst in a converter:

$$\tfrac{1}{3}SO_2 + \tfrac{2}{3}H_2S \rightarrow S + \tfrac{2}{3}H_2O$$

This is cooled to about 150°C, scrubbed to remove sulfur, reheated, and passed over through a second converter, before the tail gases are vented to the atmosphere. About $2\tfrac{1}{2}$ lb of steam are produced for each pound of sulfur. Plants ranging in capacity from 1 to 700 tons/day have been built (1965) and plants with capacities to 50 tons/day are widely used.

Instead of producing sour gas from a scrubbing process which needs further treatment, there are also a number of scrubbing processes where the recovery of the reagent simultaneously produces sulfur. Early processes of this type involved alkaline iron oxide suspensions; the basic chemistry is the following:

a. Absorption

$$H_2S + Na_2CO_3 \rightarrow NaHS + NaHCO_3$$

b. Recovery of absorption reagent

$$Fe_2O_3 \cdot 3H_2O + 3NaHS + NaHCO_3 \rightarrow Fe_2S_3 \cdot 3H_2O + 3Na_2CO_3 + 3H_2O$$

c. Recovery of oxide and production of sulfur

$$2Fe_2S_3 \cdot 3H_2O + 3O_2 \rightarrow 2Fe_2O_3 \cdot 3H_2O + 6S$$

The first two stages are carried out in an absorption tower, while the last stage is carried out in an oxidation vessel or "thionizer." A number of these processes, called the Burkheiser, Gluud, Ferrox, and Manchester have the same basic mechanisms. They differ in the number of absorption stages and the method of introducing air in the thionizer. The Ferrox process used a tall absorber with shallow thionizers and has been used to treat H_2S concentrations of 420 ppm. The Manchester process, which is similar, treated H_2S concentrations of 250 ppm, using a tall thionizer, reducing these to 5 ppm. Sulfur is skimmed from the froth on the thionizer column. The problems associated with the process were due to the formation of metal thiocyanates and thiosulfates.

More effective and economical processes have been developed using other reagents which avoid these undesirable byproducts. One uses a thioarsenate intermediate while the other uses a metavanadate complex. In the Thylox process, which was first demonstrated in 1929, H_2S is absorbed in sodium thioarsenate:

$$H_2S + Na_4As_2S_5O_2 \rightarrow Na_4As_2S_6O + H_2O$$

This step is followed by atmospheric oxidation, producing sulfur which can be skimmed from the thionizer:

$$2Na_4As_2S_6O + O_2 \rightarrow 2Na_4As_2S_5O_2 + 2S$$

Sodium carbonate (or ammonia) is added to control the pH for satisfactory operation and the product sulfur contains very little arsenic. Operating plants reduce the H_2S from 3800-ppm concentrations to 100 ppm with a primary absorber, while lower exit concentrations can be achieved with reduced input concentrations.

The recent Giammarco-Vetrocoke process (46) also uses complex arsenic compounds, but the H_2S reacts with the arsenite to form the thioarsenite, which subsequently reacts with arsenate to give the monothioarsenate. This compound decomposes to elemental sulfur and the arsenite by acidification with CO_2 under pressure. The sulfur is filtered off and the arsenite oxidized in part to arsenate by atmospheric oxygen, for recycling. The process produces a gas with less than 1 ppm H_2S and also avoids undesirable side reactions associated with the Thylox process.

The Stretford process (47–49), which has been used in numerous plants since the initial trials in 1959, avoids the difficulties in handling of the highly poisonous arsenites on a large scale. Furthermore, it is capable of reducing the H_2S in gases containing initial concentrations from 100 to 10,000 ppm to a final concentration of 1 ppm in exhaust gases. Plants with capacities up to 50 million ft^3/day have been constructed.

In the Stretford process the H_2S is absorbed by an alkaline solution (pH 8.5–9.5) containing sodium ammonium vanadate and anthraquinone 2:6 and 2:7 disulfonate (ADA); the two reagents are present in equimolar concentrations. (In earlier plants the metavanadate was used instead of the ammonium vanadate.) In addition, Rochelle Salt (potassium sodium tartrate) is added to prevent precipitation of the vanadate. The reactions may be summarized as follows:

a. Absorption:

$$H_2S + Na_2CO_3 \rightarrow NaHS + NaHCO_3$$

b. Production of sulfur:

$$2NaHS + H_2O + 4NaVO_3 \rightarrow Na_2V_4O_9 + 4NaOH + 2S$$

c. Recovery of the vanadate with ADA:

$$Na_2V_4O_9 + 2NaOH + H_2O + 2ADA \rightarrow 4NaVO_3 + 2ADA \text{ (reduced)}$$

d. The reduced ADA is oxidized with atmospheric oxygen:

$$2ADA \text{ (reduced)} + O_2 \rightarrow 2ADA + H_2O$$

The Stretford process plant is simple, consisting of an absorption column filled with conventional packing, a delay tank, and a sulfur extraction plant. The process has been used at pressures up to 20 atm (50) and so would be suitable for H_2S removal after pressure gasification or hydro-desulfurization without lowering the pressure for pipeline purposes. The sulfur produced is of comparatively high quality, with some of the reagent occluded. This can be largely removed by washing, although considerable quantities of water are needed.

There are, therefore, a number of processes available that will remove H_2S and leave only very small amounts of sulfur compounds in the resultant gas.

IV. REMOVAL OF SULFUR OXIDES

The concentration of sulfur oxides in flue gases depends on the sulfur in the fuel, the amount of air, and the amount of sulfur that is taken up by the ash. Thus, a fuel oil containing $3\frac{1}{2}\%$ sulfur, being burned with 15% excess air, will produce about 0.2% SO_2, while a 4% sulfur coal, with similar excess air, will produce 0.25% SO_2.

On the other hand, the SO_2 concentrations from the roasting of sulfide ores may be as high as 7–14%, a suitable concentration for conventional contact-process sulfuric acid manufacture. The concentration depends on the type of ore and the roasting temperature. The waste gases from such plants are generally treated to remove 96–98% of the SO_2 from the gases, so that 0.2–0.5% remains in the tail gases (2000–5000 ppm), which is the same range of concentration found in gases from combustion of medium to high sulfur content fuels (51). The Bayer double contact process, which uses intermediate absorption of acid and further catalysis stages, converts 99.8% of the SO_2 to SO_3, so that the tail gas concentrations are 100–120 ppm at normal loads, and do not exceed 240 ppm even at 20% overload.

There are a number of alkali scrubbing processes for high SO_2 content gases. These processes can reduce the SO_2 in the tail gases to 500–1000 ppm, which is probably the highest level tolerable in large volumes of these gases.

Processes for dealing with dilute SO_2 from combustion gases fall into several categories. Low-temperature scrubbing processes are effective, but leave a tail gas at ambient temperature which tends to settle at ground level near the source. Processes where the gases are treated at high temperatures are, therefore, preferable as a cold plume is not formed. This last group of processes may involve the addition of an adsorbent material with the fuel, or the passage of flue gases through an adsorbent bed. The latter processes are suitable for recovery of the sulfur as a useful byproduct.

A. Scrubbing of Combustion Gases with High Concentrations of Sulfur Dioxide

The processes, which generally are used for smelter gases, or the gases from sulfide ore roasting plant, use either aromatic amines or ammonia for the primary absorption stage. The aromatic amines successfully used in commercial plant are dimethylaniline and xylidine–water mixtures. The former being used in the ASARCO (American Smelting and Refining Co.) process, the latter in the Lurgi Sulfidine process. Absorption isotherms (Fig. 6) show that dimethylaniline is a more efficient absorbent for SO_2 concentrations above 3.5%.

In the ASARCO process (Fig. 7), SO_2 is absorbed in dimethylaniline in the lower stages of a double column. The SO_2 free gas containing dimethylaniline is scrubbed out with dilute sulfuric acid, leaving a tail gas containing about 500 ppm SO_2 and the dimethylaniline sulfate, which is fed to the regenerator in the lower section of the stripper-regenerator column. The SO_2-rich dimethylaniline is stripped with steam and the resultant SO_2 passed to storage after scrubbing and drying by washing with 98% sulfuric acid, while the dimethyl aniline is returned.

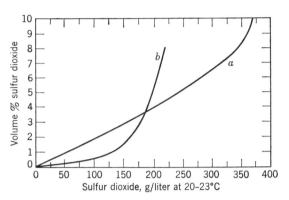

Fig. 6. Absorption isotherms for sulfur dioxide in (a) anhydrous dimethylaniline and (b) 1:1 xylidine–water mixture (42) (basis of Sulfidine process).

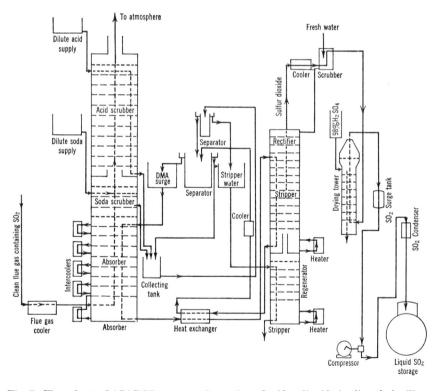

Fig. 7. Flow sheet of ASARCO process: absorption of sulfur dioxide in dimethylaniline solution (42).

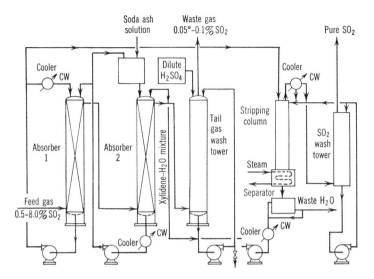

Fig. 8. Flow sheet of Sulfidine process (42): absorption of sulfur dioxide in xylidine–water mixtures.

Water from this section and sodium sulfide are added to the dimethylaniline sulfate in the lower section, which produces dimethylaniline and SO_2 (passed up to the stripping section) and sodium sulfate solution.

The Lurgi Sulfidine process (Fig. 8) is similar in principle, but a water-soluble xylidine sulfate compound is formed in the two-stage absorber. The stripping in this case is carried out by indirect heating of the rich liquors, and the tail gases average 500–1000 ppm SO_2.

The use of aqueous ammonia was considered even in the last century for SO_2 removal, but a commercial plant has only operated since 1936, at the Consolidated Mining and Smelting Co. Ltd. at Trail, B.C. The plant is known as the Cominco Sulfur Dioxide Recovery Process (Fig. 9) and has been used for reducing the SO_2 in zinc roaster plant gases from 5.5% to less than 0.2% and the tail gases from acid plants from about 1% to about 300 ppm.

About 30% of aqueous ammonia is used as an absorbent in the first stage of a two-stage absorption, followed by water scrubbing to remove ammonia from the tail gases. The temperature must be kept below 35°C to minimize these losses. The SO_2 is converted to the hydrosulfite (NH_4HSO_3) in solution and this solution is stripped of SO_2 with the addition of 93% sulfuric acid, producing ammonium sulfate and free SO_2. The removal efficiency for SO_2 varies from 85% (on a lead sinter plant) to 97% on a zinc roaster and an acid plant.

Basic aluminum sulfate has been used by I.C.I. and the Boliden Co. for the recovery of SO_2 from copper smelter gases containing about 2.5–10% SO_2 (53). At the Imatra Smelter of the Outakumpu Copper Company, four large absorption towers were used in which the aluminum sulfate was saturated at 20–30°C. The SO_2 was then regenerated in a tower where the saturated liquors were heated to 75–80°C, the final stripping tower temperature being 90°C. The SO_2 is then cooled by water sprays, dried with sulfuric acid, and liquified to give a very pure product.

The Lurgi Sulfacid Process (83) is a catalytic oxidation process in which gases containing 0.1–1.5% SO_2 are converted to sulfuric acid in a fixed catalyst bed consisting of an activated carbon catalyst at a temperature of 60–70°C (Fig. 10). The waste gases from a roaster or other combustion processes are cooled and passed to the catalyst beds in which the conversion takes place. The beds are contained in rubber-lined cylindrical vessels supported on hurdles. The acid formed in the catalyst is washed out and concentrated with the heat from the incoming gases to 65–70%. Removal efficiencies are of the order of 95% so that the maximum SO_2 in the tail gases is below 750 ppm with $1\frac{1}{2}$% SO_2 in the feed, or 150 ppm if approximately 0.3% SO_2 is in the feed gas.

In order to ensure trouble-free operation of these processes, the incoming gas stream must be carefully cleaned of particulate matter. In the case of the Sulfacid process this is carried out in a vertical Venturi scrubber, the

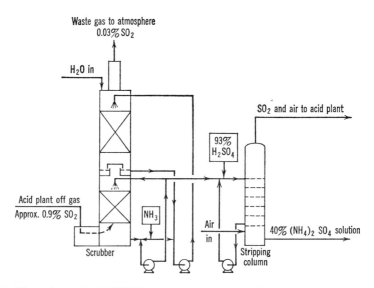

Fig. 9. Flow sheet of COMINCO process: absorption of sulfur dioxide in ammonia solution (42).

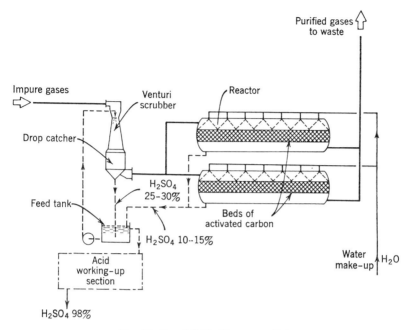

Fig. 10. Lurgi Sulfacid process (84).

Sulfidine process uses electrostatic precipitators, but the ASARCO and Cominco processes do not specify the system used in actual plants.

B. Scrubbing of Flue Gases with Low Sulfur Dioxide Concentrations

The scrubbing processes considered so far, with the exception of the Sulfacid process, are not suitable for the very low concentrations of SO_2 found in flue gases. A considerable number of scrubbing processes, using either water or aqueous solutions of alkaline materials, have been investigated during the past 30 or more years and several of these have been applied on commercial scale plant. Their wider application has been limited for a number of reasons, not the least being the low-temperature plume following a scrubbing process, which causes marked local pollution. On plants such as the Battersea and Bankside power stations it is common practice to stop scrubbing flue gases during prolonged inversions so that the plume can penetrate the inversion and disperse at higher levels.

In scrubbing (or otherwise treating) flue gases, the aim is to reduce the SO_2 concentration by about nine-tenths so that the residual SO_2 concentration is below 300 ppm. Lower levels are not readily achieved in the type and

size of plant designed for this purpose because of the comparatively low
solubility of SO_2. This should be contrasted to the processes available for
H_2S removal, where high efficiencies (better than 99.5%) are common.

The use of water for scrubbing combustion waste gases has been considered
by Spengler (2) and more recently by Germerdonk (54). Calculations by
these workers, based on absorption data for SO_2, indicate that at temperatures of 20–30°C there is a possibility of absorbing reasonable amounts of SO_2,
while above 50°C the equilibrium constant is no longer favorable. In
experiments by Germerdonk (54) gases with a SO_2 concentration of 1750 ppm
were scrubbed with water at 21.3°C in a ten-stage column. Reduction to
175 ppm was achieved if water at a rate of 11 liters/m³ flue gas (i.e., 0.087
U.S. gal/ft³) was used. This worker pointed out that, while alkaline solutions
will absorb SO_2 more readily than the recirculating acid solutions (pH 2–6) in
which some SO_2 is dissolved, these solutions will also absorb carbon dioxide,
leading to waste of reagent.

Another suggested process (2) which was patented in 1933 was the absorption of SO_2 by water at 3 atm in a column packed with charcoal. The SO_2
is stripped by air in a second column at atmospheric pressure, producing a
mixture of air with 20–25% SO_2. This can be converted to elemental sulfur
in a subsequent process. These processes have not been put into practice.

The use of naturally occurring water with slight alkalinity is the basis of
the scrubbing processes at Battersea and Bankside power stations. Thames
river water, to which some chalk slurry is added to increase alkalinity, is
used as the scrubbing liquor (55). The scrubbing towers are packed with
wooden (cedar and teak) grid packing while the whole plant is lined with
acid-resisting materials. The liquors from the scrubbing towers are oxidized
by aeration and the addition of manganese sulfate. A flow sheet is shown in
Figure 11; 95% SO_2 removal is achieved. The waste liquors are discharged
into the river. This discharge, however, increases the calcium sulfate
concentration, which can form scale in boilers and heat exchangers in plants
downstream, as well as affecting marine life in the river.

A pilot plant with 3.3 ft (1-m) diameter scrubber using natural waters
with a pH of 7 has been tested in the Kanagawa Prefecture in Japan (56).
This plant reduces the SO_2 concentration in flue gases from 1300 to 80 ppm.
The Kanagawa process appears to depend on a new type of gas–liquid
contactor.

Several processes using lime have been developed in England and in
Japan. The Howden-I.C.I. cyclic lime process was developed in 1935 and a
plant was constructed at the Fulham power station (57) (Fig. 12). A 5–10
wt% chalk or lime slurry is circulated through a wooden grid scrubbing
tower and the resultant solution is passed to a delay tank where pH is
adjusted and calcium sulfate crystallizes. The sulfate crystals are collected

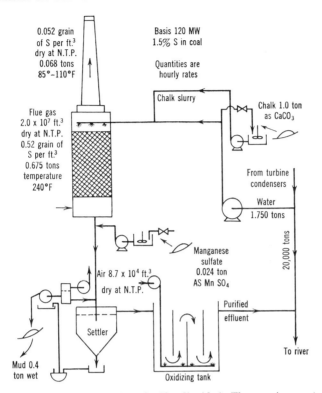

0.052 grain
of S per ft.³
dry at N.T.P.
0.068 tons
85°–110°F

Basis 120 MW
1.5% S in coal

Quantities are
hourly rates

Chalk slurry

Chalk 1.0 ton
as CaCO₃

Flue gas
2.0 x 10⁷ ft.³
dry at N.T.P.
0.52 grain of
S per ft.³
0.675 tons
temperature
240°F

From turbine
condensers

Water
1.750 tons

20,000 tons

Air 8.7 x 10⁴ ft.³
dry at N.T.P.

Manganese
sulfate
0.024 ton
AS Mn SO₄

Purified
effluent

To river

Settler

Mud 0.4
ton wet

Oxidizing tank

Fig. 11. Battersea process: absorption of sulfur dioxide in Thames river water to which an alkaline slurry has been added (42).

after settling out, but because of their degree of contamination with fly ash they cannot be used commercially.   Japan Engineering Co. have patented a process (58) (Fig. 13) which is similar in principle, but produces high-purity gypsum (CaSO₄) crystals.   A simplified process which is lower in cost has also been described, but the product from this is only suitable for landfill (58).

Ammoniacal solutions are used in some processes for flue gas treatment and can reduce the SO₂ concentration to as low as 5 ppm.   The Cominco Process has been discussed in connection with high-concentration smelter gases, but other ammoniacal liquor processes treating the dilute flue gases have been developed by Simon Carves and Mitsubishi Heavy Industries, the former producing elemental sulfur and the latter ammonium sulfate crystals.

The Fulham-Simon Carves Process (59,60) (Fig. 14) involves scrubbing of flue gases with ammoniacal liquors from coal gas plants.   The ammonium sulfite liquors from the scrubbing towers are subsequently autoclaved at 200 psi and 170°C for 3 hr producing ammonium sulfate and sulfur:

$$2NH_4HSO_3 + (NH_4)_2S_2O_3 \rightarrow 2(NH_4)_2SO_4 + 2S + H_2O$$

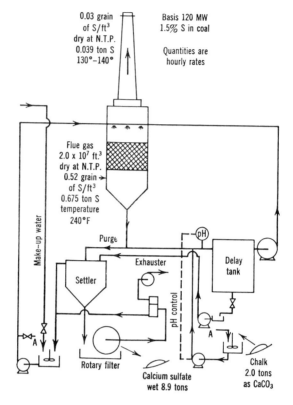

Fig. 12. Howden—I.C.I. cyclic lime process: absorption of sulfur dioxide in lime slurry liquor (42).

The first plant was installed at the Fulham power station in 1939 and a second, larger plant, treating 56,000 ft³/min, was commissioned at the North Wilford Power Station (Nottingham) in 1957 (61). This handles the combustion waste gases from 120 tons of coal with 3.0% sulfur and produces 11 tons of ammonium sulfate as well as 2000 lb elemental sulfur each day.

The Mitsubishi Ammoniacal Liquor Process (58) (Fig. 15) treats dusty flue gases and removal efficiencies of 95% are claimed for gases containing 0.1–0.2% SO₂. There is only a slight loss of ammonia. The ammonium sulfate liquors are supplied to the crystallizer in a 45 wt% solution by using a waste heat recovery process, so that no evaporator is necessary.

Another process which also produces ammonium sulfate is the Mitsubishi Manganese Oxyhydroxide Process (58) (MnO·OH). A 3% manganese oxyhydroxide slurry is used as an absorbent which is recovered by supplying ammonia and oxygen to the sulfide complex. Ammonium sulfate

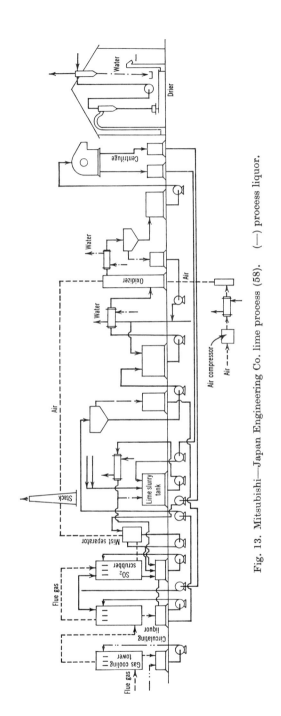

Fig. 13. Mitsubishi—Japan Engineering Co. lime process (58). (—) process liquor.

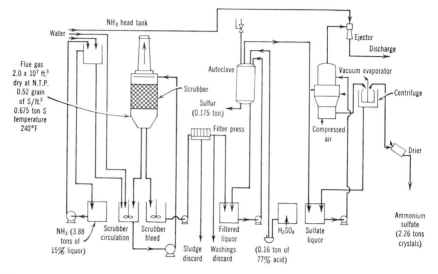

Fig. 14. Fulham-Simon Carves process: absorption of sulfur dioxide in ammonia liquor from gas works or coke ovens, with production of sulfur and ammonium sulfate (42).

is produced and the slurry is recycled. Advantages claimed for this process are that there are no losses of ammonia, and the slurry, because of the lower concentration of reagent, behaves as a solution. Furthermore, an ammonium sulfate solution with 45% weight concentration is formed so that no further evaporation is required. Efficiencies of SO_2 removal of 97% are claimed with effluent gases containing 0.1–0.2% SO_2, producing tail gas concentrations of 30–60 ppm SO_2.

One possible process, which was investigated in detail by Johnstone and Singh, involved the absorption of SO_2 by sodium sulfite solutions which has a much higher absorptive capacity than water, but not as high as the ammoniacal liquors. The process employs a reagent and an SO_2 recovery step involving zinc oxide and is therefore known as the Zinc Oxide Process (62,63). In the first stage, SO_2 is absorbed and the bisulfite formed:

$$SO_2 + Na_2SO_3 + NaHSO_3 + H_2O \rightarrow 3NaHSO_3$$

This solution is then reacted with zinc oxide, to precipitate zinc sulfite:

$$NaHSO_3 + ZnO \rightarrow ZnSO_3(ppct) + NaOH$$

The zinc sulfite is filtered, dried, and calcined in a flash calciner to give SO_2 and ZnO:

$$ZnSO_3 \rightarrow ZnO + SO_2$$

The gases containing 30% SO_2 and 70% water are dried and the SO_2 is liquefied. One difficulty with this process is the formation of sulfate as well

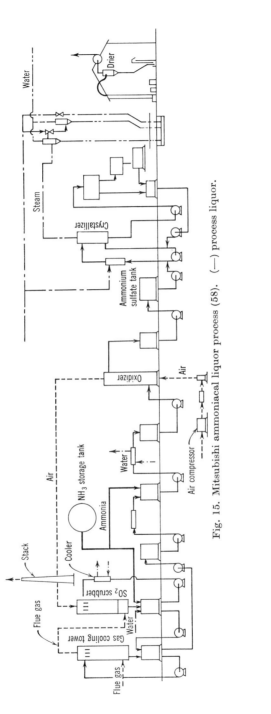

Fig. 15. Mitsubishi ammoniacal liquor process (58). (—) process liquor.

as sulfite which reduces the absorptive capacity of the solution if the sulfate concentration rises. The losses in this process amount to about 10% and so a desulfating system is employed. Lime is added, forming calcium sulfate (which is removed) and sulfite (which, after adding gaseous SO_2, forms the bisulfite). This reacts with quicklime to form a calcium sulfite precipitate and sodium sulfite liquor which returns to the absorption towers.

A recent development by Mitsubishi Heavy Industries Ltd. is the Red Mud Process (58), in which the Red Mud (also called Luxmasse) is used as an absorbent. This mud is a waste product in the extraction of alumina from bauxite and so is an extremely low-cost material. A typical composition is Al_2O_3: 18.9%; SiO_2: 17.4%; Na_2O: 8.3%; Fe_2O_3: 39.3%; TiO_2: 2.8%; ignition losses: 10.5%.

It is the complex of sodium aluminum silicate which is the active reagent in absorbing SO_2. After the absorption stage, the slurry which contains the absorbed SO_2 as sodium sulfate solution can either be disposed of as waste or it can receive further treatment to give commercial by-products. This treatment consists of removing all sodium hydroxide and sodium sulfate, and then treating the residues with sulfur dioxide. Three types of materials are formed from the residues which can be readily separated and have the following compositions:

    i. Al_2O_3: 99.2%; SiO_2: 0.8%
    ii. SiO_2: 91.9%; Al_2O_3: 8.1%
    iii. Fe_2O_3: 76.8%; TiO_2: 5.5%; Al_2O_3: 7.4%; SiO_2: 10.5%

The alumina (*i*) and silica (*ii*) separated in this way are pure white powders, free from iron oxides, and so can be used for industrial raw materials. The sulfur dioxide is used in a recirculating system, and only small losses occur.

The early absorption towers constructed at Battersea and elsewhere were all very large and filled with a wooden packing. Considerable research effort in recent years has been devoted to the development of more suitable packings and towers. A spiral wire packing has been tested with an ammoniacal liquor absorption system by Klimecek et al. (64), which has much better characteristics than conventional Raschig ring packing. This packing permits a superficial velocity five times that of the conventional packing, and the height of a transfer unit center reduces to two-thirds of the previous height. Furthermore, the cost is only 30% of the conventional packing.

A multistage turbulent contact absorption column has been tested with sodium and calcium carbonate solutions (65). The turbulent contact absorber uses a nonflooding packing consisting of low-density spheres placed between retaining grids sufficiently far apart to permit turbulent and random motion of the spheres. This arrangement gives the unit a high absorption

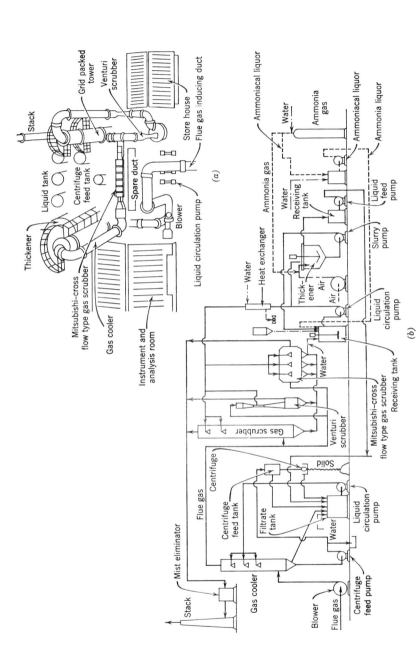

Fig. 16. Mitsubishi processes pilot plant (58). (a) Explanatory sketch of pilot plant. (b) Flow diagram of pilot plant; (—) absorbent.

rate, at high gas and liquid velocities, with small pressure drops. The equipment is of the nonclogging type and very useful in dealing with dusty gases or in cases where a solid is formed in the reaction. In the pilot plant tests on power station flue gases, where a four-stage absorber was used, sodium carbonate solutions removed 88–96% of the SO_2 while calcium carbonate removed 78–87% SO_2.

Mitsubishi have developed a cross flow scrubber (66) which consists of a series of horizontal, shallow, V-shaped trays close packed in vertical rows so that the gases penetrate the liquid surface. The absorption rate of sodium sulfite solutions was 90–95% SO_2 with inlet gas concentrations of 0.11–0.24% SO_2. Pressure drops were 2.75–6 in. WG, which is much lower than that of a Venturi scrubber with similar capacity, but higher than for a grid packing.

To overcome the cold plume problem inherent in all cold wet scrubbing processes, it has been suggested (58) that only about 30% of the flue gas should be treated by these means. The remaining 70% should be mixed directly with the treated gases, giving a slightly reduced SO_2 content in the waste gases. This method has been shown to avoid a visible white plume, as well as reducing the corrosive effect of the plume in the stack. An alternative scheme, which may prove less costly, is to treat 70% of the flue gas with a high-temperature process at reduced removal efficiencies, and, on mixing, achieve adequate lowering of SO_2 levels at elevated temperatures in the plume.

An unconventional scrubbing process, using a mixture of lithium, sodium, and potassium carbonates, in a melt at about 425°C as the scrubbing medium, is being investigated by Atomics International (104). The mixture, which melts above 400°C, is a clear, mobile liquid and readily absorbs better than 99% of the SO_2 from a gas mixture with 0.3–3% SO_2, leaving 20 ppm in the tail gases. The reactions are

$$SO_2 + M_2CO_3 \rightarrow M_2SO_3 + CO_2$$

and

$$SO_3 + M_2CO_3 \rightarrow M_2SO_4 + CO_2$$

(M represents the alkali metal ions). The regeneration system, operating at 600°C, produces H_2S in a two-stage process. First, the sulfate and sulfite are reduced to the sulfide by using producer gas $(CO + H_2)$:

$$M_2SO_3 + 2CO + H_2 \rightarrow M_2S + 2CO_2 + H_2O$$

$$M_2SO_4 + 2CO + 2H_2 \rightarrow M_2S + 2CO_2 + 2H_2O$$

Second, the sulfide is reformed to carbonate and H_2S with steam and CO_2:

$$M_2S + H_2O + CO_2 \rightarrow M_2CO_3 + H_2S$$

Standard methods of turning H_2S into elemental sulfur can be used for obtaining a useful product.   Fly ash is also retained by the melt and must be filtered out before contacting the gases in the scrubber.   If filtration of the melt is attempted, the fly ash occludes carbonate and causes serious reagent losses.   Preliminary economic assessment of the process, which is included below, shows that if the problems of handling the molten carbonates on a large scale can be overcome, then this may be a promising process.

A scrubbing process for which few details are available has been investigated on a pilot plant by Wellman-Lord Inc. at the Gannon Power Plant of the Tampa Electric Co. (106).   The process is claimed to remove 90% of the SO_2 and SO_3, together with the residual fly ash after electrostatic precipitation.   Following reaction and treatment, pure SO_2 is stripped off for use in an acid plant or for sulfur recovery.

C. Adsorption and Catalysis of SO_2 on Solids

Scrubbing processes, with the exception of the hot molten carbonate processes, are not suitable for removal of SO_2 at elevated pressures because of the cold plume problem referred to in conjunction with the Battersea and Bankside power stations.

Therefore, in recent years intensive investigations have been carried out in the United States, England, Germany, and Japan of selective adsorbents for SO_2.   The processes can be classified into those in which the SO_2 is absorbed on a very cheap material, and the resultant compound subsequently discarded, or those which use a more expensive adsorbent which can be subsequently regenerated with the production of either sulfur, sulfuric acid, or ammonium sulfate.

There are a number of systems possible for adsorption processes.   They are described below:

1. The bed is fixed and acts as a catalytic converter, in which the SO_2 gas is oxidized to SO_3, which is subsequently reacted with ammonia and/or water vapor to form either ammonium sulfate or sulfuric acid crystals.

2. The bed is fixed and acts as an absorbent, with or without the oxidizing action.   The bed is withdrawn and treated subsequently in a secondary system for recovery of the sulfur (in some form) and regeneration of the bed material.

3. The adsorbent solid is injected into the combustion system downstream from the burner in powder form and reacts as it is carried along.   The spent adsorbent is collected either on a filter or in a scrubber, where there may be a further significant scrubbing action.

4. The adsorbent solid is injected with the fuel. As it passes through the combustion zones in the flame, over a range of temperatures, it may affect the SO_2/SO_3 balance in the flames, the corrosive nature of the flue gases in the high-temperature regions of the boiler and the radiation characteristics of the flame, as well as reducing SO_2.

1. Catalytic Conversion of SO_2 in Fixed Beds

The oxidation of sulfur dioxide to the trioxide can be described by the simple equilibrium:

$$SO_2 + \tfrac{1}{2}O_2 \rightleftharpoons SO_3 + 22.6 \text{ kcal g mole}^{-1} \; (\Delta H)$$

The equilibrium constant for this reaction is

$$K_p = \frac{p_{SO_3}}{p_{SO_2} \cdot p_{O_2}^{1/2}}$$

The equilibrium constant (Table IX) shows that below 800°K the equilibrium favors the SO_3 while at flame temperatures (above 1500°K) the equilibrium favors SO_2. The equilibrium has been extensively investigated and reviewed (67) because of the commercial importance of the process for making sulfuric acid. Rate studies and catalyst investigations are also extensive.

It is important to note that the amount of sulfur trioxide formed in the flame tends to be greater than the equilibrium value for the molecular reaction equilibrium when excess oxygen is present. This increased concentration is probably due to the effect of atomic oxygen in the flame (69). Homogeneous oxidation of SO_2 by the oxides of nitrogen, without the presence of catalysts, may also play a part at the lower temperatures (900–1050°C) which can occur in the later stages in the boiler system (70). Jüngten (68) also reports that steel ducts in an experimental furnace at temperatures of 400–600°C and with excess air may act catalytically, increasing the SO_3 content to the order of 1% of the SO_2 present.

TABLE IX

Equilibrium Constant K_p for the Reaction
$SO_2 + \tfrac{1}{2}O_2 \rightleftharpoons SO_3$

Temperature		K_p	
°K	°C	Jüngten (68)	Williams (67a)
600	327	4780	4015
800	527	53.0	32.3
1000	727	3.67	1.83
1200	927	0.630	0.276
1500	1227	0.114	—

However, to use the conversion of SO_2 to SO_3 as the basis for a method for removing sulfur oxides from flue gases, it is necessary to achieve conversions in excess of 90%.   Bienstock, Field, and Myers (71) used a simulated flue gas with 0.35% SO_2 and commercial oxidation catalysts as well as one of their own composition.   The commercial catalysts, with one exception (Vanadia on silica, K_2O promoted: Harshaw catalyst, V0204) converted less than 50%.   The exception converted up to 85% at 380°C.   The special alkalized vanadia catalyst (SiO_2, 39.6%; K_2O, 16.5%; SO_3, 27.3%; V_2O, 7.1%; H_2O, 9.5%) was dried at 430°C and prepared in a granular form.   This catalyst converted 97% of the SO_2 at 365°C at flue gas compositions and would appear to give a reasonable basis for a catalytic conversion process.

A process of this type, the Penelec process (Fig. 17) was developed by Pennsylvania Electric Company in conjunction with the Air Preheater Company, the Monsanto Company, and Research-Cottrell Incorporated.

A pilot plant utilizing a commercial conversion catalyst has been built and was tested at the Pennsylvania Electric Company's Seward Station in 1961 (72).   The pilot plant had a capacity of 1500 ft³/min at 540°C.   After the boiler, the fly ash is removed in an electrostatic precipitator operating at 485°C with an efficiency of 99.9% so that the catalyst bed is not prematurely clogged or fouled.   The clean gas, containing 2000 ppm SO_2 and 20 ppm SO_3 is passed through a thin catalyst layer at 470°C with conversion efficiencies in excess of 90%.   The gases now contain 1820 ppm SO_3 and 200 ppm SO_2. The SO_3-enriched flue gases enter a rotary air preheater where their temperatures are lowered to 96°C.   A substantial amount of the acid mist formed during the cooling is deposited on the cold surfaces of the air preheater and removed.   The flue gases are then passed through an electrostatic mist

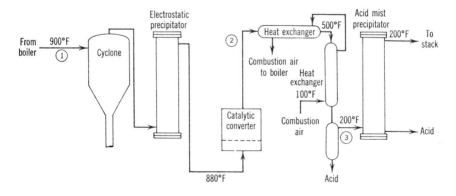

Fig. 17. Penelec catalytic oxidation process flowsheet (simplified) (72).

collector where virtually all the SO_3 is collected, and only 20 ppm SO_3 with 200 ppm SO_2 are left in the tail gases which are passed up the stack.

A second, larger prototype plant, treating 24,000 ft³/min has been tested at the 250-MW Portland No. 2 Unit of the Metropolitan Edison Company (110). The high temperature electrostatic precipitator, designed to remove 99.5% of the fly ash, was followed by the catalytic converter, tubular heat exchanger (boiler economizer), a Ljungstrom regenerative air preheater, absorption tower, and a Monsanto *Brink* mist eliminator before ejecting the waste gases via the induced draft fan to the stack. This prototype unit, which had been in operation for 4000 hours, including a continuous run of 24 days by late 1968, has demonstrated its ability to remove virtually all fly ash, as well as 90% of the SO_2, and 99.5% of the sulphuric acid produced. The acid has an average concentration of 80%, which varies slightly with stack temperature.

An alternative catalytic oxidation process in an advanced stage of development was suggested by Kiyoura (73,74) of the Tokyo Institute of Technology (Kiyoura T.I.T. Process) (Fig. 18). In this process ammonia is used, mixed with the oxidized SO_2 in the flue gases at 220–260°C to form ammonium sulfate. The air preheater is built as a two-stage unit, and the ammonia injection takes place in a Venturi mixer between these stages. After

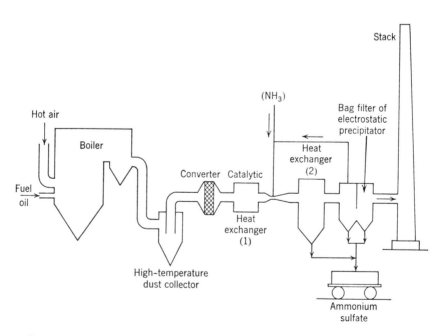

Fig. 18. Kiyoura-T.I.T. catalytic oxidation and ammonia injection process (73).

successful bench-scale tests, a 300 ft³/min pilot plant has been tested.   The plant is connected to a boiler burning a fuel oil with 3.5% sulfur and the hot flue gases are cleaned with ceramic filters before being passed to a vanadium pentoxide catalyst placed in three layers at temperatures of 420–450°C. Conversion is 91–93% at a space velocity of 50 ft³/(ft)² (min); higher velocities (up to 115 ft³/(ft)² (min)), were reported for the bench-scale plant.   The hot gases are cooled in a high-temperature heat exchanger, preheated ammonia is introduced, and further cooling in a tubular cooler to 140°C takes place. The ammonium sulfate is precipitated in an electrostatic precipitator operating at 59–63 kV with an efficiency of 97.5%.   The overall recovery is, therefore, 90%, and the purity of the ammonium sulfate is 99.2–99.6%, which approximates to reagent-grade material.   The pilot plant has operated continuously for several months.   It could appear that one advantage of this process is that the product is noncorrosive and so the corrosion problems inherent in the sulfuric acid product system are partially avoided.

The purity of the ammonium sulfate is in contrast to the predictions made earlier by Johswich (75), who felt that the impurities, particularly tarry substances, in the flue gas would act as catalyst poisons for the catalysts which operate below 300°C, as well as contaminating the product.

2. Adsorption of SO_2 in Fixed, Fluidized, and Entrained Beds

a. Dolomite

The adsorption of SO_2 in a fixed bed at high temperatures has been tested in a number of ways.   Initially, it was hoped to use a cheap, naturally occurring material such as dolomite (calcium magnesium carbonate) to adsorb SO_2.   Jones (76) and Coke (77) used beds of crushed dolomite ($\frac{1}{8}$–$\frac{1}{4}$ in.) to remove SO_2 in a flue gas at 600°C with a space velocity of 630 ft³/(ft)²(min). The efficiency of removal was 90% until the surface of the dolomite had been converted to the sulfate and 15% of the dolomite had been used.   Squires (78) discusses the reaction in some detail and indicates that calcium sulfate is formed initially and magnesium sulfate later, aided by the catalytic presence of iron oxide.   Squires proposes half calcining of the dolomite to make it porous, allowing gases to penetrate the grains of adsorbent.

To increase the surface area of the dolomite available for reaction requires crushing the material to finer particle sizes, which require an entrainment reactor rather than a fixed or fluidized bed.   Recent pilot-plant experiments by Jüngten and Peters (79) used this approach (Fig. 19).   The reactor is a tube, 20 ft long, 4-in. diameter, heated externally by gases flowing in the annulus made by a larger surrounding pipe.   The tube can be maintained at temperatures between 200 and 1100°C, and dolomite as well as limestone particles are fed in at the base of the reactor.   The reacted particles are

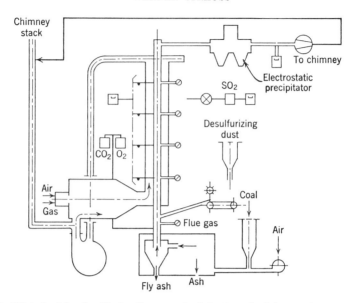

Fig. 19. Pilot plant for desulfusing flue gases by injection of calcium carbonate dust (79).

collected by an electrostatic precipitator.   The effective residence time varied between 1 and 4 sec and the dolomite and limestone were used in stoichiometric ratios from one to three times the amount required for flue gases containing 1200 ppm.   The results of the experiments at 500 and 800°C (Table X) indicate that dolomite under these conditions would not be satisfactory, but limestone, particularly precipitated and cooled calcium carbonate, might prove a suitable adsorbent.

A further process using an entrainment reactor is due to Still (87), where calcium hydroxide is introduced at temperatures below 450°C into a series of three vertical inverted U tube reactors (Fig. 20).   The gases from each inverted U tube are passed into a cyclone which removes the reagent, which is then transferred back to the first leg of the U tube.   It is expected that SO_2 removal efficiencies better than 95% will be achieved in a pilot plant currently undergoing test.

b. Alkalized Alumina

The limited success of the early dolomite adsorption experiments led the U.S. Bureau of Mines to consider other solid adsorbents.   Bienstock, Field, and Myers (71) investigated 26 different metal oxides, single and in mixtures, to test their ability to adsorb SO_2 under uniform conditions of gas flow (space velocity), grain size, gas composition, and at two temperatures, 130°C and 330°C (Table XI).   Manganese oxide and aluminum sodium oxide

TABLE X

Degrees of Sulfur Dioxide Removal Achieved in an Entrainment Reactor with Dolomite and Calcium Carbonate[a]

| Material | Chemical analysis | | Surface area, m^2/g | Degree of desulfurization stoichiometric ratio | | | | | |
| | %CaO | %CO$_2$ | | 1 | 2 | 3 | 1 | 2 | 3 |
				500°C			800°C		
Dolomite chalk hydrate	44.0	2.0	3.33	22	28	32	39	59	68
Dolomite chalk hydrate rich in Al$_2$O$_3$, SiO$_2$, ZnO	23.1	2.3	3.32	—	14	—	—	27	—
CaCO$_3$ ppct., needles	56.1	43.9	1.00	—	—	—	—	—	89
CaCO$_3$ ppct., needles coated	56.1	43.9	3.91	17	—	42	42	—	96
CaCO$_3$ ppct., spheres coated	56.1	43.9	17.68	35	—	70	45	—	90
CaCO$_3$ with alkali	45.0	34.7	1.60	25	38	43	51	68	80

[a] From ref. 79.

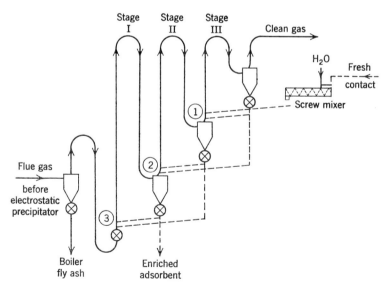

Fig. 20. Flow diagram for the Still process (87) (three stages of contact with lime adsorbent). (I–III) Reaction paths; (1–3) addition points for contacting material.

TABLE XI[a,b]

Sulfur Dioxide Absorption Tests at 330°C (625°F) (71a) (90% Removal)

Hourly space velocity of gas, 1050 hr^{-1}; mesh size of absorbent, 8–24

Absorbent	Crystalline phase (X-ray analysis)	Purity, wt %	Bulk density, g/cc	SO$_2$ absorbed, g; 100 g absorbent	Preparation
Manganese oxide	MnO$_{1.88}$	94	0.13	61	MnSO$_4$ $\xrightarrow{(NH_4)_2S_2O_8}$ precipitate washed, dried at 130°C, and heated *in vacuo* at 300–340°C for 20 hr
Manganese oxide	γ-Mn$_2$O$_3$	96	0.67	58	MnSO$_4$ $\xrightarrow{Na_2CO_3}$ precipitate washed, dried at 130°C, and heated *in vacuo* at 300–400°C for 20 hr
Hopcalite	MnO CuO^2	79 11	0.92	57	Dried *in vacuo* at 300–340°C.
Copper oxide	CuO	99	0.89	56	CuSO$_4$ $\xrightarrow{Na_2CO_3}$ precipitate washed, compressed at 4000 psi, dried at 130°C and heated *in vacuo* at 300–340°C for 20 hr
Manganese oxide	MnO$_{1.88}$	94	0.50	53	MnSO$_4$ $\xrightarrow{Electrolysis}$ precipitate washed, dried at 130°C and heated *in vacuo* at 300–340°C for 20 hr
Cobalt oxide	Co$_3$O$_4$	97	0.46	47	CoSO$_4$ $\xrightarrow{Na_2CO_3}$ precipitate washed, dried at 130°C and heated *in vacuo* at 300–340°C for 20 hr

Name	Formula	Oxide components		Composition	Value	No.	Preparation
Cobalt oxide	Co_3O_4			100	0.66	44	$CoSO_4 \xrightarrow{\text{NaOCl}/\text{NaOH}}$ precipitate washed, dried at 130°C and heated *in vacuo* at 300–340°C for 20 hr
Aluminum–sodium oxide		Al_2O_3	Na_2O	64 / 31	0.73	19	$Al_2(SO_4)_3 \xrightarrow{Na_2CO_3}$ precipitate washed, dried at 130°C and heated with H_2 at 600–640°C for 20 hr
Lead oxide	PbO			99	1.23	18	$Pb(NO_3)_2 \xrightarrow{Na_2CO_3}$ precipitate washed, dried at 130°C and heated *in vacuo* at 300–340°C for 20 hr
Chromium–sodium oxide		Cr_2O_3	Na_2O	70 / 26	0.91	16	$Cr_2(SO_4)_3 \xrightarrow{Na_2CO_3}$ precipitate washed, dried at 130°C and heated with H_2 at 600–640°C for 20 hr
Aluminum–potassium oxide		Al_2O_3	K_2O	61 / 35	0.63	15	$Al_2(SO_4)_3 \xrightarrow{K_2CO_3}$ precipitate washed, dried at 130°C, and heated with H_2 at 600–640°C for 10 hr
Sodium aluminate	$NaAlO_2$			96	0.90	10	Solution of sodium aluminate dried at 130°C and heated *in vacuo* at 300–340°C for 20 hr
Nickel oxide	NiO			91	0.74	9	$NiSO_4 \xrightarrow{Na_2CO_3}$ precipitate washed, dried at 130°C, and heated *in vacuo* at 300–340°C for 20 hr

(continued)

TABLE XI (continued)

Hourly space velocity of gas, 1050 hr^{-1}; mesh size of absorbent, 8–24

Absorbent	Crystalline phase (X-ray analysis)	Purity, wt%	Bulk density, g/cc	SO$_2$ adsorbed, g; 100 g absorbent	Preparation
Cobalt oxide	Co$_3$O$_4$	100	0.66	12	CoSO$_4$ $\xrightarrow{\text{NaOCl}/\text{NaOH}}$ precipitate washed, dried at 130°C and heated in vacuo at 300–340°C for 20 hr
Chromium–sodium oxide	Cr$_2$O$_3$ Na$_2$O	70 26	0.91	12	Cr$_2$(SO$_4$)$_3$ $\xrightarrow{\text{Na}_2\text{CO}_3}$ precipitate washed, dried at 130°C and heated with H$_2$ at 600–640°C for 20 hr
Nickel oxide	NiO	91	0.74	9	NiSO$_4$ $\xrightarrow{\text{Na}_2\text{CO}_3}$ Predipitate washed, dried at 130°C and heated in vacuo at 300–340°C for 20 hr
Nickel oxide	NiO	90	1.49	6	NiSO$_4$ $\xrightarrow{\text{NaOCl}/\text{NaOH}}$ precipitate washed, dried at 130°C and heated in vacuo at 300–340°C for 20 hr
Sodium carbonate	Na$_2$CO$_3$	99	0.98	5	Solution of sodium carbonate dried at 130°C and heated in vacuo at 300–340°C for 20 hr
Sodium stannate	Na$_2$SnO$_3$	95	0.91	4	Sodium stannate dried at 130°C and heated in vacuo at 300–340°C for 20 hr

Aluminum–lithium oxide	Li$_2$O·xAl$_2$O$_3$, Al$_2$O$_3$ Li$_2$O	87 10	0.73	4	Al$_2$(SO$_4$)$_3$ $\xrightarrow{\text{(NH}_4)_2\text{CO}_3}$ Li$_2$SO$_4$ precipitate washed, dried at 130°C and heated with H$_2$ at 600–640°C for 10 hr
Iron oxide	α-Fe$_2$O$_3$	93	0.98	3	Fe(NO$_3$)$_3$ $\xrightarrow{\text{Na}_2\text{CO}_3}$ precipitate washed, dried at 130°C and heated at 300–340°C in a stream of nitrogen for 3 hr
Sodium aluminate	NaAlO$_2$	96	0.90	3	Solution of sodium aluminate dried at 130°C and heated in vacuo at 300–340°C for 20 hr
Cadmium oxide	CdO	97	1.13	1	CdSO$_4$ $\xrightarrow{\text{Na}_2\text{CO}_3}$ precipitate washed, dried at 130°C and heated in vacuo at 370–400°C for 20 hr
Copper oxide	CuO	99	0.89	1	CuSO$_4$ $\xrightarrow{\text{Na}_2\text{CO}_3}$ precipitate washed, compressed at 4000 psi, dried at 130°C, and heated in vacuo at 300–340°C for 20 hr
Potassium carbonate		98	0.89	1	Solution of potassium carbonate dried at 130°C and heated in vacuo at 300–340°C for 20 hr

[a] Aluminum oxide (γ-Al$_2$O$_3$), bismuth oxide (α-Bi$_2$O$_3$), calcium oxide (CaO), magnesium oxide (MgO), molybdenum oxide (MoO$_3$), zinc oxide (ZnO), and potassium carbonate (K$_2$CO$_3$) absorbed less than 1 g of sulfur dioxide for 100 g of charge.

[b] Lead oxide (PbO) and calcium hydroxide (Ca(OH)$_2$) absorbed less than 1 g of sulfur dioxide also for 100 g of charge.

TABLE XI (continued)

Hourly space velocity of gas, 1050 hr^{-1}; mesh size of absorbent, 8–24

Absorbent	Crystalline phase (X-ray analysis)	Purity, wt %	Bulk density, g/cc	SO$_2$ absorbed, g; 100 g absorbent	Preparation
Nickel oxide	NiO	90	1.49	7	NiSO$_4$ $\xrightarrow{\text{NaOCl}}_{\text{NaOH}}$ precipitate washed, dried at 130°C, and heated in vacuo at 300–340°C for 20 hr
Cadmium oxide	CdO	97	1.13	5	CdSO$_4$ $\xrightarrow{\text{Na}_2\text{CO}_3}$ precipitate washed, dried at 130°C and heated in vacuo at 370–400°C for 20 hr
Sodium stannate	Na$_2$SnO$_3$	95	0.91	5	Sodium stannate dried at 130°C and heated in vacuo at 300–340°C for 20 hr
Sodium carbonate	Na$_2$CO	99	0.98	4	Solution of sodium carbonate dried at 130°C and heated in vacuo at 300–340°C for 20 hr
Aluminum–lithium oxide	Li$_2$O·xAl$_2$O$_3$, Al$_2$O$_3$, Li$_2$O	87, 10	0.73	4	Al$_2$(SO$_4$)$_3$ $\xrightarrow{\text{(NH}_4)_2\text{CO}_3}$ Li$_2$SO$_4$ precipitate washed, dried at 130°C and heated with H$_2$ at 600–640°C for 10 hr
Iron oxide	α-Fe$_2$O$_3$	93	0.98	3	Fe(NO$_3$)$_3$ $\xrightarrow{\text{Na}_2\text{CO}_3}$ precipitate washed, dried at 130°C and heated at 300–340°C in a stream of nitrogen for 3 hr

Material	Formula	Components	%	Value	No.	Preparation
Calcium hydroxide	Ca(OH)$_2$	Ca(OH)$_2$ CaO	81 19	0.36	2	Ca(NO$_3$)$_2$ $\xrightarrow{\text{NaOH}}$ precipitate washed, dried at 130°C and heated *in vacuo* at 300–340°C for 20 hr
Manganese oxide	MnO$_{1.88}$		90	0.14	33	MnSO$_4$ $\xrightarrow{\text{(NH}_4\text{)}_2\text{S}_2\text{O}_8}$ precipitate washed, and dried at 130°C.
Cobalt oxide	Co$_3$O$_4$		97	0.46	25	CoSO$_4$ $\xrightarrow{\text{Na}_2\text{CO}_3}$ precipitate washed, dried at 130°C and heated *in vacuo* at 300–340°C for 20 hr
Aluminum–sodium oxide		Al$_2$O$_3$ Na$_2$O	64 31	0.73	24	Al$_2$(SO$_4$)$_3$ $\xrightarrow{\text{Na}_2\text{CO}_3}$ precipitate washed, dried at 130°C and heated with H$_2$ at 600–640°C for 10 hr
Manganese oxide	MnO$_{1.88}$		94	0.50	23	MnSO$_4$ $\xrightarrow{\text{Electrolysis}}$ precipitate washed, dried at 130°C and heated *in vacuo* at 300–340°C for 20 hr
Manganese oxide	γ-Mn$_2$O$_3$		96	0.67	19	MnSO$_4$ $\xrightarrow{\text{Na}_2\text{CO}_3}$ precipitate washed, dried at 130°C and heated *in vacuo* at 300–340°C for 20 hr
Hopcalite		CuO MnO$_2$	11 79	0.93	13	Dried at 130°C
Aluminum–potassium oxide		Al$_2$O$_3$ K$_2$O	61 35	0.63	12	Al$_2$(SO$_4$)$_3$ $\xrightarrow{\text{K}_2\text{CO}_3}$ precipitate washed, dried at 130°C and heated with H$_2$ at 600–640°C for 10 hr

(continued)

(alkalized alumina) appeared the most promising materials, and alkalized alumina had the advantage that it was obtainable in hard, durable granules.

A cyclic adsorption process, called the Alkalized Alumina Process (71c,80,81) has been developed as a result of the preliminary studies, and two pilot plant-scale units have been investigated. The U.S. Bureau of Mines built a unit in which the adsorption was carried out in an adsorber where the particles of $\frac{1}{16}$-in. diameter were dropped counter current to rising flue gases at a temperature of 330°C (Fig. 21). The technical feasibility of the process

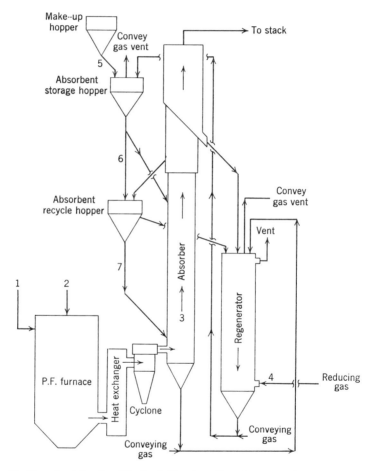

Fig. 21. Flowsheet for the Alkalized Alumina pilot plant. (*1*) Secondary air, 18,600 scfh 400°F; (*2*) primary air, 6200 scfh 140°F 150 lb/hr coal; (*3*) flue gas, 24,940 scfh; actual → 625°F (55,000 cfh); (*4*) reducing gas, 2300 scfh 1200°F, 76.7% H_2, 14.5% CO, 7.2% CO_2, 0.3% N_2, 1.3% CH_4; (*5*) absorbent, makeup 0.12 lb/hr; (*6*) absorbent, fresh feed 130 lb/hr; (*7*) absorbent, recycle 1300 lb/hr.

as been demonstrated in a 26-ft reactor, and promising rates of SO_2 removal from the flue gases of over 90% were obtained in certain experiments, using 6–28 mesh alkalized alumina pellets. Regeneration with H_2/CO mixtures at 650°C was also demonstrated, and no loss of absorptive capacity was found even after 20 cycles. There is some loss of the absorbent by attrition, approximating to 0.1 of the solids feed, but the hardness of the material can be increased by heating in air to 900°C. Different regenerating gases have been tried, and in decreasing effectiveness these were reformed natural gas, hydrogen, producer gas, and methane. Chlorine compounds from the coal are adsorbed by the alkalized alumina, and are not removed by the usual regeneration. However, they can be removed by treating the material with flue gas at 600°C. Extensive pilot plant tests are in progress (71c) and following these, full-scale application may be possible.

The U.K. Central Electricity Generating Board has a modification using a fluidized bed instead of the falling particles. The spent alumina is regenerated either with hydrogen on a small scale, reformed natural gas, or producer gas, on a large scale, at 650°C to form H_2S, CO_2, and H_2O. This gas is then fed to a Claus process plant to produce sulfur.

c. Activated Manganese Oxide

Manganese oxides, although found effective as adsorbents by Bienstock, Field, and Myers (71a) were found difficult to regenerate. The system proposed by them (Fig. 22) was similar to the Alkalized Alumina Process as regards the adsorber, but regeneration was by electrolysis. The stages suggested were

a. the reaction of Mn_2O_3 with SO_2

$$Mn_2O_3 + 2SO_2 + \tfrac{1}{2}O_2 \rightarrow 2MnSO_4$$

b. the manganese sulfate is slurried in water and sodium hydroxide is added

$$2MnSO_4 + 4NaOH + \tfrac{1}{2}O_2 \rightarrow Mn_2O_3 + 2Na_2SO_4 + 2H_2O$$

The manganese oxide is recycled to the adsorber, while the sodium sulfate solution is electrolyzed to give sodium hydroxide, which is used in the process, and sulfuric acid, which is recovered. The electrolytic recovery, however, requires large quantities of electrical energy and makes this process uneconomical.

Kun Li, Rothfus, and Adey have studied the effect of varying the surface area and the bulk density of manganese dioxide on its ability to adsorb SO_2 (109). They found, as may be expected, that increase in surface area or decrease in specific gravity among the samples tested resulted in increases

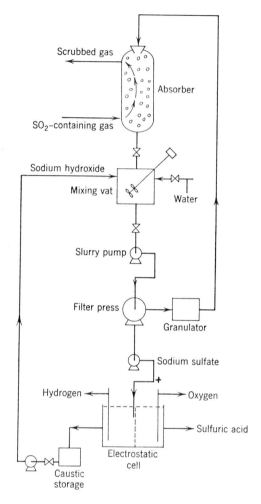

Fig. 22. Proposed scheme for scrubbing flue gas with manganic oxide (U.S. Bureau of Mines) (71a).

in adsorptive capacity. The effect of moisture was also studied, but these results were not conclusive. Correlating the oxygen content of the manganese oxides indicated that free oxygen is involved in the absorption of SO_2 to a degree depending on the oxygen deficiency in the solid to form manganese sulfate.

Mitsubishi Heavy Industries Ltd. have developed a somewhat different activated manganese oxide process, the DAP-Mn Process (82,107) (Fig. 23), where the regeneration of the adsorbent is carried out by reaction of the

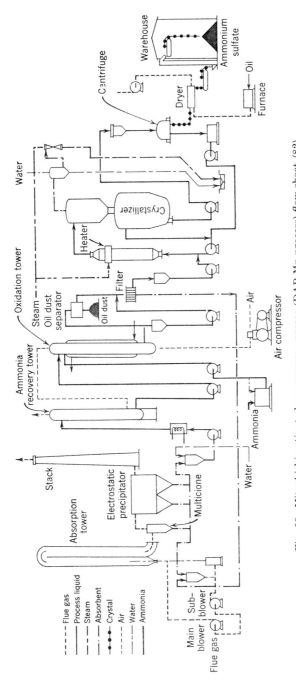

Fig. 23. Mitsubishi activated manganese process (DAP-Mn process) flow sheet (82).

manganese sulfate with air and ammonia, to produce ammonium sulfate The oxidation and subsequent regeneration takes place at room temperature the activated manganese oxide being separated by filtration. The ammonium sulfate solution is passed through a crystallizer where ammonium sulfate is recovered. If the recovery of gypsum rather than ammonium sulfate is desired, lime is added to the ammonium sulfate solution. The ammonia is then separated from solution by heating after the gypsum has been filtered off in a centrifuge.

Concentrations of manganese oxide used are of the order of 150–250 g/Nm^3 (i.e., 65–110 gr/ft^3) and temperatures of 135–160°C are used. The gas at entry contains 100 ppm, which is reduced by over 90% before exit. There is a certain amount of oil dust (0.100–0.2 g/Nm^3) in the flue gases, but this does not affect the performance of the adsorbent.

The activated manganese oxide has a composition $MnO_X \cdot YH_2O$, where X is between 1.5 and 1.8 and Y between 0.1 and 1. This composition is claimed to be different from the manganese oxides tested by Bienstock, Field, and Myers, being made by a special Mitsubishi process.

The process has so far been tested on a 1750-ft^3/min plant, but a 55-MW boiler size unit is being built at the Yokkaichi Power Station, Chubu Electric Power Company. This plant will treat 90,000 ft^3/min and will operate for one year before further larger plants are constructed.

d. Activated Carbons

The adsorbent with greatest promise of a high degree of adsorption, because of the very high surface area, is activated carbon. The use of this material at comparatively low temperatures (below 100°C), where the sulfuric acid formed is leached out, has been discussed (the Sulfacid process). Until 1957, however, it was thought that no satisfactory adsorption on carbon at higher temperatures—above 100°C—could be affected. New laboratory studies then showed that SO_2 is adsorbed and converted to H_2SO_4 and it is only possible to reverse this process above 250°C.

A process based on the use of cheap semicoke made from peat carbonized under vacuum at 600°C has been developed which is called the Reinluft Process (Fig. 24). The flue gases enter the lower section of the adsorber at about 150–200°C and part of the SO_2 is adsorbed on the carbon and is converted to SO_3. Further up the adsorber, the gases are withdrawn and then passed through a heat exchanger where they are cooled to about 110°C before being returned to the upper section of the adsorber. The gases leave the top of the adsorber at just above 100°C. The spent carbon from the bottom of the adsorber, which is saturated with sulfuric acid, is dropped into the desorber section. Here, the carbon is heated to 380–450°C with a stream of inert gas, the sulfuric acid is broken up, with the production of CO_2,

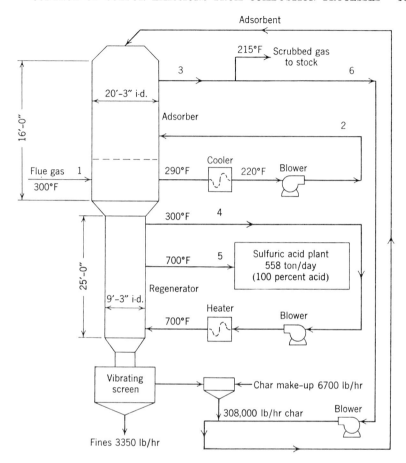

Fig. 24. Reinluft process flowsheet (dimensions and quantities for gases for 800-MW power plant) (98).

Total Gas Flow (100 units), Vol. %

Stream	1	2	3	4	5	6
N_2	76.2	76.2	76.5	28.6	28.6	76.5
O_2	3.4	3.4	3.3	1.3	1.3	3.3
CO_2	14.2	14.2	14.2	17.8	17.8	14.2
H_2O	6.0	6.0	6.0	27.3	27.3	6.0
SO_2	0.2	0.2	Trace	25.0	25.0	Trace
SO_3	Trace	—	—	—	—	—
Million scfh	87.3	91.7[a]	88.2	5.5	0.719	1.1

[a] Includes 5% leakage.

water vapor and SO_2 in concentrations of 10–15%, which is suitable as a feed gas for a contact acid plant. The char is recycled after the fines have been removed.

The first pilot plant, operated at the Wolfsburg Volkswagen Works (85), gave SO_2 removal rates varying from 45 to 96% as the gas rate was decreased from 18 ft^3/(ft)^2(min) to 4.7 ft^3/(ft)^2(min). Extensive tests were also carried out at the U.K. Ministry of Technology Warren Spring Laboratory (86) where efficiencies of over 90% were obtained with similar conditions. However, carbon losses due to attrition and chemical decomposition, serious corrosion problems, and the danger of combustion of the reactivated carbon indicated that serious problems would be faced by a commercial-scale plant. A more recent plant has been built to clean 20,000 ft^3/min (NTP) following the electrostatic precipitator of a 150-MW boiler at the STEAG—Kraftwerk Kellermann in Lünen (87). The flue gas temperatures vary between 95 and 125°C with an SO_2 content of 700–1400 ppm. The plant was in operation for 26 weeks during 1966, and was able to achieve a degree of SO_2 removal of 65–70%, which was lower than the design value of 75%.

To improve the process requires a cheap carbon which is much harder than the semicoke used in the Reinluft Process and not so combustible. Dratwa and Jüngten (88) have investigated a number of carbons made from black coal with 50% oxidation and of oxycokes made by the coking of air-oxidized hard coals. These carbons, as well as peat carbons, were investigated by Jüngten for their physical properties and ability to adsorb SO_2. The oxycoke proved to be very hard, even after repeated cycling in the adsorption system, unlike the peat coke or activated black coal carbon. The oxycokes also had an extensive fine-pore structure, which was a function of the pyrolysis temperature (89). This coke was able to adsorb much greater quantities of SO_2 with lower residence time than the peat carbons (Table XII).

As a result of the favorable laboratory tests on the oxycoke formed from preoxidized black coal, pilot plant experiments have been undertaken in the experimental boiler plant at Mathias Stinnes A.G., Bottrop.

Another activated carbon, which may also have similar properties, has been suggested by Strauss (90). This is made by extruding flame carbons into pellets. The flame carbons were formed from furnace oils to which activating additives had been added before burning under controlled conditions.

e. Activated Silica Gel

Another material which has been found suitable for the high-temperature adsorption of SO_2, is silica gel which has been treated with iron salts (91). The treatment consists of saturating the silica gel with iron salt solutions and

TABLE XII

Degree of SO_2 Removal from Simulated Combustion Gases by Different
Carbons after Passage of 12,700 ft³ (NTP) (88) (11,700 ft³ Gas (NTP)/ft³
of Adsorbent)

Carbon	Temperature °C	Residence time, sec	% SO_2 removed
Activated carbon from	80	13	60
black coal with	120	13	45
50 % oxidation	160	13	35
	80	5.6	30
	120	5.6	20
	160	5.6	15
	80	3.8	
	120	3.8	10–5
	160	3.8	
Peat coke	80	13	70
(3 regenerations)	120	13	65
	160	13	57
after 2 regenerations	80	13	100
	120	13	100
	160	13	100
Oxycoke	80	5.6	97
	120	5.6	94
	160	5.6	75
	80	3.8	90
	120	3.8	73
	160	3.8	65

heating this to 600°C. The treated gel is suitable as an adsorbent at
temperatures of 350–400°C. Regeneration of sulfur trioxide is carried out
at 700°C. The proposed plant consists of two adsorption stages: in the first,
the flue gases pass up through a bed of the silica gel, while above this, the
granules are allowed to fall freely counter-current to the gas stream.

3. Adsorption of SO_2 on Material Introduced into the Combustion Chamber

The problem of corrosion in the boiler combustion chamber, rather than
SO_2 removal, prompted a number of workers, particularly Wickert (92), to
introduce materials such as dolomite into the combustion chamber. As well
as preventing both high-temperature and low-temperature corrosion and
marked improvement in the maintenance requirements of boilers (93,94),
the addition of these materials was also found to reduce the emissions of SO_2.
There is considerable evidence that for the adsorption on the dolomite or
limestone to be effective, the material has to be calcined before use, and this

reaction takes place when the material is introduced into the boiler furnace near the fuel burners at temperatures above 1100°C. Calculations have shown that at this temperature it takes 0.5 sec for a 20-μm granule to be calcined (95). Wickert calcined the material used in pilot plant work at 1250°C and he obtained results for lime (CaO) and calcined dolomite (CaO: MgO). He further found that adding 1–2% Fe_2O_3 to the dolomite catalyzes the reaction:

$$2MgO + 2SO_2 + O_2 \xrightarrow{(Fe_2O_3)} 2MgSO_4$$

This catalyst has no effect in the case of CaO.

Pilot plant-scale experiments have been carried out by a number of workers. Early experiments in the refinery boilers at the Mobil Bremen Refinery (92) and at Wolfsburg (92,94) used a system which introduced the dolomite powder (optimum particle size 10–15 μm) with an air blast across the firing chamber above the burners. High degrees of sulfur removal were reported at low SO_2 concentrations, for example 72% removal at 250 ppm; 85% removal at 136 ppm.

The most extensive large-scale tests have been carried out in an ash slagging boiler with a capacity of 220,000 lb/hr. Dolomite was blown in at five points, as shown in Figure 25a (denoted 1–5). The temperature at the first point was approximately 1500°C, while at 5 it was approximately 900°C. As shown in Figure 25b, the maximum effectiveness was obtained at point 3, where the temperature was 1150°C, which is in agreement with laboratory experiments. The dolomite quantity was 2.5 times the stoichiometric amount needed for combination with the sulfur in the coal. Here point 3 represents a compromise between suitable temperature and residence time. Further tests were carried out, and these are reported in some detail (87) with hydrated lime, powdered limestone, and caustic potash as well as with dolomite. Hydrated lime gave the best results, which were somewhat better than those shown for dolomite.

Pollock, Tomany, and Frieling (65) investigated the addition of limestone (10% of the coal fired), the material being fused together with the coal. The sulfur dioxide reduction from flue gases containing 550–890 ppm was between 40 and 80%. The material here was recovered in the scrubber described earlier in conjunction with the scrubber investigations by these workers. It was noted that lower concentrations of limestone had no effect.

Far more successful than the completely dry Wickert process has been an extension of the earlier work by Pollock, Tomany, and Frieling, by a group from the Detroit Edison Company and Combustion Engineering Inc. (96): Dolomite injected in the corners above the coal burners, by injectors diverging from the direction of coal injection and tilted 15% upward, the coal burners being tilted similarly downward. The dolomite injection was at rates of approximately stoichiometric to 1½ times stoichiometric for the SO_2

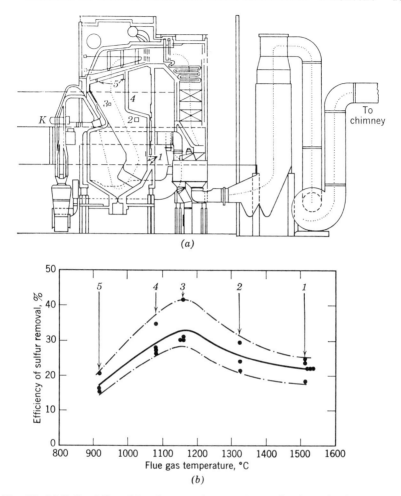

Fig. 25. (a) Boiler followed by electrostatic precipitator showing injection points for dolomite (1–5) (87) (Wickert Process); (b) Degree of sulfur removal obtained by dolomite injection in boiler at points 1–5 (87) (Wickert Process).

present. The flue gases were then scrubbed in a high-efficiency scrubber containing a fluidized bed of $\frac{5}{8}$-in. glass marbles. Removal efficiencies of 90–99.5% were obtained. These high efficiencies needed slurry flow rates of 8 gal/min for 1000 ft³/min of gas.

V. THE ECONOMICS OF SULFUR OXIDE CONTROL PROCESSES

While the technical feasibility of many of the processes discussed in the preceding sections is well-established by laboratory and pilot-plant experimentation, it is the most economical processes which will eventually be used.

What constitutes the most economical process depends to a large exten
on both the availability of suitable raw materials for processing, as well a⊃
on the markets for the possible byproducts of sulfur oxide control: sulfur
sulfuric acid, ammonium sulfate, and calcium sulfate.

Thus sulfur, although the most difficult to produce, has the most ready
market, while ammonium sulfate and calcium sulfate have limited markets
and although they may be suitable byproducts in Japan, this is not likely
to be the case in the United States. Sulfuric acid is not easily transported
over long distances, but if it can be used close to the power plant for pickling
liquors for steel works, or for superphosphate fertilizer, it may be a worthwhile byproduct.

In order to make a realistic comparison between the relative economic
feasibility of the different processes, it becomes necessary to reduce them to
common bases of estimation. This proves rather difficult for processes from
different countries. Nonetheless, estimates for a number of these processes
have been made, and it is possible to obtain some idea of the relative cost per
unit of power produced and for the unit of fuel consumed.

TABLE XIII

Summary of Costs of Scrubbing SO_2 from Flue Gases[a] of a 120-MW Power Plant by the
(1) I.C.I. Howden Cyclic Lime Process, (2) the Fulham-Simon Carves Ammonia Process,
and (3) the Johnstone and Singh Sodium Sulfite Process

| Process | SO_2 content of flue gases before scrubbing (ppm) | | | | | |
	830, 90% removal		3000, 90% removal		3000, 70% removal	
Capital[b]						
1	$1,646,750		$1,992,200		$1,747,200	
2	$3,221,100		$4,445,400		$4,055,400	
3	$2,512,160		$3,170,810		$2,687,490	
	No credit for products	Credit[d] for products	No credit for products	Credit[d] for products	No credit for products	Credit[d] for products
Operating costs,[c] $/ton coal						
1	1.24	--	1.93	—	1.69	—
2	2.96	1.59	6.46	1.48	5.32	1.44
3	2.15	2.00	3.18	2.22	2.66	1.90
Operating costs, mills/kWhr						
1	0.62	—	0.97	—	0.85	—
2	1.48	0.85	3.23	0.74	2.66	0.72
3	1.08	1.00	1.59	1.11	1.33	0.95

[a] 475,000 tons of coal with 1.5 and 5% sulfur.
[b] Includes working capital at 10% of fixed capital.
[c] No provision for return on investment, only 5% on fixed capital.
[d] Byproduct prices: sulfur, $28/ton; ammonium sulfate, $32/ton; sulfur dioxide, $14/ton.
[e] Anhydrous ammonia charged at $100/ton.

The U.S. Bureau of Mines has studied scrubbing processes (1959) and more recently (1964) the high-temperature processes for SO_2 removal from flue gases. In the survey of scrubbing processes for 120-MW stations (47), 1.5 and 5% sulfur coals were considered, which produce flue gases at a rate of 20 m/ft³/hr with 0.083% (i.e., 830 ppm) and 0.30% (3000 ppm) SO_2. The degree of removal was 70 and 90%, respectively. The processes considered were the Howden-I.C.I. Cyclic Lime process (57), the Fulham-Simon Carves process (59–61) and the Johnstone and Singh Regenerative Sodium Sulfite process (62,63), which are described in Section IV-B. Table XIII shows the comparative costs for these on a 1958 basis.

A more recent survey of costs for the Fulham-Simon Carves process based on an actual operating plant at North Wilford has been carried out by Wood (61) for an oil-burning, 275-MW power station burning 1600 tons of 3.25% sulfur oil as well as for a 400-MW plant burning a 4.8% sulfur oil. An attempt has been made here to place the data for the 275-MW station on a common cost basis, and extend this to an 800-MW power station. The extension has been made assuming directly proportional cost increases in all costs except for the operating labor, which is assumed to remain at the same numbers independent of plant size (Table XIV). In this calculation, the price of the raw materials (particularly ammonia) and of the byproducts (sulfur, ammonium sulfate) are critical factors in the financial feasibility of the process.

The Sulfacid process data given by the manufacturer for a 120-MW power station scrubbing plant (84) can be treated in a similar way to the Fulham-Simon Carves process (Table XV).

Recent research in high-temperature gas cleaning processes has led to a detailed costing of three of these processes by Katell et al. (81,98–100). These processes were the Reinluft process (75), the Penelec Catalytic Oxidation process (72), and the Alkalized Alumina process (80). The costing was made on the basis of coal-fired, 800-MW power stations, burning 3% sulfur coal with 20% excess air. The calculations involve the removal of 90% SO_2 in the flue gases and the production of either sulfuric acid (Reinluft and Penelec processes) or sulfur (Alkalized Alumina process). The detailed costing for these processes is shown in Tables XVI, XVII, and XVIII, and the effect of varying the price of the byproduct sulfur or sulfuric acid in Figures 26 and 27. The production of superphosphate from the acid produced in the Penelec process is also being considered (100) (Fig. 28).

The alternative to the Penelec process by Kiyoura, in which ammonia is added to the oxidized SO_3 in gases to give ammonium sulfate has been costed on a similar basis to the work by Katell et al., but 600 MW were considered as the unit size (Table XIX).

The economics of desulfurization with byproduct sulfur production has

TABLE XIV

Cost Estimates for the Fulham-Simon Carves Process on a 24-hr Basis[a]

	Quantity used	275 MW Original cost, $	800 MW Original cost	Standard cost, $
Capital investment		5.76m.	16.8m.	16.8m.
Materials				
Ammonia[b]	53 tons	2,600	7,600	15,500
78 % H_2SO_4	25.4 tons	496	1,450	1,450
Services				
Steam	500 tons	675	1,970	1,970
Power[c]	43,600 kW	382	1,110	1,020
Water[d]	1,470 tons	58	170	118
Operating labor		219	219	219
Maintenance		315	920	920
Capital charges[e]		1,660	4,840	7,040
		6,405	18,279	28,237
Credits				
Sulfur[f]	20.2	510	1,490	1,720
Ammonium sulfate	196.6	7,650	22,300	18,400
		8,160	23,790	20,120
Net cost		−1,755	−5,511	8,117
Oil consumed tons/day		1,600	4,560	4,560
Cost/ton ($)		−1.09	−1.19	1.74
Cost: mills/kWh.	No credit	0.97	0.95	1.47
	Credit for products	−0.265	−0.278	0.425

[a] $US 2.4 = one pound sterling has been used in the conversion.

[b] Ammonia: original cost, $49/ton; standardized cost, $100/ton.

[c] Power: original cost, 8.75 mills/kWhr; standardized cost, 8.0 mills/kWhr.

[d] Water: original cost, 14,4 cents/1000 US. gal; standardized cost, 10 cents/1000 U.S. gal.

[e] Capital charges: original cost as given; standardized cost, 15 % of capital.

[f] Sulfur: original cost, $25.20/ton; standardized cost, $28.00/ton.

[g] Ammonium sulfate: original cost, $39.00/ton; standardized cost, $32.00/ton.

TABLE XV

Cost Estimate for the Sulfacid Process (84), 1-hr Basis

	120 MW		800 MW,
	Quantity used	Original cost, $	Standard cost, $
Capital		2.45m.	16.5m.
Services: scrubbing			
Power[a]	700 kWh	8.75	37.60
Water[b]	66,000 gal	0.61	4.40
Labor	1 man	3.00	3.00
Maintenance		4.40	29.50
Capital charges[c]	13%	28.50	206.00
Services: acid recovery			
Power	181 kWh	2.26	9.70
Oil[d]	0.67 tons	13.40	90.00
Water (clean)	33,000 gal	3.04	22.20
Water (boiler feed)[e]	740 gal	0.28	1.85
Labor	1 man	3.00	3.00
Maintenance		1.75	11.80
Capital charges[c]	13%	11.40	82.00
		35.13	220.55
Credits			
98% Sulfuric acid[f]	1.8 tons	36.00	240.00
Steam (350 psia)[g]	2.8 tons	7.00	47.00
		43.00	297.00
Oil burned		29 tons	195 tons
Cost: $/ton		1.29	1.09
Mills/kWhr	No credits	0.67	0.43
	With credits	0.31	0.27

[a] Power: original cost, 12.5 mills/kWhr; standardized cost, 8 mills/kWhr.

[b] Water: original cost, 9.2 cents/1000 gal; standardized cost, 10 cents/1000 gal.

[c] Capital charges: original cost, 13%; standardized cost, 14%.

[d] Fuel oil: original cost, $20.00/ton; standardized cost, $20.00/ton.

[e] Boiler feed water: original cost, 38 cents/1000 gal; standardized cost, 38 cents/1000 gal.

[f] 98% sulfuric acid: original cost, $20/ton; standardized cost, $20/ton.

[g] Steam: original cost, $2.50/ton; standardized cost, $2.50/ton.

TABLE XVI

Total Estimated Capital Requirements and Operating Personnel (98)

	Reinluft process			Alkalized alumina process			Catalytic oxidation process		
	cost, $	%	Opr/shift	cost, $	%	Opr/shift	cost, $	%	Opr/shift
Absorber-regenerator	9,629,300	67.7	1	4,171,500	49.0	1	—	—	—
SO$_2$ removal equipment	—	—	—	—	—	—	13,727,600	80.8	2
Sulfuric acid plant	1,600,000	11.3	1	—	—	—	—	—	—
Gas producer	—	—	—	645,000	7.6	—	—	—	—
Sulfur recovery	—	—	—	1,780,000	20.9	1	—	—	—
Plant facilities	561,500	3.9	1	329,800	3.9	½	686,400	4.0	½
Plant utilities	825,400	5.8	1	484,800	5.7	1½	1,009,000	5.9	½
Total construction	12,616,200	88.7	4	7,411,100	87.1	3	15,423,000	90.7	3
Initial catalyst requirements	465,600	3.3		341,600	4.0		699,300	4.1	
Total plant cost (insurance and tax bases)	13,081,800	92.0		7,752,700	91.1		16,122,300	94.8	
Interest during construction	327,000	2.3		193,800	2.3		403,100	2.4	
Subtotal for depreciation	13,408,800	94.3		7,946,500	93.4		16,525,400	97.2	
Working capital	808,200	5.7		563,500	6.6		473,600	2.8	
Total investment	14,217,000	100.0		8,510,000	100.0		16,999,000	100.0	

TABLE XVII

Estimated Working Capital (98)

		Per cent
Reinluft process		
Activated char, 60 days ($80/ton)	$385,900	47.7
Direct labor, 3 months (direct + maintenance)	114,100	14.1
18.5% payroll overhead, 3 months	21,100	2.6
Operating supplies, 3 months	24,500	3.0
Indirect cost, 4 months	116,400	14.4
Fixed cost, 0.5% of insurance base	65,400	9.1
Spare parts	35,000	4.3
Miscellaneous expense	45,800	4.8
Total	808,200	100.0
Alkalized alumina process		
Coal supply for producer, 60 days ($4/ton)	$86,400	15.3
Absorbent makeup, 60 days ($0.25/lb)	164,000	29.1
Direct labor, 3 months (direct + maintenance)	62,200	11.0
18.5% payroll overhead, 3 months	11,500	2.0
Operating supplies, 3 months	11,700	2.1
Indirect cost, 4 months	60,800	10.8
Fixed cost, 0.5% of insurance base	38,800	6.9
Spare parts	60,000	10.7
Miscellaneous expense	68,100	12.1
Total	563,500	100.0
Catalytic oxidation process		
Catalyst supply, 60 days	$25,400	5.4
Direct labor, 3 months (direct + maintenance)	110,800	23.4
18.5% payroll overhead, 3 months	20,500	4.3
Operating supplies, 3 months	25,500	5.4
Indirect cost, 4 months	115,900	24.4
Fixed cost, 0.5% of insurance base	80,600	17.5
Spare parts	45,000	9.5
Miscellaneous expense	49,900	10.1
Total	473,600	100.0

TABLE XVIII

Estimated Annual Operating Cost (98)

Reinluft process:

Direct cost:

Raw materials and utilities:

Credit-steam: 51.1 M lb/hr × 7,920 hr/yr × $0.50/M lb	$202,400
Heat: 53,322,000 Btu/hr × 7,920 hr/yr × $0.50/MM Btu	211,200
Power: 1,473 kWhr/hr × 7,920 hr/yr × $0.006/kWhr	70,000
Char: 6,699 lb/hr × 7,920 hr/yr × $80/ton	2,122,200
Raw water: 135.7 M gph × 7,920 hr/yr × $0.10/M gal	107,500
	$2,308,500

Direct labor:

96 man-hr/day: $2.75/man-hr × 365 day/yr	96,400
Supervision: 15% of labor	14,500
	110,900

Plant maintenance:

48 mean: $6,000/yr	288,000
Supervision: 20% of maintenance labor	57,600
Material	144,000
	489,600

Payroll overhead: 18.5% of payroll	84,500
Operating supplies: 20% of plant maintenance	97,900
Total direct cost	$3,091,400
Indirect cost (administration and general overhead)—50% labor, maintenance, and supplies	349,200
Total capital charges	1,990,400
Gross operating cost	$5,431,000

Alkalized alumina process:

Direct cost:

Raw materials and utilities:

Absorbent makeup	$901,700
Coal: 15 tph × 7,920 hr/yr × $4/ton	475,200
Power: 1,900 kWhr/hr × 7,920 hr/yr × $0.006/kWhr	90,300
Heat: 78,560,000 Btu/hr × 7,920 hr/yr × $0.70/MM Btu	435,500
Water: 33 M gph × 7,920 hr/yr × $0.10/M gal	26,100
Credit-heat: 78,560,000 Btu/hr × 7,920 hr/yr × $0.50/MM Btu	−311,100
	$1,617,700

Direct labor:

72 man-hr/day: $2.75/man-hr × 365 day/yr	72,300	
Supervision: 15% of labor	10,800	83,100

Plant maintenance:

23 men: $6,000/yr	138,000	
Supervision: 20% of maintenance labor	27,600	
Material	69,000	234,600
Payroll overhead: 18.5% of payroll		46,000
Operating supplies: 20% of plant maintenance		46,900
Total direct cost		$2,028,300
Indirect cost (administration and general overhead)—50% labor, maintenance, and supplies		182,300
Total capital charges: 14% of total investment		1,191,400
Gross operating cost		$3,402,000

Catalytic oxidation process:

Direct cost:

Raw materials and utilities:

Power: 3,067 kWhr/hr × 7,920 hr/yr × $0.006/kWhr	$145,700	
Steam: 110 M lb/day × 330 day/yr × $0.50/M lb	18,200	
Water: 2,200 M gal/day × 330 day/yr × $0.10/M gal	72,600	
Catalyst	139,900	$376,400

Direct labor:

72 man-hr/day: $2.75/man-hr × 365 day/yr	72,300	
Supervision: 15% of labor	10,800	83,100

Plant maintenance:

50 men: $6,000/yr	300,000	
Supervision: 20% of maintenance labor	60,000	
Material	150,000	510,000
Payroll overhead: 18.5% of payroll		82,000
Operation supplies: 20% of plant maintenance		102,000
Total direct cost		$1,153,500
Indirect cost (administration and general overhead)—50% labor, maintenance, and supplies		347,600
Total capital charges: 14% of total investment		2,379,900
Gross operating cost		$3,881,000

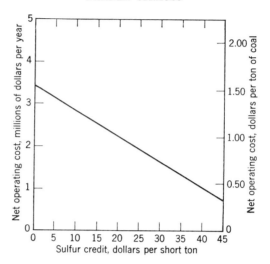

Fig. 26. Relationship of net operating cost and sulfur byproduct credit—Alkalized Alumina process (99).

been summarized in a nomogram (Fig. 29) by Pirsh, Rusanowsky, and Young (108) of Babcock and Wilcox. The nomogram has been worked out on the basis that with byproduct recovery there is no increase in the cost of generating power. Thus by combining installed cost, operating cost, and byproduct credit these must net to zero. In the example shown in Figure 29, it can be deduced that the permissible operating cost of the cleanup system can be up to 0.08 mill/kWhr if the installed cost is $11 per kW and also if a 3% sulfur coal is used, with a sulfur price of $35 per (long) ton. The byproduct credit in this case is slightly over 0.3 mill/kWhr.

Processes using relatively cheap additives, without recovery of worthwhile byproducts, have also been costed. Haley (101) has calculated the cost of adding dolomite to the flame in sufficient quantity (assuming 30% effective conversion to the sulfate) to remove different proportions of the SO_2 in the flue gases, viz. 34, 67, and 90% (Table XX). His estimates, like those of Katell et al., are based on an 800-MW power plant, where limestone can be obtained at a very economical rate of $2/ton. Haley's costs of 0.27 mills/ kWhr are therefore lower than the values given by Zentgraf (87) for the Wickert and Still processes, which use much more expensive adsorbents (Table XXI). In all these processes, the used reagent is collected by electrostatic precipitators and not by scrubbers.

As scrubbers have also been tested in the dolomite and limestone addition processes, both for large-scale (96) and medium-scale (102) application, comparative costs in these cases are included.

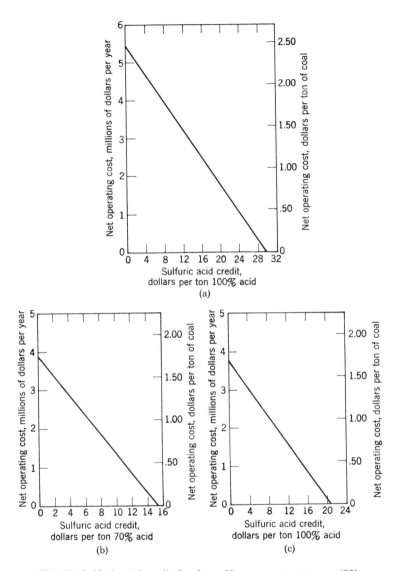

Fig. 27. Sulfuric acid credit for dry sulfur recovery processes (99).

(a) Reinluft process: (b) Catalytic oxydation process (Penelec or
Cat-Ox process): (c) Alkalized Alumina process.

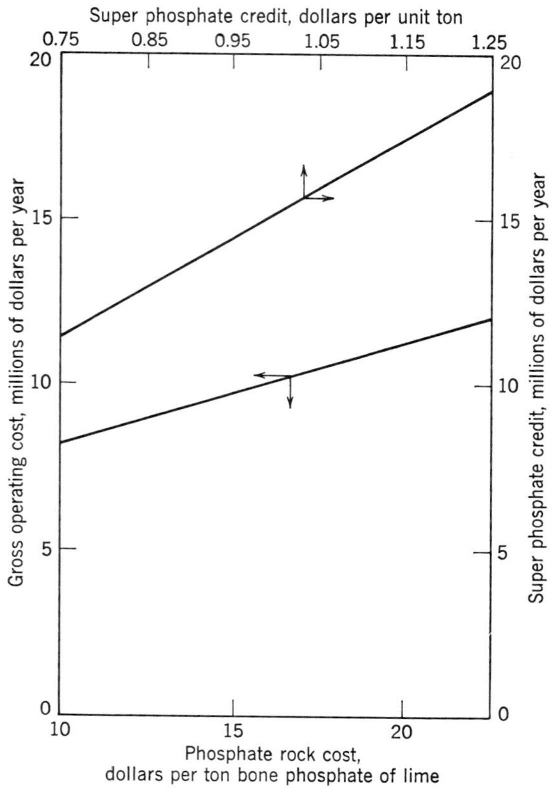

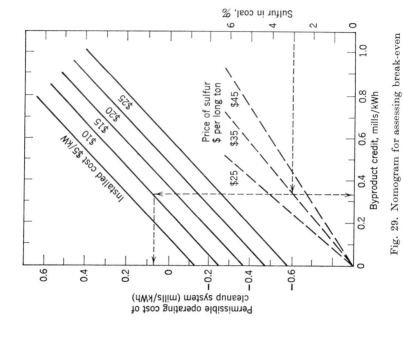

Fig. 28. The effect of superphosphate production, gas–solid reaction, on operating costs when removing sulfur from flue gas by catalytic oxidation (Penelec process) (100).

Fig. 29. Nomogram for assessing break-even costs for desulfurization—sulfur recovery systems (108).

TABLE XIX

Estimated Operating Costs for Kiyoura-T.I.T. Process (Based on 600 and 800-MW Plants Burning 3.6% S Fuel Oil)

	Kiyoura costs, 600 MW	Standardized costs, 800 MW
Total capital investment	$6,671,000	$8,900,000
Direct costs	2,129,600	4,445,000
Ammonia: Kiyoura: $56.18/ton		
St. Cost: $100/ton	1,518,700	3,600,000
Fuel: Kiyoura: $19/ton		
St. Cost: $20/ton	418,900	589,000
Power: Kiyoura: 8 mills/kWhr		
standard: 8 mills/kWhr	192,000	256,000
Direct labor	83,000	83,000
Plant maintenance and material	234,600	312,000
Payroll overhead (18.5% of payroll)	46,000	73,200
Operating supplies (20% of maintenance)	46,900	62,400
Total direct costs	2,540,200	4,975,600
Total capital charges (14% of invest.)	933,940	1,247,000
Indirect costs (admin. overhead)	182,300	182,300
Gross operating costs	3,656,440	6,404,900
Credit—sale of ammonium sulfate	3,220,000	4,270,000
Kiyoura: $32.20/ton		
St. Cost: $32.00/ton		
Net operating cost	436,440	2,134,900
Net operating cost: $/ton fuel oil	0.44	1.60
mills/kWhr—without credits	0.85_0	1.14_5
with credits	0.10_4	0.38_1

Large-scale studies lead to SO_2 removal efficiencies as high as 99% with dolomite concentrations only 10% above the stoichiometric requirement, and cost estimates for a 500-MW plant are shown in Table XXII. The cost estimates for a 500-MW plant are here considerably lower than for the Cyclic Lime Process.

The cost of the Mitsubishi Manganese Oxide Process has been prepared for a 750-MW power station, and costs approximately $1.10 per ton of fuel oil burned (100), which is similar in cost to the Sulfacid Process. The different processes are compared in Table XXIII.

It can be concluded that the cost of desulfurizing a high-sulfur residual oil is similar in cost to the most efficient methods of removing sulfur dioxide from the flue gases. In the case of coal combustion, it appears likely that if low-cost dolomite is available and comparatively low degrees of sulfur removal are adequate (i.e., of the order of 60%) dolomite addition followed by electrostatic precipitation may be the most economical. On the other

TABLE XX

Economic Appraisal of Dry Limestone Process (800-MW Plant)[a]

	$/year @ 90% load factor
Direct cost:	
Raw materials and utilities	
Limestone (for 3% sulfur coal) @ $2/ton delivered (30% chemical utilization)	$925,000
Grinding costs	113,000
Heat loss allowance (raise limestone temperature)	50,000
Water	5,000
Waste disposal and hauling @ $0.80/ton net	411,000
	$1,504,000
Direct labor:	
24-man-hour day @ 2.75/hr × 355 days/yr = $24,100	
Supervision @ 15% = 3,600	$27,700
Maintenance @ 3% investment	24,000
Payroll overhead @ 18.5% payroll	5,100
Supplies	10,000
Sub-total	$1,570,800
Indirect cost: (administration and general overhead)	26,400
Total capital charges: 14% investment	112,000
Net operating cost	$1,709,210
(0.27 mills/kWhr; 3.0 ¢/MM But; $0.78/ton)	

[a] Functionally comparable to comparisons by Bienstock and Field at the 58th Annual Meeting, Air Pollution Control Assn. Capital cost allowance: complete facilities $800,000. Operating cost: estimated annual cost for removing 2/3 sulfur in coal.

Removal of SO_2 in stack would enable boiler to operate at lower stack temperatures (340–260°F), increasing efficiency about 2%. This would result in a gross annual fuel saving of about $230,000 (based on 10¢/$10^6$ Btu credit). No credit recorded here for this factor.

At 30% conversion of limestone to sulfate and using a 3% sulfur, 12.7% ash coal; daily requirements for two-thirds SO_2 removal are: Raw materials: coal, 6,710 tons/day; limestone, 1,400 tons/day. Waste: ash, 725 tons/day (assuming 85% of ash in stack); limestone–calcium sulfate mixture, 1,557 tons/day. Total waste: 2,282 tons/day.

TABLE XXI

Comparison of Dolomite, Limestone, and Lime Addition

	A. D. Little (101), Dolomite	Wickert process (87)		Still process (87), CaO from brown coal ash
		Limestone	Slaked lime	
Cost of Adsorbent	$2/ton	$7.55/ton	$14.20/ton	n.a.
mills/kWhr	0.27	0.65	1.33	0.50
Effectiveness of SO_2 removal (%)	67	20–30	50–70	65–70
Capital per kW installed	1.00	n.a.	n.a.	n.a.

TABLE XXII

Cost Estimates for Dolomite and Soda Ash Addition in the Boiler, Followed by Scrubbing

	500-MW Plant (96) with dolomite (110%) addition 3% S coal	20 MW Plant (102) with soda ash addition 2.1% S coal
Capital	$1.11m.	$125,000
Operating Costs	¢/ton	¢/ton
Dolomite ($2/ton) or soda ash	24.9	200
Extra disposal cost	12.8	—
Calcining dolomite	3.0	—
Power (mills/kWhr—8)[a]	14.5	21
Charges, labor, etc.	13.4	7
	68.6	228
Credits (savings in corrosion, precipitator, stacks, etc.)		
Capital cost savings	$16.5m.	—
Operating ¢/ton	27.0	300[b]
Costs: mills/kWhr, no credits	0.257	1.14
credits	0.145	0.36

[a] The original paper uses 5 mills/kWhr, but the cost estimates here uses 8 mills/kWhr.
[b] Saving due to reduced fuel costs of $3/ton because high sulfur coal used.

TABLE XXIII

Cost Comparison of Sulfur Oxide Removal Processes

Process	Efficiency of SO_2 removal	Capital, $/kWhr installed	without credits mills/kWhr	without credits $/ton	with credits mills/kWhr	with credits $/ton	
Penelec[a]	90%	21.2	0.613	1.75	−0.183	−0.52	Coal-fired 3% S coal
Kiyoura T.I T.[b]	90%	11.1	1.145	4.80	0.381	1.60	Oil-fired 3.6% S oil
Reinluft[c]	90%	17.8	0.857	2.45	0.061	0.174	Coal-fired 3% S coal
Alkalized alumina[d]	90%	10.6	0.537	1.54	0.271	0.78	Coal-fired 3% S coal
Dry limestone[e]	90%	1.0	0.39	1.13	—	—	Coal-fired 3% S coal
	67%	1.0	0.27	0.78	—	—	Coal-fired 3% S coal
	34%	1.0	0.15	0.43	—	—	Coal-fired 3% S coal
Sulfacid[f]	95%	20.6	0.43	2.57	0.27	1.09	Oil-fired 2.1% S oil
Fulham-Simon Carves[g]	95%	21.0	1.47	6.07	0.43	1.74	Oil-fired 3.2% S oil
A.I. Molten Carbonate[h]	90%	10.2	0.454	1.30	—	—	Coal-fired 3% S coal

[a] Sulfur acid credited at $19.40 for 70% acid; break-even price: $14.80.

[b] Ammonia charged at $100/ton (Kiyoura uses $56.18/ton). Ammonium sulfate credited at $32.20/ton.

[c] Absorbent char at $80/ton. Sulfuric acid credited at $27.65/ton 100% acid; break-even price: $29.70/ton.

[d] Sulfur credit price $28/ton; break-even price: $56.90/ton.

[e] Dolomite cost: $2/ton.

[f] Fuel oil $20/ton; 98% sulfuric acid: $20/ton credit.

[g] Ammonium sulfate: $32/ton; sulfur: $28/ton.

[h] Laboratory-scale preliminary investigation indicates that sulfur credited at $48/ton represents a break-even price for the process.

hand, following the dolomite addition by scrubbing gives very high degrees of sulfur removal but no useful byproduct.

If there is the possibility of byproduct sales of either sulfur, sulfuric acid, superphosphate, or ammonium sulfate, then one of the catalytic conversion or alkalized alumina adsorption systems may be more desirable. It can, furthermore, be concluded that in most cases the recovery of the byproducts will not result in a profit for the operation, but in general the removal of SO_2 from combustion gases is associated with a minimum increase in the cost of power generated of the order of 0.3 mill/kWhr.

References

1a. F. A. Rohrman and J. H. Ludwig, *Chem. Eng. Progr.*, **61** (9), 59 (1965).

1b. F. A. Rohrman, J. H. Ludwig, and B. J. Steigerwald, "SO_2 Emissions to U.S. (1966)," in *Control Technology R & D Programs*, National Center for Air Pollution Control, Cincinnati, Ohio.

2. G. Spengler and G. Michalczyk, *Die Schwefeloxyde in Rauchgasen und in der Atmosphäre*, V.D.I., Verlag, Düsseldorf, 1964, pp. 21–22.

3. J. H. Ludwig and P. W. Spaite, *Chem. Eng. Progr.*, **63** (6), 82 (1967).

4. F. A. Rohrman, B. J. Steigerwald, and J. H. Ludwig, *Power*, 82 (May 1967).

5. F. E. Ireland, *One Hundred and Third Annual Report on Alkali etc. Works*, the Chief Inspectors, H.M.S.O., London, 1967, p. 40.

6. G. Scurfield, "Air Pollution and Tree Growth," *Forestry Abstr.*, No. 29, **21**, 1960.

7. Anon., *Environ. Sci. Technol.*, **1**, 282 (1967).

8. M. D. Thomas, in *Air Pollution*, W.H.O., Geneva, 1961, pp. 233–278.

9. I. J. Hindawi, "Injury by Sulfur Dioxide, Hydrogen Fluoride and Chlorine," 67–159, presented at 60th Annual Meeting of the Air Pollution Control Association, Cleveland, Ohio, June 1967.

10. A. Parker, in *Air Pollution*, W.H.O., Geneva, 1961, p. 374.

11. F. E. Ireland, *One Hundred and Third Annual Report on Alkali etc. Works*, Chief Inspectors, H.M.S.O., London, 1967, p. 57.

12. G. Chedd, *New Scientist*, **36**, 281 (1967).

13. F. A. Rohrman, J. H. Ludwig, and B. J. Steigerwald, *Coal*, (April 1965) (reprint).

14. H. R. Brown and H. Berry, in *Coal in Australia*, Vol. 6, 5th Empire Min. and Met. Congr. A.N.Z., 1953, pp. 93–125.

15. W. S. Smith, *Atmospheric Emissions from Fuel Oil Combustion*, U.S. Dept. H.E.W., Divn. of Air Pollution, 1962.

16. A. B. Edwards, in *Brown Coal*, P. L. Henderson, Ed., Melb. U.P., 1953, p. 22.

17a. M. Kemezys and G. H. Taylor, in *Symposium on Inorganic Constituents of Fuel*, Melbourne University, 1964, pp. 45–56.

17b. W. S. Smith and C. W. Gruber, *Atmospheric Emissions from Coal Combustion*, U.S. Dept. H.E.W., April 1966, p. 16.

18a. L. S. Zarabin, *Coke Chem. (U.S.S.R.)*, (*English Transl.*), **7**, 8 (1963).

18b. R. E. Zimmerman, *Chem. Eng. Progr.*, **62** (10), 61 (1966).

19. A. Z. Yurovsky and I. D. Remesnikov, *Coke Chem. (U.S.S.R.)*, (*English Transl.*), December, 1958, pp. 8–13.

20. T. T. Frankenberg, "Removal of Sulfur from Fuels and Products of Combustion," A.S.M.E. Winter Meeting, 1964, paper 64WA/APC-2.

21. M. P. Silverman, M. H. Rogoff, and I. Wender, *Appl. Microbiol.*, **9**, 491 (1961).
22. R. M. Jimeson, *Chem. Eng. Progr.*, **62** (10), 53 (1966).
23. "NBS Commercial Standards Specifications for Fuel Oils, CS12-48," Sept. 25, 1948, quoted by W. S. Smith in *Atmospheric Emissions from Fuel Oil Combustion*, U.S. Dept. H.E.W., 1962.
24. H. C. Carpenter and P. L. Cottingham, "A Survey of Methods for Desulfurizing Residual Fuel Oils," U.S. Bur. Min. Inform. Circ. 8156, 1963.
25. H. D. Hartough in *Advances in Petroleum Chemistry and Refining*, Vol. 3, K. A. Kobe and J. J. McKetta, Eds., Interscience, New York, 1960, pp. 419–481.
26. V. A. Kalichesvsky and K. A. Kobe, *Petroleum Refining with Chemicals*, Elsevier, Amsterdam, 1956, pp. 259–271.
27. A. Szayna, Sulfur Absorbent, Method of Regeneration, Treatment of Hydrocarbons, U.S. Patents 2,273,297–99, Feb. 17, 1942.
28. L. S. Galstaun, B. J. Steigerwald, J. H. Ludwig, and H. R. Garrison, *Chem. Eng. Progr.*, **61** (9), 49 (1965).
29. H. H. Meredith, "Desulfurization of Caribbean Fuel," Air Pollution Control Assoc. Annual Conf., Paper 67-163, 1967.
30. A. M. Squires, *Chem. Eng. Progr.*, **74** (26) (1967).
31. C. G. Fredersdorff and M. A. Elliott, "Coal Gasification," in *Chemistry of Coal Utilization*, H. H. Lowry, Ed., Wiley, New York, 1963, pp. 892–1022, ch. 20.
32. M. A. Elliott and H. R. Linden, "Manufactured Gas," in *Kirk-Othmer Encyclopedia of Chemical Technology*, Vol. 10, 2nd ed., H. F. Mark et al., Eds., Interscience, New York, 1966, pp. 353–442
33. *Gas Making*, 2nd ed., Technical Services Branch, B.P. Co., London, 1965.
34. A. R. Powell, in *Chemistry of Coal Utilization*, Vol. 2, H. H. Lowry, Ed., Wiley, 1945, p. 921.
35. L. Reeve, *J. Inst. Fuel*, **31**, 319 (1958).
36. B. R. Kean, *J. Inst. Fuel*, **28**, 155 (1965).
37. A. M. Squires, *Gasifying Oil for Clean Power*, Am. Inst. Chem. Engrg., 60th Annual Meeting, November 26–30, 1967, preprint 37E.
38. A. M. Squires, "Cyclic Use of Calcined Dolomite," in *Advances in Chemistry*, 69, Fuel Gasification (ACS), 1967, pp. 205–229.
39. A. V. Bureau and J. F. Olden, *The Chemical Engineer*, **206**, CE55-62, March (1967).
40. J. E. Munroe and E. G. Masdin, *Brit. Chem. Eng.*, **12** (3) 369 (1967).
41. A. L. Kohl and F. C. Riesenfeld, *Gas Purification*, McGraw-Hill, New York, 1960, pp. 18–152.
42. W. Strauss, *Industrial Gas Cleaning*, Pergamon Press, New York, 1966, pp. 96–103
43. C. J. Wandt and L. W. Dailey, *Hydrocarbon Process.*, **46** (10) 155 (1967).
44. Anon., *Chem. Eng.*, **59**, 210 (1952).
45. H. Grekel, L. V. Kunkel, and R. McGalliard, *Chem. Eng. Progr.*, **61** (9), 70 (1965).
46. G. Giammarco, Italian Patents, 537564, 1955; 560161, 1956; 565320, 1957.
47. T. Nicklin and B. H. Holland, *Cleaning of Coke Oven Gas by the Stretford Process*, Coke Oven Mgrs. Assn., Cardiff, January 1963, pp. 1–8.
48. T. Nicklin and B. H. Holland, "Removal of Hydrogen Sulfide from Coke Oven Gas by the Stretford Process," Symposium: Cleaning Coke Oven Gas, Saarbrücken, March 1963, pp. 1–20.
49. Anon., *Chem. Eng.*, **71**, 128 (July 20, 1964).
50. C. Ryder and A. V. Smith, *Inst. Gas Eng. Conm.*, **624**, 1, (1962).
51. W. Möller and K. Winkler, *Double Contact Process for Sulfuric Acid Production*, Air Pollution Control Assoc. Annual Meeting, Paper 67-115, Cleveland, June 1967.

52. E. P. Fleming and T. C. Fitt, *Ind. Eng. Chem.*, **42**, 2253 (1950).

53. M. P. Appleby, *Trans. J. Soc. Chem. Ind.*, **56**, 139 (1937).

54. R. Germerdonk, *Chem.-Ing. Tech.*, **37** (11), 1136 (1965).

55. R. L. Rees, *J. Inst. Fuel*, **25**, 350 (1953); also Institution of Mechanical Engineers Conference on the *Mechanical Engineers Contribution to Clear Air*, 1957, p. 34.

56. Anon., *Chem. Eng.*, **74**, 43 (March, 1967).

57. J. L. Pearson, G. Nonhebel, and P. H. N. Ulander, *J. Inst. Fuel*, **8**, 119 (1935).

58. M. Atsukawa, Y. Mishimoto, and K. Matsumoto, *Mitsubishi Heavy Industries Ltd. Technical Review*, **2**, 134 (1965).

59. T. Kennaway, *Iron and Steel Inst. Special Report*, **61** (1958); *J. Air Pollution Control Assoc.*, **7**, 266 (1958).

60. E. Wallis *Brit. Chem. Eng.*, **7**, 833 (1962).

61. C. W. Wood, *Trans. Inst. Chem. Eng.*, **38**, 54, London (1960).

62. H. F. Johnstone and A. D. Singh, *Ind. Eng. Chem.*, **29**, 286 (1937).

63. H. F. Johnstone and A. D. Singh, *Ind. Eng. Chem.*, **32**, 1037 (1940).

64. R. Klimecek, J. Skrivanek, and J. Bettelheim, *Staub-Reinhalt. Luft*, **26**, 235 (1966).

65. W. A. Pollock, J. P. Tomany, and G. Frieling, Paper 66-WA/CD-4, A.S.M.E. Winter Annual Meeting, New York, 1966.

66. K. Matsumoto and Y. Shiraishi, *Mitsubishi Heavy Industries Ltd. Technical Review*, **5**, 59 (1968).

67. D. J. Williams, *Coal Res. C.S.I.R.O.*, **23**, 7 (July 1964).

67a. Average of Equilibrium Constants of Bodenstein and Pohl, Kapustinsky and Shamovsky, and Evans and Wagman, quoted by Williams.

68. H. Jüngten, *Erdoel Kohle*, **16**, 119 (1963).

69. A. Hedley, "A Kinetic Study of SO_3 Formation in a Pilot Scale Furnace," in *Mechanism of Corrosion by Fuel Impurities*, H. R. Johnson and D. J. Littler, Eds., Butterworth, 1964, pp. 204–215.

70. C. F. Cullis, R. M. Henson, and D. L. Trim, *Proc. Roy. Soc.*, *(London), Ser A*, **295**, 72 (1966).

71. (a) J. H. Field, D. Bienstock, and J. G. Myers, "Process Development in Removing SO_2 from Hot Flue Gases," Pt. I, *U.S. Bur. Mines Report of Invest.*, **5735** (1961); (b) Pt. II (with R. C. Kurtzrock), *U.S. Bur. Mines Report of Invest.*, **6037** (1963); (c) Pt. III, *U.S. Bur. Mines Report of Invest.*, **7021** (1967).

72. R. M. Bouvier, *Proc. Am. Power Conf.*, **24**, 138 (1964).

73. R. Kiyoura, *J. Air Pollution Control Assoc.*, **16**, 488 (1966); *Staub-Reinhalt. Luft*, **26**, 524 (1966).

74. R. Kiyoura, "Studies on the Removal of SO_2 from Hot Flue Gases, II," *60th Ann. Meeting Air Pollution Control Assoc.*, Cleveland, No. 67–91 (1967).

75. F. Johswich, *Brennstoff-Wärme-Kraft*, **14**, 105 (1962).

76. N. E. Jones, Ph.D. thesis, University of Sheffield, 1960.

77. J. R. Coke, Ph.D. thesis, University of Sheffield, 1963.

78. A. M. Squires, *Chem. Eng.*, **74**, 133 (Nov. 20, 1967).

79. H. Jüngten and W. Peters, *Staub-Reinhalt. Luft*, **28**, 89 (1968).

80. D. Bienstock, J. H. Field, and J. G. Myers *J. Eng. Power, Trans. A.S.M.E.*, *Ser. A*, **86** (3) 353 (1964).

81. D. Bienstock, J. H. Field, S. Katell, and K. D. Plants, *J. Air Pollution Control Assoc.*, **15**, 459 (1965).

82. M. Atsukawa, Y. Mishimoto, and K. Matsumoto, *Mitsubishi Heavy Industries Ltd. Technical Review*, **4**, 33 (1967).

83. K. Gasierowski, *Mitt. Ver. Grosskesselbesitzer*, **83**, 83 (1963).

84. Lurgi Apparatebau G.m.b.H., *Sulfacid Process*, pp. 1–9.

85. E. Wahnschaffe, *Mitt. Ver. Grosskesselbesitzer*, **83**, 72 (1963).

86. L. E. Reed, P. R. Trott, and S. Sutton, "Removal of Sulfur Oxides from Flue Gas: the Reinluft Pilot Plant," Report No. LR 15 (AP), September 1965.

87. K. M. Zentgraf, *Staub-Reinhalt. Luft*, **28**, 94 (1968).

88. H. Dratwa and H. Jüngten, *Staub-Reinhalt. Luft*, **27**, 301 (1967).

89. H. Jüngten and J. Karweil, *Erdoel Kohle*, **15**, 898, 985 (1962).

90. W. Strauss, Aust. patent application 41506/68, August 1, 1968.

91. H. G. Heitmann and J. Sieth, *Mitt. Ver. Grosskesselbesitzer*, **83**, 82 (1963); German patent 1217535, May 26, 1962.

92. K. Wickert, *Mitt. Ver. Grosskesselbesitzer*, **83**, 74 (1963).

93. H. Ulrich, *Mitt. Ver. Grosskesselbesitzer*, **81**, 413 (1962).

94. E. Schneider, *Mitt. Ver. Grosskesselbesitzer*, **80**, 354 (1962).

95. A. E. Potter, R. E. Harrington, and P. W. Spaite, "Limestone Dolomite Processes for Flue Gas Desulfurization," presented at the Am. Chem. Soc. Meeting, Chicago, Ill, September 11, 1967.

96. A. L. Plumley, O. D. Whiddon, F. W. Shutko, and J. Jonakin, "Removal of SO_2 and Dust from Stack Gases," Am. Power Conf., Chicago, April 25–27, 1967.

97. J. H. Field, L. W. Brunn, W. P. Haynes, and H. E. Benson, "Cost Estimates of Liquid Scrubbing Processes for Removing Sulfur Dioxide from Flue Gases," *U.S. Bur. Mines Rep. Invest.*, **5469**, 1959.

98. S. Katell, "An Evaluation of Dry Processes for the Removal of SO_2 from Power Plant Flue Gases," *Symp. on Economics of Air Pollution Control*, Am. Inst. Chem. Eng. 59th National Meeting, Columbus, Ohio, May 15, 1966.

99. S. Katell, *Chem. Eng. Progr.*, **62** (10), 67 (1966).

100. S. Katell and K. D. Plants, *Hydrocarbon Process.*, **46** (7) 161 (1967).

101. H. E. Haley, *Electrical World*, **167** (20) 71 (May 15, 1967).

102. R. Kopita and T. G. Gleason, *Chem. Eng. Progr.*, **64**, 74 (1968).

103. Solvay Process Corp., Tech. Service Report No. 6.61, Syracuse, N.Y., March 30, 1961.

104. R. D. Oldenkamp and D. E. McKenzie, "The Molten Carbonate Process for Control of Sulfur Oxide Emissions," Air Pollution Control Assoc. Meeting, St. Paul, June, 1968.

105. R. D. Harris, "The Occurrence of Sulfur in Bituminous Coals and Methods of Removal," presented at Tech. Sales Conf., Nat. Coal Assn., Pittsburgh, Pa., September 16, 1965.

106. Anon., *Electrical World*, 29 (October 9, 1967).

107. S. Ludwig, *Chem. Eng.*, **75**, 70 (Jan. 29, 1968).

108. E. A. Pirsh, N. P. Rusanowsky, and N. W. Young, "An Appraisal of Air Pollution in the Power Industry," paper presented to the American Power Conference, Chicago, April 1968.

109. Kun Li, R. R. Rothfus, and A. H. Adey, *Environ. Sci. Tech.*, **2**, 619 (1968).

110. J. G. Stites, W. R. Horlacker, J. L. C. Bachofer, and J. S. Bartman, "The Catalytic-Oxidation System for Removing SO_2 from Flue Gas," Monsanto report, St. Louis, 1969.

Control of Internal Combustion Engines

R. G. TEMPLE

*Department of Chemical Engineering, University of Aston,
Birmingham, England*

I. INTRODUCTION

In general, abatement of atmospheric pollution follows an intense local problem. The general public is usually more readily aroused by the sight or smell of pollutants, and visible but frequently less harmful pollution is often the subject of legislation long before invisible yet more serious pollutants are considered. Little attention was paid to pollution from motor vehicles until the peculiar climatic conditions of Los Angeles led to the formation of a photochemical smog which was found to be related to the rapidly rising concentration of motor vehicles, and the Californian Motor Vehicle Pollution Control Board (MVPCB) was set up in 1960. Its main abatement proposals culminated in the incorporation in gasoline engine vehicles of crankcase and exhaust emission-control devices and systems for minimizing evaporative emissions.

The California problem naturally drew attention from the U.S. Federal authorities and from other countries, particularly those with an interest in exporting motor cars or accessories to the United States.

These widespread and varied interests have encouraged urgent and intensive research over the whole field of automotive engine emissions leading, in a relatively short period, to major contributions to our knowledge of the chemical reactions involved. One reason is that legislation in this field depends on accurate measurement of pollutants and so the development of legislative controls has been accompanied by considerable research into analytical techniques.

Diesel engine pollution at a serious level becomes evident as black smoke. Legislative control has been developed in many user countries, and since diesel pollution is markedly visible, control has relied to some extent on visual assessment. However, in the absolute measurement of smoke density it is essential to eliminate individual and environmental variations; such measurements have been incorporated in control legislation.*

* Author's Note:

Rapid developments in exhaust pollution control have taken place during the preparation of this material. The normal interval between submission and publication has

II. GASOLINE ENGINE EXHAUST EMISSIONS

A. Constituents

Gasoline engine exhaust is likely to consist of the following gases in varying concentrations, according to driving conditions: carbon monoxide, hydrocarbons, oxides of nitrogen, sulfur dioxide, aldehydes, carbon dioxide, hydrogen, oxygen, water vapor, and nitrogen. Carbon monoxide, hydrocarbons, and oxides of nitrogen are recognized as the most serious pollutants; the considerable variation in their exhaust concentrations with driving conditions is shown in Table I.

TABLE I

Typical Exhaust Gas Constituents

Pollutant	Idling	Acceleration	Cruising	Deceleration
Carbon monoxide %	4–9	0–8	1–7	2–9
Hydrocarbons				
(as hexane) ppm	500–1000	50–800	200–800	3000–12000
Oxides of nitrogen ppm	10–50	1000–4000	1000–3000	5–50

Engine conditions (speed, fuel–air ratio, manifold vacuum, etc.) exert a significant influence on the exhaust composition, and these conditions are largely determined by the various driving modes. For example, optimum fuel–air ratio depends on the varying demands of the several driving modes. Maximum power requires a mixture 10–15% richer than stoichiometric, while cruising is normally best (i.e., most economical) at up to 10% below stoichiometric. At idle, most engines require even richer mixtures to compensate for residual combustion products in the cylinder. At acceleration, induction manifold pressure suddenly falls, resulting in condensation of fuel on the manifold walls and temporarily weakening the mixture entering the engine. Various enrichment devices are used in the carburetor to prevent stalling at this condition. On deceleration, throttle closure results in high induction manifold vacuum. This draws exhaust gas back into the

---

therefore seen further developments which could not be adequately described without unduly delaying publication.

The pace of such development, for example in exhaust emission control, is made clear by the extent of the list of control systems submitted for approval in 1968, compared with the previous list. By that time (1968) most major world automobile manufacturers developed, and received approval for, systems for control of exhaust, crankcase and evaporative emissions.

cylinders, reducing the air intake, and producing exceedingly rich mixtures. When the pressure ratio across the throttle exceeds 2:1, a standing shock wave is set up, resisting forward flow.

While modern carburetors are designed to cope with these varying conditions, the main pollutant concentrations are significantly affected. Hydrocarbons, as seen in Table I, are most prominent during deceleration. Nonflammable mixtures are present in parts of the cylinder encouraged by residual exhaust and poor mixing at low inlet mixture velocities. At other driving conditions, combustion is more complete and hydrocarbon emissions correspondingly lower. Figures 1 and 2 show effects of air fuel ratio and spark timing on exhaust hydrocarbon emissions (1). Interpretation of emission data may be confused unless a clear distinction is made between emissions expressed as mass rate (lb/hr) or as a fraction of exhaust volume, in parts per million (ppm) (Fig. 2).

Hydrocarbon emissions are seen to increase in a near-linear fashion when expressed on a weight (lb/hr) basis but to decrease significantly when volumes (ppm) are related to air flow. While the statutory limits are in ppm (the original studies being related to atmospheric calculations) the California test cycle is weighted according to a "lb/hr" basis and currently standards on this basis are favored.

Carbon monoxide, due to its toxicity, has been for many years the subject of exhaust control studies. Since the reduction of carbon monoxide is also in the interests of high efficiency, improved engine design had gone some way to reducing exhaust concentration before the overall exhaust problem was recognized. The reduction, however, was not sufficient to counteract the increase in the number of vehicles, and by 1960 was still above the 1–2% generally agreed to be desirable. Figure 1 shows the direct relation between carbon monoxide emission and air–fuel ratio. As with hydrocarbon emissions, leaner mixtures lead to lower pollutant concentrations.

Nitrogen oxides (NO_x) were early to be recognized as one of the principal precursors of photochemical smog. Approximately 99% of the oxides of nitrogen present in exhaust emissions have been shown to be in the form of nitric oxide (2). While the literature concerning exhaust hydrocarbon and carbon monoxide content is quite extensive, it is far less so in respect to nitric oxide. This is mainly due to the fact that exhaust nitric oxide measurement is a more cumbersome and lengthy matter. Check techniques have been devised to allow use of nondispersive infrared analysis (3).

As with hydrocarbon and carbon monoxide emission, nitric oxide concentration shows considerable variation with fuel–air ratio. Figure 3 shows clearly the peak values reached at near stoichiometric mixtures for steady running. The insignificant emissions of NO_x during idling are also clearly brought out in that diagram; these low values are related to the high

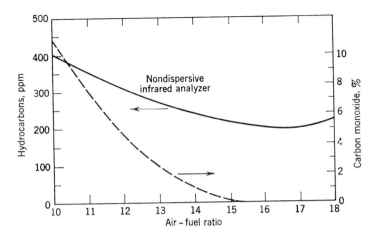

Fig. 1. Effect of air–fuel ratio on exhaust emissions.   (—) hydrocarbons; (– – –) carbon
monoxide.

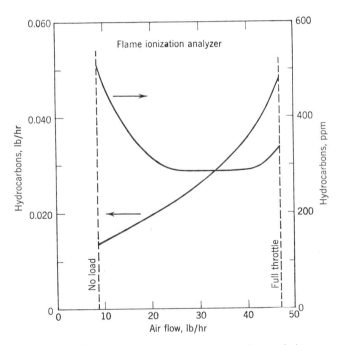

Fig. 2. Effect of engine air flow on hydrocarbon emissions.

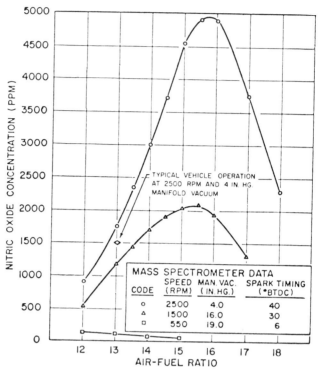

Fig. 3. Effect of air–fuel ratio on exhaust NO concentrations for various speed–load combinations.

manifold vacuum which increases residual gas dilution and lowers charge density, thus reducing flame speed and reaction temperature. The effects of other engine variables on nitric oxide concentrations are by no means simple. Engine speed, air–fuel ratio, distribution, and spark timing are all likely to have significant effects on cylinder combustion conditions. Owing to the interaction of the many process variables, it is difficult to isolate the effects completely but the experimental work bears out the view that NO_x concentration is dependent on peak combustion-gas temperature and available oxygen.

Aldehydes were identified at an early stage in Los Angeles as end products of the photochemical smog-forming reactions. While many are known to be powerful irritants, aldehydes in general have received less attention than carbon monoxide and hydrocarbon in recent literature. Their importance as exhaust constituents lies in the fact that they are major end products of the photochemical reaction involving reactive hydrocarbons and oxides of nitrogen (4,5). Thus their identification and measurement could well be a vital step in the establishment of pollution control methods. Emissions

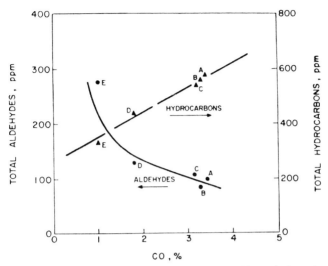

Fig. 4. Aldehyde and hydrocarbon versus carbon monoxide emissions (various cars, California standard cycle, hot).

ranging from 10 ppm to several hundred ppm are reported (6,7). However, in contrast to hydrocarbon emissions, aldehydes show considerable increase at high air–fuel ratios. For several multicylinder vehicles a clear relation between aldehyde and carbon monoxide emissions has been shown (7), and since carbon monoxide content in the range shown is a measure of air–fuel ratio, it is clear that increasing the air–fuel ratio has a marked deleterious effect on total aldehydes (Fig. 4). This work indicated that extreme leaning of the mixture to reduce hydrocarbons may result in an *increase* in pollution problems.

Polynuclear aromatics such as 3.4-benzpyrene have received less attention than other exhaust pollutants whose effects are more directly related to a particular atmospheric irritant or nuisance. However, the danger to health from such recognized carcinogens is now well established and with the rapid rise in the incidence of lung cancer, all possible sources of these materials are receiving much greater scrutiny. Published data on the emissions of 3.4-benzpyrene from petrol engines vary widely, but it has been shown that a significant proportion of the total 3.4-benzpyrene in city air can come from automobiles (8). Further studies are required to establish the mode of dispersion of this substance in the atmosphere. Fuel consumption (particularly aromatic content) is found to exert a marked influence on polynuclear aromatic emissions. Concentrations ranging from 0.5 gr "tar"/1000 ft³ exhaust for isoctane or di-isobutylene, to 2.6 gr "tar"/1000 ft³ for commercial gasoline (36% aromatic) are reported (8); 3.4-benzpyrene content of the tar showed similar variation, with values of 100 ppm for commercial gasoline.

B. Regulations

1. Limits

Statutory limits for hydrocarbon and carbon monoxide concentrations were defined for all 1967 models on sale within the state of California and for all imported vehicles for the 1968 model season. The standards for exhaust emissions were:

Hydrocarbons   270 ppm by volume (as hexane)
Carbon monoxide   1.5% by volume

Further emission standards were laid down in 1965 by the California Legislature to apply to all 1970 models. The 1970 limits are:

Hydrocarbons   180 ppm by volume (as hexane)
Carbon monoxide   1% by volume
Oxides of nitrogen   350 ppm by volume (as NO_2)

Exhaust gas concentrations are to be adjusted to a dry exhaust volume containing 15% ($CO + CO_2$). 1970 California standards call for adjustment of the exhaust concentrations by the ratio $15/(\frac{1}{2}CO + CO_2 + 10 \times HC)$. Federal regulations impose similar standards: $15/(CO + CO_2)$ for load and idle mode, $15/(6HC + CO + CO_2)$ for the deceleration mode.

Meanwhile the Federal Government proposed that the California 1967 limits should apply to all 1968 models for sale in the United States. After representations by foreign vehicle manufacturers, concessions were allowed which recognized the difficulty of pollution control for smaller engines. California also accepted these concessions and the following limits became obligatory for 1968 models (9):

Engine capacity	Hydrocarbons ppm by volume (as C_6H_{14})	Carbon monoxide, % by volume
Over 140 in.³	275	1.5
100–140 in.³	350	2.0
50–100 in.³	410	2.3
Under 50 in.³	No limit	

The standards refer to a composite sample for a representative driving cycle shown in Sect. B.2 below.

While the U.S. and California limits are of great concern to other countries with large motor manufacturing interests, only in Japan have statutory limits been laid down.   These are for carbon monoxide, and are as follows:

| 1967 models: | 3% by volume |
| 1968 models: | 2.5% by volume |

Since there is little likelihood of photochemical smog in Europe it is anticipated that legislation in European countries would be mainly concerned with
carbon monoxide levels.

2. Test Procedures

A truly representative sample of exhaust gases can only be obtained if the
individual variations in driving habits can be allowed for in a representative
driving pattern.   To establish such a pattern the American Automobile
Manufacturers' Association (A.M.A.) set up a Traffic Survey Panel in 1956.
This panel organized a detailed study of driving conditions in Los Angeles.
A representative test route was selected and 135 test runs were made,
reflecting variations in rush-hour and off-peak traffic conditions, in manual
and automatic transmission vehicles, and in "driver" habits.   Continuous
recordings of speed, speed change, manifold vacuum, and fuel consumption
were made (10).

From the results, an 11-mode test cycle was proposed which would simulate
a 20-min trip through Metropolitan Los Angeles, starting with a cold engine.
After further investigation the California MVPCB laid down a 7-mode test
cycle for chassis dynamometer operation.   To comply with state regulations
emission limits must be achieved in engines operated over this 7-mode cycle
seven times from a cold engine start.   The test cycle is shown in Table II;
gear changes are indicated in Figure 10.

The first four 7-mode cycles are classed as "warm-up" cycles and emissions
from these are weighted 35%.   The fifth cycle is run but not assessed.
The sixth and seventh cycles are classed as "hot" cycles and emissions
weighted 65%.

The test is run on a chassis dynamometer loaded so that manifold vacuum
at 50 mph is equal to that of the car previously driven on a straight road in
top gear.   Inertial flywheels of a prescribed weight are to be attached to the
dynamometer according to the weight of the vehicle and gear changes
throughout the cycle are prescribed for each type of manual and automatic
gear-box.   The temperature in the test room must be between 68 and 86°F
and the car must be stored for 12 hr at a temperature not falling below 60°F.
A deterioration factor is calculated on the assumption that deterioration is
linear after the first few thousand miles.   Emissions after 50,000 miles are
taken as the lifetime average (assuming lifetime of 100,000 miles).   The
50,000-mile level of emissions is determined by measuring emissions at
4000-mile intervals for a smaller sample of "durability" cars over an approved

TABLE II

Seven-Mode Driving Cycle

| Sequence | | Acceleration | Time in, | Cumulative | Weighting |
No.	Mode	mph/sec	sec	time, sec	factor
1	Idle[a]		20	20	0.042
2	(0–25)	2.2	11.5	31.5	0.244
	( )		)14		
3	(25–30)	2.2	2.5	34	Data not read
4	30		15	49	0.118
5	30–15	−1.4	11	60	0.062
6	15		15	75	0.050
7	(15–30)	1.2	12.5)	87.5	0.455
	( )		)29		
8	(30–50)	1.2	16.5	104	Data not read
9	50–20	−1.2	25	129	0.029
10	20–0	−2.5	8	137	Data not read

[a] On first cycle only idle engine in neutral at 1000–1200 rpm for 40 sec. All subsequent idle periods will be as specified: in gear, at normal speed, and for 20 sec. Gear changes are indicated in Figure 10.

route of 50,000 miles. A deterioration factor: emissions at 50,000 miles/ emissions at 4000 miles is then determined and applied to vehicles under test which have completed 4000 miles and are classified as "emission data" vehicles. The number of durability and emission data cars is decided by sales forecasts, etc.

In this way, the effects of the many interacting "driving pattern" influences on emission levels are accounted for, and a composite sample obtained which is reasonably representative of a mean driving pattern. The exhaust gas concentration determined in this sample must then be adjusted to a dry exhaust volume containing 15% by volume of carbon monoxide.

3. Analytical Methods

a. Hydrocarbons

The California MVPCB (and U.S. Federal) standards define hydrocarbons as "the organic constituents of vehicle exhausts as measured by a hexane-sensitized nondispersive infrared (n.d.i.r.) analyzer or equivalent method." In the absence of general agreement on alternatives, n.d.i.r. analysis is accepted, although its limitations for hydrocarbon determinations are recognized, and development of alternative analytical methods is the subject of much research effort (11,12).

b. Carbon Monoxide

California and Federal legislation similarly requires that carbon monoxide shall be measured by a nondispersive infrared analyzer or by an equivalent method.

c. Oxides of Nitrogen

California 1970 standards require that oxides of nitrogen shall be measured by the phenol-disulphonic acid or equivalent method.

4. Future Developments

a: Definition and Rating of "Reactive" Hydrocarbons

While limitation of total hydrocarbon emissions is an obvious step in the direction of pollution reduction, it is recognized that certain hydrocarbons are more "reactive" than others; in this context, this signifies their participation in the photochemical reactions which produce smog. It is obviously desirable to establish relative weightings for individual hydrocarbons and to date there is considerable conflict of opinion as to what this weighting should be.

The California Department of Public Health has therefore proposed the following definition of Reactive Organic Compounds and has invited comments and suggestions on analytical procedures.

b. Proposed List of Reactive Compounds

i. Olefins

ii. Aromatics except benzene

iii. Higher paraffins (C_6 and over)

iv. Aldehydes

The proposals include aldehydes which are the major constituents of the oxygenated compounds in exhaust gases   It is acknowledged that measurement of total aldehyde emissions would be preferable to procedures for *all* the oxygenated compounds and suggestions on analytical procedures are invited.

5. Certification of Control Devices

The 1959 California Legislation provided that the MVPCB notify the Department of Motor Vehicles when two or more approved control devices had been certified as acceptable. After this certification date approved devices became mandatory according to the following timetable:

1. One year later on all new vehicles and all used vehicles on transfer of registration.

2. On the second December 31st after the certification date on all commercial vehicles registered in California.

3. On the third December 31st on all vehicles registered.

The above timetable commenced in August, 1964, with the certification of four exhaust emission control devices for new cars, and one for used cars. Additional systems have since been added and details of some of the principal devices are outlined below.

C. Control Methods

Early attempts to reduce hydrocarbon and carbon monoxide emissions involved catalytic oxidation devices and thermal afterburners. Carburetor and ignition modifications were naturally the subject of intense study and Chrysler Corporation devised a system ("the Cleaner Air Package") which met Californian requirements by engine modifications alone. Other large automobile manufacturers, for example, General Motors, Ford, American Motors, and International Harvester, have preferred systems which achieve their objective by oxidation of combustibles in the exhaust ports; air is injected close to the back of the exhaust valves in these Manifold Air Oxidation ("Man-Air-OX") Systems. Other control methods under development include modification to combustion chamber design, inlet air climatic control, fuel injection systems, and exhaust heating.

The two systems presently in use to meet California requirements are the Cleaner Air Package system of engine modification and various forms of the "Man-Air-OX" systems.

1. Engine Modification Systems

a. The Chrysler "Cleaner Air Package"

This was the first system to comply with California regulations merely by engine modifications (13). The present Cleaner Air System has evolved from the Cleaner Air Package and produces satisfactory emission levels mainly by means of leaner fuel/air mixtures, adjusted ignition timing, higher engine speed and air flow at idle, and special choke calibration. The main items of the Cleaner Air Package are shown in Figure 5. Other modifications are made peculiar to each engine. These include redesigned combustion chambers, improved manifold heat transfer facility, and new carburetor metering systems to improve carburation and cylinder distribution. Most of the modifications are directed toward the idle, acceleration, and deceleration modes in which the bulk of emissions occur.

Increased idle speed (with appropriate air flow) with the leaner mixture ensures that engine conditions are similar to cruising. The idle mixture is

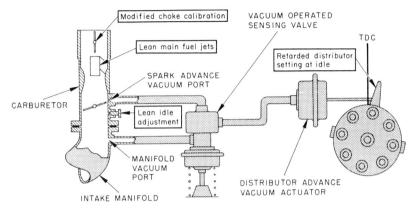

Fig. 5. Chrysler cleaner air package, schematic.

restricted by a device limiting the screw adjustment and this ensures that emissions remain within the acceptable range. Spark timing is retarded at idle (Fig. 6.) and this increases exhaust gas temperatures with consequent better combustion of the hydrocarbon and carbon monoxide.

During acceleration modes, emissions are low due to the leaner mixture; a special choke calibration deals with the richer mixtures needed for cold starting. The modifications to cylinder distribution and manifold heat exchange permit the choke to open sooner than in unmodified engines.

Deceleration is characterized by high concentrations of exhaust emissions and if it occurs from high engine speeds a considerable contribution to total emissions results. High intake manifold vacuum encourages dilution of the incoming charge with exhaust gases drawn back into the cylinder before the exhaust valves are fully closed; the ensuing combustion is thus far from complete, causing high emissions. The Cleaner Air System reduces these

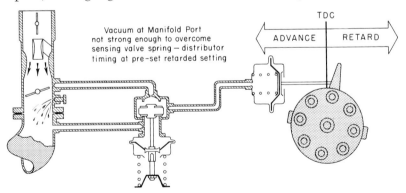

Fig. 6. Chrysler cleaner air package at idle condition.

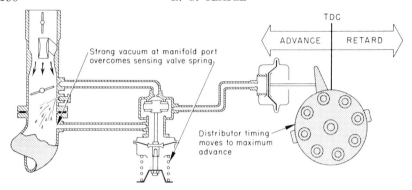

Fig. 7. Chrysler cleaner air package at deceleration condition.

emissions first by means of the increased air flow allowed in the closed position of the throttle blade; this reduces manifold vacuum. A further modification improving combustion during deceleration is advanced spark timing. This is accomplished by the vacuum control sensing valve which routes manifold vacuum direct to the spark advance unit (Fig. 7).

Reference to Figure 6 shows the isolation of the manifold connection at idle (so that the timing unit moves to retard). When accelerating, carburetor vacuum acts directly on the spark advance unit (sensing valve also as in Fig. 6.)

Chrysler cars fitted with the appropriate Cleaner Air System modifications meet California and Federal requirements for hydrocarbons and carbon monoxide. Practically no additional maintenance is called for beyond normal requirements.

2. The Zenith Duplex Induction System

This system incorporates, in parallel with the induction manifold, a primary feed pipe of smaller cross section whereby the mixture is taken through an exhaust-heated chamber (14,15).

Flow in the primary feed-pipe (Fig. 8) is controlled by a separate throttle (C) the main larger section being closed during part-throttle running by the "secondary" throttle valve (D). Figure 9 shows alternative positions for the throttles in the primary (lower) and secondary (upper) manifolds. The secondary throttle is not brought into action until near full throttle operation. Thus at acceleration, deceleration, and idling the primary heated manifold alone is in use. Some variation in the road speed achieved with this manifold is inevitable with different vehicles; however, in the development models the primary throttle alone achieved acceleration up to 50 mph and beyond, and thus covered the whole of the Californian test cycle (see Fig. 10).

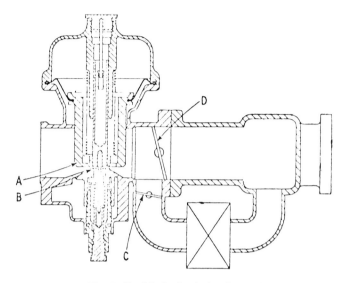

Fig. 8. Zenith duplex induction system.

Carburetor features are naturally also of primary importance in mixture control and a special thermostatic valve is incorporated to offset the richening effect as the engine warms up. This valve bleeds air downstream of the carburetor and maintains a leaner mixture. In addition a diaphragm bypass valve is incorporated which short-circuits the primary throttle at a selected vacuum, thus lowering the manifold depression and supplying extra mixture to support combustion. Ignition is retarded at idle and deceleration by means of a small valve, operated by lost motion on the acceleration linkage, which feeds vacuum to a retard capsule on the distributor.

The high standard of mixture distribution in all driving modes allows higher air-fuel ratios to be used, and emissions quoted are well within U.S. Federal standards for the range of engine capacities.

3. Volvo Dual Manifold Emission Control System

In order to comply with U.S. 1968 exhaust standards inlet manifold modifications were made to the Volvo B18B twin-carburetor engine. The dual manifold emission control system was chosen in favor of an afterburner system when it was found that the latter required accurate balancing of the twin carburetor and after a comparative evaluation of cost and performance of both systems.

The emission control system (16) relies on lean carburation, retarded spark timing at idle, increased idle speed, and throttle bypass valve for deceleration. The modified induction system incorporates a heat transfer and turbulence

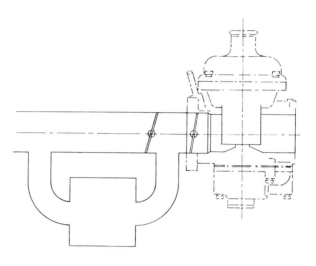

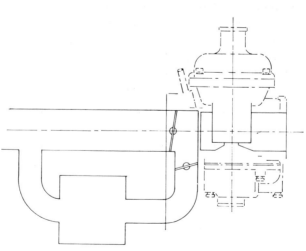

Fig. 9 . Alternative throttle arrangements in Zenith duplex induction system.

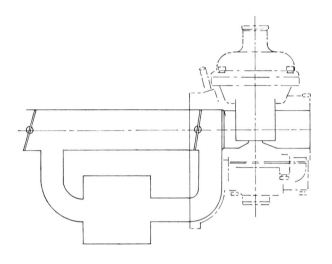

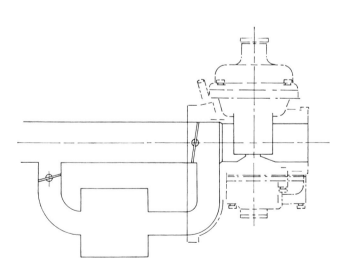

Fig. 9 *(Continued)*

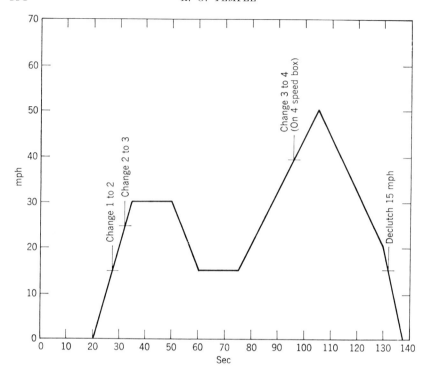

Fig. 10. California emission test cycle.

chamber and is shown in Figure 11 compared with the conventional twin carburetor layout of the unmodified engine (Fig. 12). The carburetor throttle and manifold throttle are linked as shown in Figure 13, the opening point of the manifold throttle corresponding to a cruising speed around 75 mph, covering the whole of the California and other driving cycles.

A static spark timing of 5° before top dead center is used, this relatively moderate retard being chosen to avoid "running on" and overload of the cooling system. The deceleration valve is a simple vacuum-operated poppet valve which claimed to show only moderate limitation of the engine braking effect.

50,000-mile durability tests of cars incorporating these engine modifications have shown that hydrocarbon and carbon monoxide emissions are well within prescribed limits.

4. Man-Air-OX Systems

a. General Motors Air Injection Reaction (A.I.R.) System

This system is basically an air injection system combined with engine modifications to increase its effectiveness (17). Figure 14 shows the

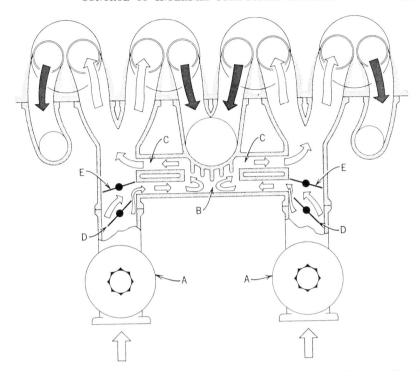

Fig. 11. Induction system for Volvo emission control engine. *A*, carburetor; *B* and *C*, cross flow pipes; *D*, carburetor throttle; *E*, manifold throttle.

belt-driven air pump and air distribution lines leading to the exhaust ports. The pump is a Saginaw positive displacement, nonlubricated, vane pump (Fig. 15), capable of delivery against 16.4 in. Hg (18).

While this air distribution system is common to all General Motor vehicles, the necessary engine modifications vary to some extent for particular engine–carburetor–transmission combinations. Modification of the standard carburetor flow characteristics is necessary to achieve close control of mixture ratio in the low flow region. Retarded ignition timing at idle effectively contributes to the reduction of emissions without reducing fuel economy or performance. The effectiveness of the system is demonstrated in Figure 16.

An antibackfire gulp valve is incorporated which permits a controlled amount of air to enter the intake manifold on throttle closure and thus prevents buildup of unburnt fuel in the exhaust. Suitable pressure relief and check valves are also incorporated for protection of the air injection system.

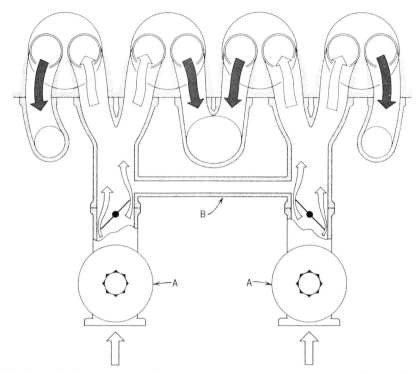

Fig. 12. Induction system for Volvo noncontrol engine. *A*, carburetor; *B*, cross flow pipe.

b. The Ford Thermactor Exhaust Emission Control System (19)

This system comprises an air pump, with distribution manifolds and necessary relief and check valves (Fig. 17) and, for some engines, modified distributors and carburetors.

A semiarticulated, vane-type, positive displacement pump was selected after extensive tests with many designs. A pressure relief valve, while

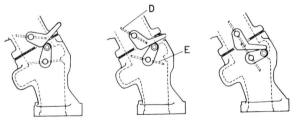

Fig. 13. Throttle positions in Volvo dual manifold system. Left to right: 25% load, 50% load, 100% load. *D*, carburetor throttle; *E*, manifold throttle.

Fig. 14. General Motors air injection reactor system.

protecting the pump against excessive back pressure, also limits exhaust temperatures at highest engine speeds. At this condition excessive air injection could cause exhaust temperatures to exceed working limits of standard materials and necessitate special materials and heat shielding of other components. A further advantage of the relief valve is to minimize horsepower losses resulting from excessive air injection into the exhaust system. The ideal relief valve setting varies for each engine–vehicle combination according to the back pressure developed in the exhaust system. However, for manufacturing simplicity two pressure settings were chosen: for six cylinder, and for eight cylinder engines, respectively. A "nonreturn" check valve (to protect the air pump against drive-belt failure), and anti-backfire valve are key features as with other Man-Air-OX systems. The original "gulp" type antibackfire valve was found, on smaller engines, to have undesirable side effects such as engine speed-up in gear changes especially under choked cold start conditions, and a sluggishness in engine response on deceleration. A bypass valve was therefore developed which momentarily reduces the air supply to the exhaust ports following each throttle closure.

c. Other Systems

Other U.S. motor manufacturers (American Motors, International Harvester, Kaiser, Jeep, and Checker) introduced various combinations of air

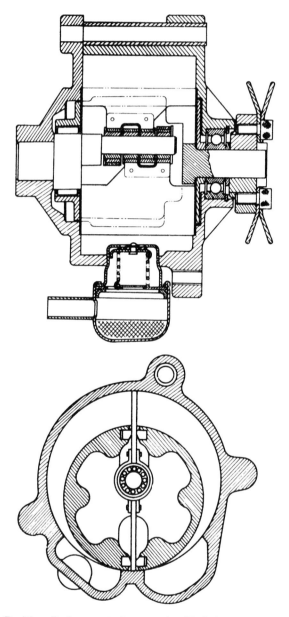

Fig. 15. Positive displacement air pump for Air Injector Reactor System.

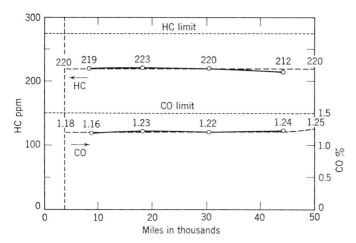

Fig. 16. Average exhaust emissions from nine test cars fitted with General Motors air injection reactor system.

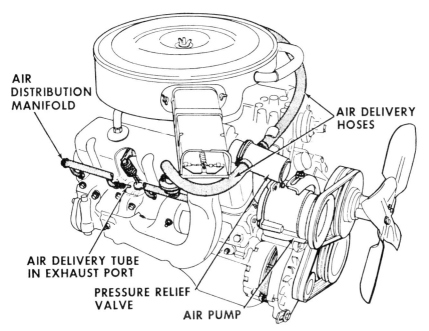

Fig. 17. Ford Thermactor exhaust emission control system.

pump and air distribution manifold together with carburetor and timing modifications where necessary for 1967 models to be licensed in California. Three manufacturers, Ford Motor, American Motors, and General Motors introduced both air-injection and engine modification systems to cover their whole range of models.

For 1968 models most of the major world automobile manufacturers had approval for their exhaust emission control systems. Air injection systems were approved for Adam-Opel, Bavarian Motor Works, British Motor Corporation, Citroen, Daimler-Benz, Ford of Great Britain, Isuzu, Porsche, Rolls Royce, S.S. Automobiles, Toyota Motor Co., and Volkswagen.

Engine modification systems were approved for Chrysler Corporation, Cord, Fiat, Jaguar, Lotus, Peuguot, Rover Co., Renault, SAAB, Simca, Standard Motor, and Volvo. The remaining automobile companies appearing in the 1968 list of approved exhaust systems, viz. American Motors Corporation, Checker, Ford Motor Co.,(USA), General Motors Corporation, International Harvester, Kaiser, Jeep, Nissan Motor Co., and Shelby specified air injection systems for certain models and engine modification for others.

III. CRANKCASE EMISSIONS

Crankcase emissions arising from gases which escape past the pistons ("blowby" gases) represent a considerable proportion of potential hydrocarbon emissions. They were recognized as serious pollutants by 1959 and were the subject of early legislative control in California. Positive crankcase ventilation to the inlet manifold enables the emissions to be dealt with in the cylinders or by exhaust-control devices.

A. Constituents

Analysis has shown that crankcase emissions are principally unburned fuel/air mixture and thus hydrocarbon concentrations are in the region of 10,000–20,000 ppm (as hexane). The olefin content of the blowby emissions is of particular importance since higher olefins are known to be major contributors in the development of photochemical smog. Ethylene, in contrast, shows extremely low photochemical reaction rate (21). It is shown (22) that olefins excluding ethylene represent a much greater proportion of blowby gases than of exhaust. Thus crankcase emissions, if allowed to escape direct to the atmosphere, would be extremely serious sources of hydrocarbon pollutants.

A smaller proportion (20–30%) of the crankcase emissions consists of combustion products and thus carbon monoxide concentrations are less than 10% of typical exhaust emissions. Other pollutant combustion products will arise in similar proportions to the main exhaust.

B. Regulations

1. Limits

In 1960 the California legislature adopted a standard limit for crankcase hydrocarbon emissions of not more than 0.15% by weight of the fuel supplied, under the test conditions outlined in Sect. B.2 below. Control devices were obligatory on all cars by the end of 1965 and a new standard of 0.10% of fuel supplied was issued for California in that year.

The U.S. Federal Government in 1966 made no provision for crankcase emission testing but stated that no discharge to atmosphere was permissible (23).

France imposed regulations on crankcase emissions and specified test conditions similar to those of California (1963) in 1965 adopting the hydrocarbon limit of 0.15% by weight of fuel supplied.

2. Test Procedures

The 1963 California test procedure for "Positive Crankcase Ventilation" (PCV) devices is designed to check reliability and establish ability to comply with the standard limits over the following engine cycle:

Mode	Weighting
Idle	25% total time
30 mph at 16 in. Hg manifold vacuum	25% total time
30 mph at 10 in. Hg manifold vacuum	50% total time

In 1965 the weighting factors were modified to 19% at idle, 37% at 30 mph 10 in. Hg manifold vacuum, 11% at 30 mph 16 in. Hg manifold vacuum, 8% at 30 mph 2 in. Hg manifold vacuum, 25% at deceleration (taken as zero blowby).

C. Control Methods

1. General Requirements

Procedure for approval and certification of PCV devices was laid down for California in 1963. In addition to the hydrocarbon emission limit, approved devices had to comply with the following conditions.

1. No adverse effect on engine operation or performance.

2. Safe operation.

3. Durability, permitting efficient operation for 12,000 miles with normal maintenance.

4. No excessive heat, noise, or odor.

5. Not unduly costly.

6. Device must not create or contribute toward noxious or toxic content of ambient air.

In addition, protection against carburetor backfire was required and the applicant for approval was asked to show that the device would not draw oil from the crankcase. In order to comply with criterion *5* above it was necessary to demonstrate that the average air/fuel ratio was not reduced by more than 1% nor increased by more than 4%.

Although the alteration of air/fuel ratio can be readily allowed for in factory-installed devices, the above restriction provides a reasonable design objective for devices intended for used vehicles (24).

2. Positive Crankcase Ventilation (PCV) Devices

PCV systems installed to meet the 1963 California requirements included tube-to-air cleaner devices, tube-to-intake manifold devices with automatic metering through a ventilation valve, and combinations of these "split-flow" devices. Such devices have been incorporated by most major world motor manufacturers for vehicles to be sold in countries or states where legislation applies.

3. Tube-to-Air Cleaner Devices

Such a device is the American Motors Crankcase-to-Air-Cleaner system shown in Figure 18. Adequate positive ventilation is achieved at all loads

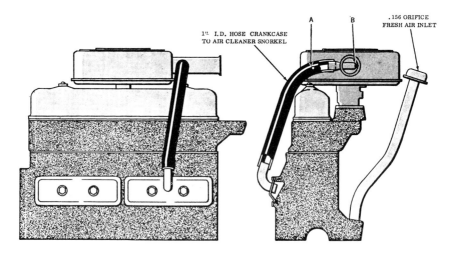

Fig. 18. American Motors crankcase-to-air cleaner crankcase ventilation system schematic.

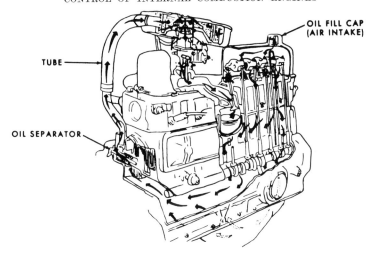

Fig. 19. Installation of air cleaner crankcase emission control system on a Ford 2231CID engine.

by means of the single ventilation tube which is introduced at right angles into the air cleaner snorkel (25). Its end is beveled at 45°, the bevel facing downstream so that a pressure differential is set up by air flow in the air cleaner. Suction is thus created in the vent tube and this is found to be adequate even during idle and deceleration modes.

The Ford Air Cleaner Emission Control System (26) similarly uses a single tube between crankcase and air cleaner (see Fig. 19). In both systems air inlet to the crankcase is restricted by means of an orifice in the filler cap or breather, thus avoiding oil carry-over into the induction system at high engine speeds. In addition to filtering incoming air the air cleaner element serves to remove low volatile contaminants such as gums and tars which would interfere with carburetor and induction valve functions.

4. Tube-to-Intake Manifold Devices

These systems depend on the operation of a ventilation valve for automatic metering of gas flow from crankcase to inlet manifold. A typical design is shown in Figure 20. Such a valve incorporates a fixed orifice at highest vacuum (as shown) while a variable area orifice (annulus) is additionally provided by spring action when manifold vacuum falls (27). This and similar valves are designed so as to modulate ventilation with manifold vacuum. Furthermore, such a valve must serve as an antibackfire valve isolating the crankcase immediately when a pressure wave arises.

Later valve designs incorporate a self-cleaning plunger to avoid clogging the ventilation tube (Fig. 21). Such valves were found to provide

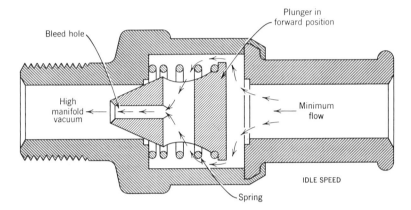

Fig. 20. Crankcase ventilation valve.

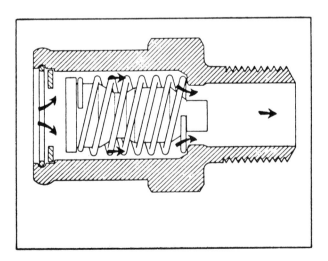

Fig. 21. Self-cleaning crankcase ventilation control valve. Movement of loose-fitting plunger tends to keep port free from deposits.

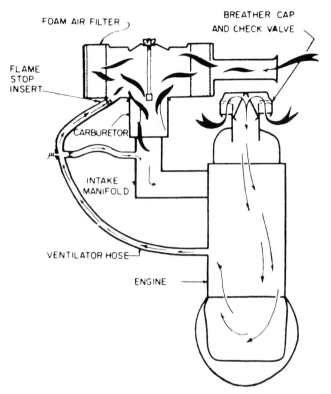

Fig. 22. Split-flow crankcase ventilation system.

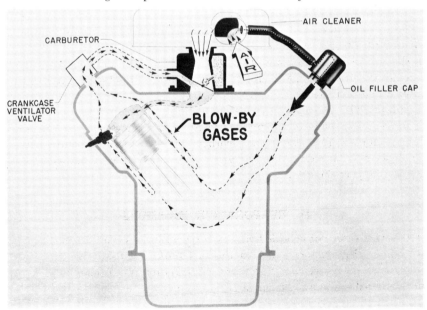

Fig. 23. Crankcase ventilation control system.

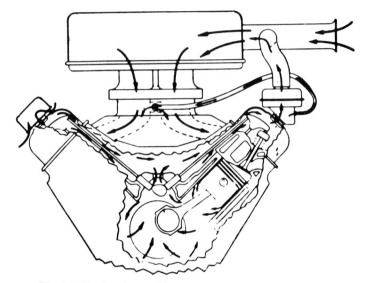

Fig. 24. Dual action positive crankcase ventilation system.

satisfactory flow characteristics while maintaining adequate crankcase ventilation at all driving modes (28).

5. *Combination ("Split-Flow") Ventilation Systems*

In these systems crankcase ventilation is achieved through connections to both air cleaner and intake manifold. Typical systems are shown in Figures 22, 23, and 24. Such systems incorporate an orifice to restrict flow in the intake manifold connection so that almost all emissions are drawn through this route at speeds up to about 35 mph at road load. Above this value air cleaner suction takes effect ensuring adequate ventilation under all conditions (29). A check valve may be installed in the air cleaner connection to prevent bypassing of the filter element at low engine speeds. In some designs an extra filter element is incorporated for this purpose but a restriction may still be required in the air cleaner connection to act as a flame arrester (30); cf. Figure 22.

IV. EVAPORATIVE EMISSIONS

A further source of volatile hydrocarbon emissions is their direct escape to atmosphere by evaporation from the fuel tank and from the carburetor. Such emissions represent a significant contribution to atmospheric pollution and their reduction was one of the first objectives of legislative action.

A. Constituents

The hydrocarbons are naturally the "light ends" of the gasoline and the proportions of each present will be subject to considerable daily and seasonal variation with ambient temperature. Such variation is also affected by certain changes in fuel specification, particularly those which are made to improve cold start characteristics in cold weather.

In general vapor losses will mainly consist of C_3 to C_7 hydrocarbons with 70–75% in the butane–pentane region (31,32). The amount of evaporation from the petrol tank depends on fuel volatility, ambient temperature, and the liquid surface area. Evaporation from carburetor vents depends on location and method of venting, fuel volatility, internal carburetor dimensions, and ambient (i.e., under the hood) temperatures.

In view of these many variables, actual losses of fuel by evaporation show a wide variation. In total they were shown to reach values of 7–9% by weight of fuel used for U.S. 1956–1957 cars (prior to control legislation). The relative importance of the four major emission sources is generally in the order of: exhaust, 65%; crankcase, 20%; carburetor, 9%; and fuel tank, 6%.

B. Regulations

1. Limits

California State Department of Public Health proposed standards for evaporative emissions in 1964 and published interim test procedures with the following limits in November, 1966 (33).

Fuel tank emissions: 6 g hydrocarbon/"standard" 24-hr day (see below for test schedule)

Carburetor hot soak emissions: 2 g hydrocarbon/soak-engine coolant at 180°F minimum and ambient air at 85–95°F min. Soak duration of 1 hr.

2. Test Procedure

The California test schedule commences at 6 A.M. on a sunny day when minimum temperature is between 60 and 70°F and predicted maximum 90°F; at 7.30 A.M. vehicle to be driven for 20 min over a $7\frac{1}{2}$ mile urban route, then parked on asphalt for hot soak until the ambient temperature reaches 90°F; after further 20-min trip, vehicle parked for 1-hr soak; wind speeds during soak periods held below 5 mph by wind shield.

Hot soak carburetor losses are determined by measuring change in density of the fuel in the carburetor before and after the soak period.

Fuel tank losses are measured from 6 A.M. to end of second soak—assumed to be equivalent to 24 hr. Tank emissions are collected and weighed in charcoal absorption traps with precautions against atmospheric moisture.

All above tests are carried out using a summer-grade fuel of specified volatility range.

C. Control Methods

Control methods now being designed for 1970 models rely on storage of the evaporative emissions in a reservoir during the "soak period." The reservoir can be either a charcoal canister designed for this purpose or the crankcase itself.

The basic Evaporative Loss Control Device (ELCD) based on adsorption in charcoal is shown in Figure 25 (34). Three essential components are involved: the adsorbent canister, a pressure-balancing valve, and a purge-control valve. The canister traps hydrocarbon vapors at any parts of the engine cycle when they cannot be fed to the engine. They must then be retained until they can be released to the intake manifold (Fig. 26). A cross section of such a canister is shown in Figure 27. It consists of a bed of suitably sized charcoal held between retaining screens.

The function of the pressure-balancing valve is twofold: (a) during engine operation to maintain metering pressures in the carburetor at design levels so as to prevent interference with its normal efficacy; (b) while the engine is shut down to close all external vents and route hydrocarbon vapors to the reservoir canister (see Fig. 25).

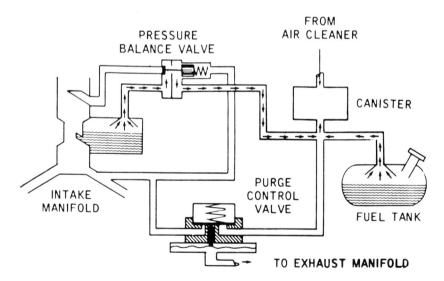

Fig. 25. Basic evaporative loss control device—hot soak condition.

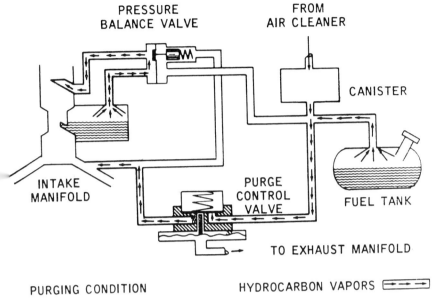

Fig. 26. Basic evaporative loss control device—purging condition.

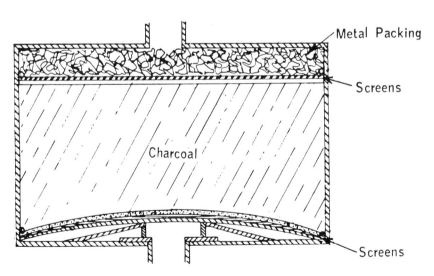

Fig. 27. Cross section of ELCD canister.

The pressure-balancing valve is actuated by intake manifold vacuum against a spring which closes the normal route for the carburetted mixture immediately when this vacuum falls to zero.

In contrast the purge-control valve is actuated by exhaust back pressure. When unseated by this pressure the canister is directly connected with the intake manifold. Air is drawn through the canister, picking up hydrocarbon vapors and going forward directly to the engine under the influence of the manifold vacuum (Fig. 26). The purge-control valve is so designed that the stripping of the canister is precisely phased to engine operating mode, the quantity of stripped hydrocarbon–air mixture being kept in step with engine throughput since this is exponentially related to the exhaust backpressure actuating the valve.

In this way the ELCD system deals with all types of evaporative emissions except carburetor running losses. These are dealt with by elimination of the usual external vents and the use of a throttle-actuated poppet, leading hydrocarbon vapors evolved at idle to the canister. The system has been shown to conform with the proposed California requirements regarding evaporative loss control. It is, of course, also vitally necessary that exhaust emissions should not be adversely affected; with this system changes in hydrocarbon and carbon monoxide emissions were quite insignificant.

An alternative scheme for evaporative control involves the use of the crankcase as the reservoir for these emissions (35). Figure 28 shows the routes

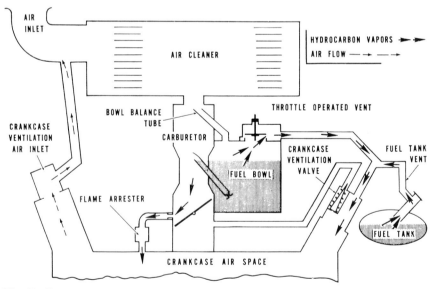

Fig. 28. Evaporative loss control via crankcase air space (shut down and soak condition).

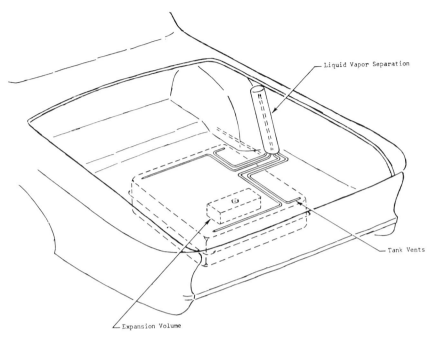

Fig. 29. Fuel tank venting system for evaporative loss control.

of hydrocarbon vapors during shut down and hot soak.   When the engine is restarted these vapors are drawn in to the engine by inlet manifold vacuum. Increase in hydrocarbon and carbon monoxide emissions is slight, so that the overall effect of the controls is a considerable reduction in pollution.

Fuel tank ventilation is complicated by the need to remove hydrocarbons in all parking attitudes.   Vents at all corners and an expansion volume in the tank are essential (Fig. 29).   The vapor–liquid separator must be included since the loss control scheme would otherwise be a closed system, introducing syphon possibilities between tank and crankcase.

Various other means of containing evaporative emissions are under active consideration.   These include means of draining the float chamber to the fuel tank on switch-off, evaporation from the latter being controlled.   (An electric pump is then required to refill the carburetor at switch-on.)   Fuel tank insulation to reduce heat absorption has been suggested but this would be of limited advantage for commuter parking where long periods of hot-soak conditions are experienced at the highest daytime temperatures.   Improved heat insulation of the float chamber is of special advantage where frequent intermittent parking occurs; overall reduction of temperature under the hood is equally important in this respect.

V. DIESEL ENGINE EXHAUST EMISSIONS

On account of visible smoke emissions, the diesel engine is widely blamed for much atmospheric pollution. However, the medically harmful pollutants such as carbon monoxide, benzpyrene, and aldehydes are emitted only in low concentrations, while the oxides of nitrogen, though by no means negligible, are present in much lower proportions than in gasoline engines. There is thus less concern on the part of legislators to set limits to the gaseous constituents of diesel exhaust. However, black diesel exhaust smoke is readily noticeable and is a potential safety hazard. Many countries have therefore introduced legislation to limit such smoke, an objective readily achieved by proper engine maintenance and adjustment. Diesel exhaust odor, a further sign of malfunctioning, is now also attracting the attention of pollution control legislators.

A. Constituents

1. Black Smoke—Unburned Carbon Particles

Unburned carbon, appearing as visible black smoke, is a clear indication of inefficient operation; as such, its elimination is a matter of personal as well as public interest to diesel vehicle operators. The composition of exhaust smoke has been variously reported between 75 and 95% carbon (36) showing significant variation with engine loading (37). Particle size varies in the 0.1–0.3 μm range with smaller particles predominating.

2. "White" Smoke

A fine mist of partly vaporized fuel and water droplets is often produced in "cold-start" conditions or on misfire (38). This is "white" smoke and is a powerful irritant due, in part, to accompanying aldehydes in the exhaust gases. Fortunately it is of short duration and is of little importance in normal driving schedules. Many other workers distinguish between "hot" smoke (black) and "cold" smoke (white).

3. Blue Smoke

Although "white" smoke and (particularly) "black" smoke have attracted wide attention by many workers, less is known about "blue" smoke (38). This does not become visible until several feet from the exhaust and is probably the result of a cooling and (ultimately) condensation process. Precipitation of the droplets in blue smoke yields a dark amber liquid of the viscosity of light lube oil. Mass spectrometric analysis has shown this to be a mixture of hydrocarbons.

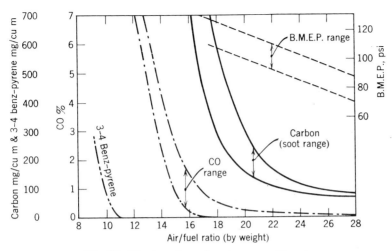

Fig. 30. Diesel engine exhaust gas constituents.

Blue smoke droplets are of much smaller diameter than those of white smoke or the particles in black smoke. They represent a particular fraction of the unburned fuel in the exhaust, viz. that fraction which will condense in the colder conditions some feet away from the exhaust pipe. The production of blue smoke is thus a function both of engine conditions and of fuel specification. It is heaviest at medium load, the maximum emission occurring at 40% rated load with straight-run fuels, at 60% with cracked fuel. At higher engine speeds the maximum blue smoke emission occurs at lower loads, this shift being related to exhaust temperature.

4. Odor

Diesel exhaust odor, although somewhat unpleasant, is not of itself dangerous, except in confined situations (e.g., test houses, stationary engine facilities, fleet garages). However, as symptomatic of some pollution, "odor" is now regulated by California legislation and other authorities will, no doubt, follow this lead. There seems to be no direct correlation between odor and pollutants; thus odor and irritant intensity have to be assessed by panel estimates. In comparative studies it has been assumed that aldehydes and oxides of nitrogen are the most probably odoriferous constituents. Minor sources of odor, such as organic peroxides and acids are unlikely to be present in sufficient quantity to contribute to noticeable levels.

5. Other Pollutants

The pollutants arousing most concern in spark-ignition engines have been shown by many workers to be present in relatively insignificant quantities in the case of diesel engines. Figure 30 shows the concentration of some pollutants related to air/fuel ratio (39). Carbon content is seen to increase rapidly at higher air/fuel ratio than carbon monoxide. (The range of values shown covers a wide range of production model engines.) Furthermore concentration of 3.4-benzpyrene is of negligible importance at acceptable air/fuel ratios, and measurable quantities of nitrogen oxides are not detected until fuel delivery rates are nearly twice normal values, so that pollution from either source is unlikely to be important. Carbon, emitted as black smoke, remains the most serious pollutant.

B. Smoke Measurement

Since a well-tuned engine in good condition is the best means of minimizing pollution, control depends primarily on the ability to measure black smoke density. Smoke meters for this purpose have most commonly been based on either filter discoloration or on light absorption, using photocell techniques. Both types are based on the Brown formula relating smoke density to light absorption:

$$D \propto \frac{1}{K} \log \frac{I_0}{I}$$

where D is smoke density, I_0 is intensity of incident light, I is intensity of light measured, and K is the instrument characteristic (for direct measurement, K is the length of column of smoke in the instrument; for filtration method, $K = V/A$, where V is column of gas sample and A is filter area).

1. Direct Light Absorption Meters

Early diesel smokemeters developed from the simple arrangement of a photocell and a collimated light source in the chimney stacks of power stations or large boiler or furnace installations. For exhaust systems, overheating and fouling of the photocell and source must be prevented and an early form of direct meter designed by Bokemueller in 1939 (40) went some way to eliminating these problems (Fig. 31). This type of "full flow" meter has several obvious limitations and in general "sampling" type meters have been found much more convenient for comparative purposes. A typical basic design of sampling meter is shown in Figure 32. Several modifications were made to this design to eliminate errors due to condensation in the exhaust, soot deposition on the light source and photocell lens, and to manifold pressure variations which alter the density of the exhaust gas sample. Preheating devices, readily cleaned windows to protect the photocell assembly, and a pressure-regulating valve have been incorporated to achieve greater reproducibility. A most satisfactory meter was ultimately developed by British Petroleum Co. (41) and, after modifications, was marketed by

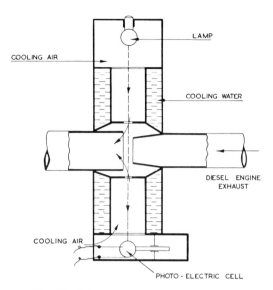

Fig. 31. Bokemueller full-flow smokemeter.

Leslie Hartridge Ltd.   The now well-known Hartridge Smokemeter (Fig. 33)
uses two identical measuring tubes, one filled with a sample of exhaust
smoke, the other a reference tube, scavenged with clean air, the photocell
and lamp being moved from one to the other by a simple mechanism.
Clearances between the lamp and photocell lens and the ends of the smoke
tube are designed to ensure that the former are swept by clean air to prevent
fouling.   Condensation troubles are reduced by the continuous flow of
exhaust, and the internal surface of the tubes is of matte black with lateral
fins to eliminate reflected light from water droplets.   Further modifications
were introduced following test experience in Belgium and in its final form the
Hartridge meter has been very widely used.

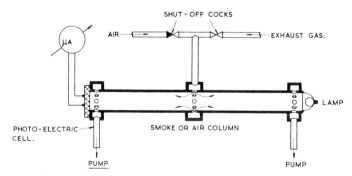

Fig. 32. Sampling smokemeter (light absorption type).

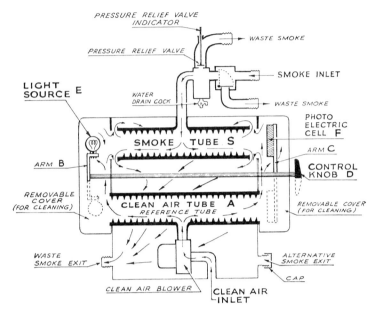

Fig. 33. Hartridge smokemeter.

2. Filter-Type Sampling Meters

The simplest form of filter type meter is a simple handpump by which the gas sample is drawn through the filter paper which is then compared visually with a standard scale of ten shades ranging from white to black. The Bacharach Instrument is typical and its Oil Burner Smoke Scale is widely used as a standard of comparison. Manual operation inevitably involves variations in rate of pumping and thus of total sample volume, both of which lead to poor repeatability. The Saurer Smokemeter was devised to eliminate these variations by means of a simple double aspirator with sliding gas and water valves (Fig. 34). The instrument is perhaps more sensitive to exhaust pressure and relies on visual comparison of the soiled filter paper. However, a similar instrument has been proposed as a standard in Japan. A fundamental limitation is that this is not a readily portable instrument; a laboratory or test-bench environment is essential for accurate use.

The problem of variable sampling rate and volume is overcome in a modification of the Bacharach pump instrument, in which the sample is drawn in by means of a motor-driven, rotary, suction pump set to run for one minute during which time the vacuum is manually controlled. Variations in volume are not entirely eliminated due to changes in filter texture (36). A further improvement is achieved by the Von Brand smokemeter in which a continuous recording is obtained by drawing a spool of filter paper through an

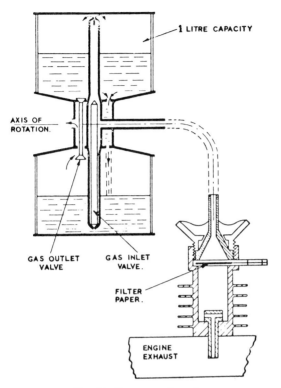

Fig. 34. Saurer smokemeter.

electrically heated filter head. This meter was originally devised for industrial chimneys but has been adapted to diesel engine testing.

A sampling meter which has achieved wide acceptance in diesel engine smoke-metering is the Bosch Smokemeter. In this instrument the sample is drawn in by means of a spring-operated plunger which is held in its minimum position and released remotely at will by pneumatic operation of a rubber diaphragm. The operating principles are illustrated in Figure 35, which includes modifications introduced by the U.K. manufacturing licensees— Dunedin Engineering Co. Close specification of cylinder and sampling pipe volumes, of spring loads and sampling rate, and of filter paper area and quality, enable high reproducibility to be obtained. Furthermore, a photo-cell reflectometer unit is used to give a precise assessment of the darkening of the filter paper (Fig. 36).

In addition to its wide acceptance in Europe the Bosch Smokemeter is recommended, in addition to the J.S.A.E. meter, for legislative standards in Japan. It combines reliability with robust construction and is yet readily

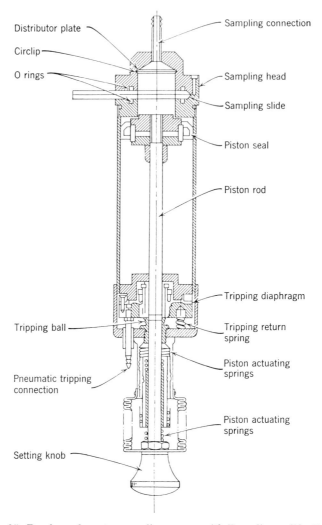

Distributor plate

Circlip

O rings

Sampling connection

Sampling head

Sampling slide

Piston seal

Piston rod

Tripping diaphragm

Tripping ball

Tripping return spring

Pneumatic tripping connection

Piston actuating springs

Piston actuating springs

Setting knob

Fig. 35. Bosch smokemeter-sampling pump, with Dunedin modifications.

portable and accurate, the photocell assessment unit rendering skilled visual comparison unnecessary.

With all smokemeters a representative sample is essential and the type of probe needed to obtain such a sample is to some extent dependent on the characteristics of the smokemeter, particularly with regard to the need in some types of meter for positive pressure in the sample line.

Consistent metering also depends on acceptance of standard procedures

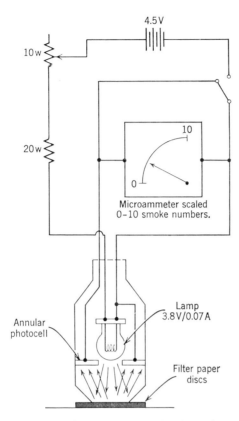

Fig. 36. Bosch smokemeter evaluating unit.

for installation, calibration, measurement, and maintenance. Such procedures for the Hartridge and Bosch/Dunedin smokemeters are included in a
proposed British Standard "Diesel Engines for Road Vehicles."

Smoke measurement under road conditions is fraught with difficulties
and in this respect also elaborate standardization procedures are essential if
reasonable correlation with steady-state testing is to be achieved.

C. Regulations

Increasing numbers of diesel-powered commercial vehicles have focussed
public attention on diesel exhaust smoke and complaints against the offensive
minority of vehicles emitting black smoke have led to legislation in many
countries. Initially based on subjective assessment, such legislation is
increasingly based on smoke meter readings. As with spark ignition engines,
vehicle smoke measurement must be carried out under prescribed driving

conditions. In the case of diesel engines the primary concern is to establish that acceptable smoke limits are not exceeded under the most conducive conditions (this is in noteworthy contrast to the need to monitor gaseous pollutants from gasoline engines throughout a standard driving schedule so that resultant concentrations may be used to give some indication of smog-forming potential).

For diesel engine tests, a choice must be made between several possible driving modes at which to carry out smoke determinations:

1. Under full load at various constant speeds
2. Under full load at low acceleration rate
3. Under full load at low deceleration
4. Under "free acceleration" in neutral gear

Legislation has been in force for some years in several European countries and maximum smoke levels are generally quoted for the Hartridge and/or Bosch smokemeters.

In France smoke limits were decreed in 1964 for new vehicles. Graduated limits were specified, light trucks being required to meet a lower smoke level (40 Hartridge units) than heavy trucks (upper limit 60 Hartridge units). This discrimination is made on the grounds that the smaller vehicles operate to a far greater extent in built up areas where pollution is more serious, whereas heavy trucks often bypass such areas. In West Germany, on the other hand, although no legislation appears to be in force, maximum smoke levels (recommended in V.D.I. Directive No. 2281) are related to rated horsepower so that the heaviest vehicles must have lower smoke emissions than light trucks and cars.

In January 1965 all Belgian vehicles were required to conform to maximum smoke level equivalent to 70 Hartridge units, a limit which has since been reduced to 50 Hartridge units. An opacimeter with photoelectric cell is specified. In both Belgium and France smoke is to be measured under "free acceleration" conditions. (Full throttle in neutral gear to give three successive similar maximum smoke readings.)   Norway has also adopted this test, with a maximum of 70 Hartridge units or a Bosch Smoke No. of 5.5.

Specialists in other countries have suggested that the "Free Acceleration test" may lead to an unfairly high rating and argue the case for steady load testing as being more indicative of actual performance in service. In Sweden, for example, a low rate of acceleration at full load is decreed for road checks. Smoke level is recorded at the midpoint of a specified speed range with test limit of 3.5 B.S.N. (Bosch Smoke No.) and service limit 4.5 B.S.N. Finland has similar legislation with maximum fixed at 70 Hartridge Units.

In the United Kingdom, a considerable amount of work has gone into establishing an objective smoke limit, that is, an upper limit of acceptability of carbon concentration in exhaust.   Bench tests at steady state confirmed that the relationship between rate of exhaust discharge (G) and solids concentrations (C) can be reasonably closely expressed by the equation

$$CG^n = K$$

the constant K depending on the level of acceptance required.   Defining the exhaust discharge rate in terms of cylinder volume and engine speed it was found that experimental curves for the above relationship indicated values of $n = \frac{1}{2}$.

Extensive tests with a panel of 48 observers were carried out to establish the relation between acceptability limits and smoke meter readings (42).

After consideration of the results and of the need to provide a margin for production variations and deterioration in service the British Standards Institute subcommittee on "Diesel Engines for Road Vehicles" has accepted, for a trial period of 3 years after implementation, a smoke density standard based on a curve $C\sqrt{G} = 2.3$ (the 50% objector level).   Service tests of eleven engines of British manufacture showed this to be a realistic limit and to comply with the "free acceleration" Belgian test and the Swedish "low acceleration" test.

Reference to particular smokemeters is avoided but calibration curves (of carbon content) for the Bosch and Hartridge meters are included and the use of these, subject to standardized procedures, is approved.

D. Reduction of Smoke

A well-designed diesel engine, regularly maintained and sensibly operated without overloading will produce very little smoke and thus it may be contended there is no compelling incentive to seek other means of smoke reduction.   However, there are cases where conscientious fleet operators derate their vehicles to reduce emissions and any sound means to limit or eliminate such derating would find ready acceptance.

1. Fuel Quality

Fuel properties and constituents probably provide the greatest opportunities for smoke reduction.   Higher volatility undoubtedly leads to reduced black smoke emissions, but to some extent conflicts with the need for high specific gravity and viscosity to ensure adequate power and good fuel consumption.   Furthermore, the use of a more volatile fuel to reduce smoke may lose any apparent advantage since reduced power available may make it necessary to adjust the fuel stop, leading to poorer combustion.

In addition to volatility, thermal stability of the fuel is of prime concern. Unstable diesel oils quickly give rise to injector fouling with consequent sharp increase in smoke emission. The fouling takes the form of (a) lacquer-like carbonaceous deposits in the injector body causing decreased fuel flow or seizure, and (b) coked deposits on injector sprays, leading to uneven spray characteristics, smoke, and power loss.

Great improvements in thermal stability are reported by Irish and Wiseman (43). Thermally unstable nitrogen and oxygen impurities, together with most of the sulfur, were removed by the Unifining process (hydrogenation). The hydrogen-treated fuel gave an injector life more than double that obtainable with straight-run commercial fuels.

2. Fuel Additives

In recent years various additives have been tried for suppressing diesel smoke. Ashless surfactant agents for maintaining clean fuel injectors showed only limited value in smoke limitation. Organometallic compounds such as calcium sulfonate showed more promise but were disappointing in commercial trials (44). However, in the last four years a group of really effective antismoke additives have been developed. These are barium-organic compounds and have been found effective at very low concentrations (0.2–0.25% by volume). Prolonged tests have shown that no deleterious effects on engine performance accrue from their use. Smoke tests with one such additive are reported by Golothan (45). Average smoke reduction at 25 mph in top gear was in the order of 30 Hartridge units and no serious decline in performance was noted even after 35,000 miles. While combustion chamber deposits were somewhat increased these did not cause any problems although some smoke-level fluctuations arose from the alternate buildup and dispersal of "trumpet" deposits on injectors. Fuel filters of the fine paper type were blocked more quickly but this again did not prove a serious problem. Marketing experience in Belgium has shown average black smoke reduction of 15–30 Hartridge units in the "free acceleration" test and no fuel filter blockage was reported.

Similar strikingly beneficial effects are reported in service tests in the United States of a barium-containing additive. Instantaneous smoke was markedly suppressed and significant improvement in previously fouled injectors noted in many cases (43). The use of the additive enabled vehicles to achieve a marked recovery in "smoke-limited horsepower."

The cost of the additive is no minor concern in the highly competitive market in which bulk suppliers operate today. Cost of blending at depots or of packaging as "shots" for filling stations must also be included on the debit side of additive usage. An overall increased cost of 1–2 pence (i.e.,

1–2¢) per gallon is quoted on U.K. prices. The only technical difficulty which is reported is a tendency to facilitate emulsification of the diesel fuel with water.

VI. FUTURE DEVELOPMENTS IN POLLUTION CONTROL

As has been shown, the attention which has been focussed on motor vehicle pollution has resulted in widespread investigations leading to legislative control of the most serious pollutants. Future developments are likely to involve attention to other pollutant vapors and particulate emissions, apart from continuing investigations to maintain established air cleanliness levels. More sophisticated analytical techniques will be sought in those areas where uncertainties persist so that the effects of existing and proposed control techniques may be more accurately monitored. Main concern in connection with particulate emissions relates to lead, but a recent report of the World Health Organization indicates that the levels of lead in blood and urine have not risen in 30 years and that the incidence of highest lead levels is by no means related to urban conditions. It is thus more likely that lead emissions will be of greater concern if catalytic exhaust control is further developed, and in connection with existing hydrocarbon control methods. Gagliardi has drawn attention to this problem in a recent paper (46).

Increasing attention will undoubtedly be given to a careful monitoring of the effects of the control legislation. Several papers reviewing the progress made to date have already appeared (e.g., reference 47), but the assessment of progress in this respect is extremely difficult. Significant improvements may be hard to establish, at the levels of dilution involved.

The inexorable increase in car population complicates the picture but at the same time renders exact monitoring the more essential. The attention already focussed on the problem and the achievements to date should ensure that motor vehicle emissions, though usually unseen, will never again be unnoticed.

Acknowledgement

Permission to include copyright material has been kindly granted by the following:

Ethyl Corporation (H. J. Gibson) for Figures 2 and 10
Ford Motor Co. (H. A. Nickol) for Figure 3
Sun Oil Co. (P. E. Oberdorfer) for Figure 4
Chrysler Corporation (C. M. Heinen) for Figures 5, 6, 7, 23, 28, and 29
Zenith Carburetter Co. (U.K.) for Figures 8, 9, and 10
Volvo A.B. (Sweden) (A. Larborn) for Figures 11, 12, and 13
General Motors Co. (G. W. Niepoth) for Figures 14, 15, and 16
Ford Motor Co. (J. M. Chandler) for Figure 17
American Motors Corporation (L. R. Hamkins) for Figure 18

Ford Motor Co. (W. R. Was) for Figure 19
General Motors A.C. Division (J. T. Rausch) for Figures 20 and 24
Walker Manufacturing Co. (W. M. Carpenter) for Figure 22
Esso Research & Eng. Co. (P. J. Clarke) for Figures 25, 26, and 27
Perkins Engine Co. (U.K.) (M. Vulliamy) for Figures 30, 31, 32, and 34
Leslie Hartridge Ltd. (U.K.) for Figure 33
Robert Bosch GmbH (Germany) for Figure 35

References

1. H. J. Gibson, Paper A9, 11th Automobile Technology Congress, Munich, June, 1966.
2. R. M. Campau and J. C. Neerman, S.A.E. Paper 660116, Automotive Engineering Congress, Detroit, January 1966.
3. T. A. Huls and H. S. Nickol, S.A.E. Paper 670485, Detroit, May 1967.
4. J. D. Caplan, S.A.E. Paper No 650641, International West Coast Meeting, Vancouver, August 1965.
5. A. P. Altschuller, *J. Air Pollution Control Assoc.*, **16**, 257 (1956).
6. R. E. Pegg and A. W. Ramsden, Paper V1/1, International Clean Air Conference, London, September 1966.
7. P. E. Oberdorfer, S.A.E. Paper 670123, Automotive Engineering Congress, Detroit, January 1967.
8. C. R. Begeman, S.A.E. Paper 440C, Automotive Engineering Congress, Detroit, January 1962.
9. U.S. Federal Register, Vol. 31, No 61, Pt II, March 1966.
10. D. M. Teague, S.A.E. Paper 171, National West Coast Meeting, August, 1957.
11. M. W. Jackson, *J. Air Pollution Control Assoc.*, **11** (12), 697–702 (1966).
12. J. L. Jones, E. A. Schuck, R. W. Eldridge, N. Endow, and F. W. Cranz, *J. Air Pollution Control Assoc.*, **13** (2), 73–77 (1968).
13. E. W. Beckman, W. S. Fagley, and J. O. Sarto, S.A.E. Paper 660107, Automotive Engineering Congress, Detroit, January 1966.
14. G. Lawrence, J. Buttivant, and T. O'Neill, S.A.E. Paper 670484, Detroit, January 1967.
15. *Automobile Engineer*, pp. 96–99, March 1967.
16. A. O. J. Larborn and F. E. S. Zackrisson, SAE paper 680108, Automotive Engineering Congress, Detroit, January 1968.
17. W. K. Steinhagen, G. W. Niepoth, and S. H. Mick, S.A.E. Paper 660106, Automotive Engineering Congress, Detroit, January 1966.
18. W. B. Thompson, S.A.E. Paper 660108, Automotive Engineering Congress, Detroit, January 1966.
19. J. M. Chandler, J. H. Strick, and W. J. Voorhies, S.A.E. Paper 660163, Automotive Engineering Congress, Detroit, January 1966.
20. S. Toyoda, K. Nakajima, and T. Toda, S.A.E. paper 670687, Detroit, June 1967.
21. E. A. Schuck and G. J. Doyle, Report No. 29, Air Pollution Foundation, California, October 1959.
22. P. A. Bennett, C. K. Murphy, M. W. Jackson, and R. A. Randall, Paper No. 142A, S.A.E. Annual Meeting, Detroit, January 1960.
23. U.S. Federal Register, Vol. 31, No 61, Pt II, March 1966.

24. J. T. Middleton. *J. Air Pollution Control Assoc.*, **13** (2), 78–80 February (1963).

25. R. T. Van Derveer and D. L. Hittler, S.A.E. Paper 648B, Automotive Engineering Congress, Detroit, January 1963.

26. W. L. Was and A. E. Stanyar, S.A.E. Paper 648E, Automotive Engineering Congress and Exposition, Detroit, January 1963.

27. L. Raymond, *S.A.E. J.*, **72** (8), 48–54 (1964).

28. C. M. Heinen, "The Car and Air Control" (Fig. 4), paper presented to National Pollution Control Exposition and Conference, Houston, April 1968.

29. G. R. Fitzgerald and D. B. Lewis, S.A.E. Paper 648A, Automotive Engineering Congress, Detroit, January 1963.

30. E. C. Lentz, S.A.E. Paper 648C, Automotive Engineering Congress, Detroit, January 1963.

31. "Evaporation Loss of Petroleum from Storage Tanks Part 1," *A.P.I. Proc.*, **32** (Sec. 1) 220 (1952).

32. J. T. Wentworth, "Carburettor Evaporation Losses," S.A.E. Paper 12B, Annual Meeting, Detroit, January 1958.

33a. California Standards for Motor Vehicle Emissions, Revised March 1967. State of California, Dept. of Public Health, Bureau of Air Sanitation; (b) California Test Procedure and Criteria for Motor Vehicle Exhaust Emission Control, State of California Air Resources Board (Revised Sept. 1967).

34. P. J. Clarke, J. E. Garrard, C. W. Skarstrom, J. Vardi, and D. T. Wade, S.A.E. Paper 670127, Automotive Engineering Congress, Detroit, January 1967.

35. C. M. Heinen, "The Car and Air Control," paper presented to National Pollution Control Exposition and Conference, Houston, April 1968, p. 11.

36. R. A. C. Fosberry and D. E. Gee, Motor Industry Research Association Report No. 1961/5, July 1961.

37. H. Stott and H. Bauer, *M.T.Z.*, **18** (5), 127, May 1967.

38. J. B. Durant and L. Eltinge, S.A.E. Paper No. 3R, Annual Meeting, January 1959.

39. M. Vulliamy and J. Spiers, S.A.E. Paper No 670090, Automotive Engineering Congress, Detroit, January 1967.

40. A. Bokemueller, *Jahrbuch der Brennkrafttechnischen Gesellschaft, e.v.*, Vol. 20, 1939.

41. A. L. Wachal, *Automobile Engr.*, 303–305, July 1953.

42. A. E. Dodd and L. E. Reed, Motor Industry Research Association, Report No 1964/12, April 1964.

43. G. E. Irish and E. L. Wiseman, A.S.T.M. Symposium on Diesel Fuel Oils, A.S.T.M. Tech. Publ. No. 413, July 1966, p. 84.

44. R. E. Pegg and A. W. Ramsden, Paper V1/1, International Clean Air Conference, London, September 1966.

45. D. W. Golothan, Paper V1/3, International Clean Air Conference, London, September 1966.

46. J. C. Gagliardi, S.A.E. Paper 670128, Automotive Engineering Congress, Detroit, January 1967.

47. M. L. Brubacher, California MVPCB publication, January 18, 1967.

Electrostatic Precipitation

Myron Robinson[*]

Research-Cottrell, Inc., Bound Brook, New Jersey

[*] Present address: Health and Safety Laboratory, U.S. Atomic Energy Commission, New York, New York.

I. INTRODUCTION

A. Precipitator Fundamentals

The electrostatic or electrical precipitator is less frequently known as a treater, a Cottrell, or an electrofilter. The last designation is in agreement with the German *Elektrofilter*, the French *électrofiltre*, and the Russian *elektrofil'tr*. It is commonly found in one of two basic forms (Fig. 1). In the simpler the precipitator comprises a grounded cylinder designated, in accordance with its function, the collecting or passive electrode, and coaxial with it a high-potential wire called the corona-discharge or active electrode. An alternative basic design consists of two grounded parallel plates (the collecting electrodes) together with an array of parallel discharge wires mounted in a plane midway between the plates. Gas with suspended solid or liquid particles is passed either through the tube or between the plates. If a sufficient difference of potential exists between the discharge and collecting electrodes, a corona discharge will form about the wire(s). The corona serves as a copious source of unipolar ions of the same sign as that of the discharge electrode(s). The ions, in migrating across the interelectrode space under the action of the impressed electric field, in part attach themselves to aerosol particles moving with the gas through the system. The charged particles, in turn, are attracted to the collecting surface, adhere, and so are extracted from the gas stream. Solid particles build up a layer on the collecting surface, from which the accumulated deposit is dislodged periodically by rapping or flushing and allowed to fall into a hopper or sump for subsequent removal. Liquid particles (droplets) form a film on the collecting surface, the precipitated liquid then dripping off into a sump. The cleaned gas is discharged at the outlet of the precipitator.

Various modifications of the "classic" designs are possible. The two-stage precipitator, for example, which finds its principal use in domestic and commercial air cleaning for air conditioning and aerosol-sampling applications, divides the particle-charging and precipitating functions into adjacent segments of the apparatus (Fig. 2). Other precipitator variations also exist, or have been proposed, in which the corona charging mechanism is replaced by ionizing radiation from a radioactive or other source. It has been further suggested that the externally applied collecting field in the two-stage precipitator be discarded in favor of the precipitating action of the aerosol's own space-charge field. However, single-stage models of the varieties illustrated in Figure 1 have so far found almost universal application in the cleaning of contaminated industrial gases by electrostatic precipitation.

B. Previous Studies of Precipitation

There have been several comprehensive reviews of the theory and practice of electrostatic precipitation. Foremost among these contributions may be mentioned the monographs of White (1), Rose and Wood (2), and Strauss (3) in English, of Uzhov* (4,5) in Russian, and of Lutyński (8) in Polish. These authors have done much to systematize the data available from widely scattered sources. Lodge's slim volume (9), published in 1925, is today only of historical interest.

In addition to these comprehensive treatments, basic information is available in a number of briefer articles, e.g., Lowe and Lucas (10), Roberts (11), Roberts and Walker (12), Heinrich and Anderson (13), Meldau (14), Underwood (15), Uzhov (6), Gottschlich (16,17), Ertl (18,19), Simon (20), Pauthenier (21), American Industrial Hygiene Association (22), Pennsylvania State University seminars (23), Adrian (two-stage precipitators) (24), and Jones (25) and Lippmann (26) (air-sampling precipitators). A market research report of the air pollution control industry in the United States—including electrostatic precipitation—has been made by Predicasts (27). A forthcoming precipitation study sponsored by the National Center for Air Pollution Control (U.S.A.) will include an extensive bibliography, a statistical review of precipitator utilization, operating characteristics, and economics in all applications, as well as a critique of the state of the art (28).

---

* Uzhov (6) cites an additional Russian work (7) otherwise unknown to the present writer.

Briefly annotated bibliographies of the precipitator literature through 1955
have been prepared by Strehlow (29) and Scheffy (30).

An all-inclusive account of the physics of corona discharge at atmospheric
and lower pressures is given by Loeb (31). More general treatments of
electrical breakdown in gases are given by Meek and Craggs (32), Cobine (33),
von Engel (34), Loeb (35–37), and Llewellyn Jones (38).

This chapter aims at updating prior precipitator surveys by describing the

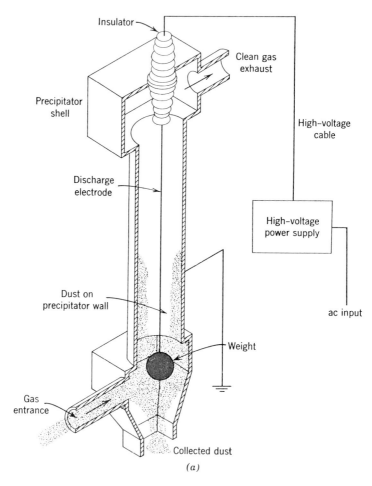

(a)

Fig. 1. Single-stage electrostatic precipitators: (a) tubular (1 μm), (b) duct type.
In both cases the corona discharge and precipitating field extend over the full length
of the apparatus. (Figure 1a used with permission of Addison-Wesley Publishing
Company.)

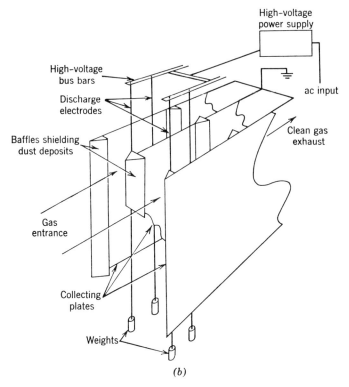

High-voltage
power supply

High-voltage
bus bars

Discharge
electrodes

ac input

Baffles shielding
dust deposits

Clean gas
exhaust

Gas
entrance

Collecting
plates

Weights

(b)

Fig.1 *(continued)*

results of subsequent studies, some not hitherto published, as well as earlier work, which for circumstances of publication or reason of language, were not previously included. However, to provide a more unified picture of the subject and avoid a disconnected summary of unrelated developments, substantial material has been incorporated from the older literature. The emphasis throughout the chapter is on physical principles and predictive methods rather than equipment and empirical descriptions; on the science of electrostatic precipitation rather than the art. Special attention is given to new and growing areas of application, especially with reference to gas composition and extremes of temperature and pressure.

C. Historical Origins

The existence of electrostatic forces of attraction had been known to classical antiquity, but the earliest reported observations in aerosol electrostatics date from about 1600. Closely associated with studies of smoke and

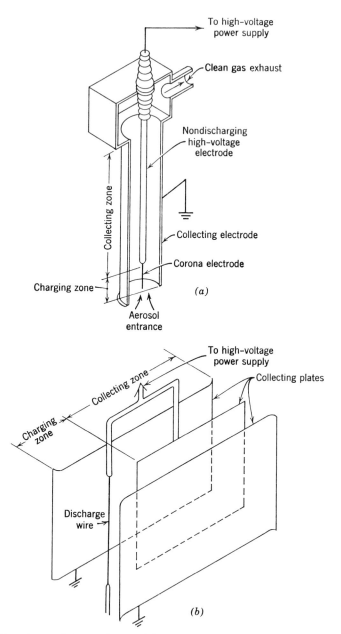

Fig. 2. Two-stage electrostatic precipitators: (a) tubular, (b) duct type. The charging zone in each model is confined to the region about the corona wire. The downstream collecting zone provides a pure electrostatic field in which the previously charged particles are precipitated on the collecting surface.

TABLE I

Origins of Electrostatic Precipitation

1600	William Gilbert observes that frictionally charged dielectrics "entice smoke sent out by an extinguished light" (45).
1667	Savants of the Academy del Cimento report that a stream of smoke rising near a piece of rubbed amber will in part "bend · · · and be arrested by the Amber and [in] part · · · mount upwards" (46).
1672	Otto von Guericke constructs an electrostatic generator consisting of a frictionally charged sulfur sphere which glows in the dark and produces "roarings and crashings." He also discovers the effectiveness of pointed conductors in attracting charged bodies (47).
1675	Robert Boyle notes that "A chafed Diamond · · · [will] · · · make ascending smoak deviate from its former line · · · and · · · act at a considerable distance" (48). He further describes "a curious diamond · · · [that] gave light in the dark when it was rubbed · · · and · · · became electrical" (49).
1709	Francis Hauksbee discovers the electric wind. "The Force of the Effluvia · · · was · · · felt upon the Face, if · · · [a] rubb'd [glass] Tube were held near it". He finds that "Dust and Powders, the streams of Liquids, · · · and · · · Smoke · · · immediately adhered to the surface of the Tube" (50).
1718	Isaac Newton discussed the corona glow and the electric wind but without reference to particle behavior (51).
1746	Jean Nollet describes the point corona in noting that electrified points display "brushes of light" (52).
1747	Benjamin Franklin writes about "the wonderful effect of pointed bodies, both in throwing off and drawing off the electrical fire" (53).
1756	William Watson finds that a charge is acquired by an insulated conductor placed in a column of smoke rising out of an electrified vessel of burning turpentine (54).
1771	Giambatista Becarria observes that "the movements that the electric wind can excite · · · repel the hot smoke of a candle · · · and the flame will · · · be repelled by the wind" (55).
1777	Tiberius Cavallo originates the concept of particle charging time. "Smoak · · · attracted by · · · [an] electrified body · · · is not immediately repelled because it is a bad Conductor and acquires Electricity very slowly" (56).
1785	Charles de Coulomb investigates loss of charge from an insulated conductor and explains leakage to occur in part by convection of charged particles through the air (57).
1820	M. Hohlfeld demonstrates the first electrodepositional device consisting of a clear-cut corona electrode and particle-collecting surface (58).
1838	Michael Faraday remarks that "the power of particles of dust to carry off electricity in cases of high tension is well known." (59a). He extends the study of the electric wind to dielectric liquids and builds an ion-drag liquid pump (59b).
1850	G. F. Guitard independently repeats Hohlfeld's experiment and precipitates tobacco smoke in a bottle (60).
1883	J. A. Clark and Oliver Lodge rediscover electrostatic precipitation (61).
1884	Lodge and Alfred Walker unsuccessfully attempt the precipitation of lead-oxide fume on a commercial scale (62).
1906	Frederick G. Cottrell independently rediscovers electrostatic precipitation and successfully applies it to the collection of sulfuric-acid mist (39).

other particulate suspensions in an electric field was the discovery of the man-made corona discharge after 1670. Among the early investigators (and frequent rediscoverers) of electroseparation, electrodeposition, and the high-voltage discharge are to be counted some of the foremost names in the history of science. As Table I shows, the essential phenomena of fine-particle electrostatics had been demonstrated in principle long before an advancing technology provided the accessory equipment required for successful com-mercial application.

Modern electrostatic precipitation is closely connected with the name of Frederick G. Cottrell. In 1907, by adapting the then newly developed mechanical rectifier and high-voltage transformer to his needs, Cottrell became the first to build a workable electrostatic precipitator (39). With Walter Schmidt's development of the fine-wire discharge electrode in 1912, Cottrell's demonstration of the general superiority of negative corona over positive in 1913 (40), and continuing improvements in rectifying equipment to give high voltages at reasonable currents, the three basic ideas to dominate the art, even today, were established.

The early background of electrostatic precipitation is described elsewhere in greater detail by Robinson (41,42). Accounts of the precipitator's more recent history are given by Cottrell (39), Schmidt and Anderson (43), and White (1a,44).

II. MECHANISMS OF CORONA FORMATION (31-38)

A. Initiation and Maintenance of the Discharge

Consider a system of two juxtaposed electrodes immersed in a gas such as atmospheric air, one electrode having a much smaller radius of curvature than the other, the gap between the two electrodes being large compared to the radius of the smaller. Typical examples of such a configuration are a point opposite a plane, a coaxial wire and pipe, or a wire parallel to a plane. As the potential difference between the electrodes is raised, it is found that the gas near the more sharply curved electrode breaks down at a voltage less than the spark-breakdown value for the gap length in question. This incomplete breakdown, called corona, appears in air as a highly active region of glow, bluish white or possibly reddish in color, extending into the gas a short distance beyond the discharge electrode surface. The corona on a positive wire has the appearance of a more or less uniform sheath covering the wire surface facing the opposite electrode. On a negative wire the cor-ona is concentrated in tufts of glowing gas spaced at intervals along the wire.

The initiation of the corona discharge requires the availability of free electrons in the gas in the region of the intense electric field surrounding the discharge electrode. Since the supply of these electrons depends on the random action of ionizing sources such as natural radioactivity and cosmic rays, the corona discharge is intermittent at voltages very near the corona threshold.

In the case of a negative discharge wire, free electrons in the high-field zone near the wire gain energy from the field to produce positive ions and other electrons by collision. These new electrons are, in turn, accelerated and produce further ionization, thus giving rise to the cumulative process termed an electron avalanche. The positive ions formed in this process are accelerated toward the wire. By bombarding the negative wire and giving up relatively high energies in the process, the positive ions cause the ejection from the wire surface of secondary electrons necessary for maintaining the discharge. In addition, high-frequency radiation originating in the excited gas molecules within the corona envelope may photoionize surrounding gas molecules, likewise contributing to the supply of secondary electrons. Electrons of whatever provenance are attracted toward the anode and, as they move into the weaker electric field away from the wire, tend to form negative ions by attachment to neutral oxygen molecules. These ions form a dense unipolar cloud filling by far most of the interelectrode volume, and constitute the only current in the entire space outside the region of corona glow. The effect of this space charge is to retard the further emission of negative charge from the corona, and, in so doing, limit the ionizing field near the wire and stabilize the discharge. However, as the voltage is progressively raised, complete breakdown of the gas dielectric, that is, sparkover, eventually occurs.

Should the discharge electrode be positive, the physical mechanisms are quite different. In this case, electrons formed by chance ionizing events in the high-field space in the neighborhood of the wire establish electron avalanches which move in toward the wire. As with the negative discharge, these avalanches are responsible for the visible, highly ionized state of the gas close to the wire. Positive ions formed in the avalanches are left behind by the electrons and drift from the region of the wire into a field of progressively decreasing intensity. The ions are thus incapable of acquiring sufficient energy between molecular collisions to produce either significant ionization in the gas or electron ejection at the cathode surface. Photons originating in the corona glow presumably give rise to ionization in the gas and perhaps cause photoemission at the cathode. In this manner the secondary electrons upon which a self-sustaining discharge depends are generated. Nevertheless, virtually all the current in the space external to the corona envelope is carried by the positive ions.

When the corona-starting voltage is reached in either polarity, the current increases slowly at first and then more rapidly with increasing voltage. As sparkover is approached, relatively small increments of voltage give sizable increases in current, as shown, for example, in Figure 3.

A negative discharge under conditions prevailing in electrostatic precipitators, generally (though not invariably) yields a higher current at a given voltage than positive, and the sparkover voltage, which sets the upper limit to the operating potential of the precipitator, is also usually greater. The

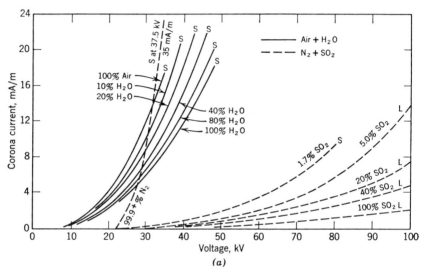

Fig. 3. Negative corona current–voltage characteristics of wire-tube systems. *L* designates the limiting voltage of the power supply and *S* sparkover. (*a*) The effect of adding to the gas a constituent of relatively low ion mobility. Solid curves: water vapor in dry air (tube diameter, 7.6 cm; wire diameter, 0.25 mm; temperature, 200°C; pressure, 1 atm). Broken curves: sulfur dioxide in nitrogen (tube diameter, 15.2 cm; wire diameter, 2.8 mm; room temperature and pressure). In fly-ash precipitation, the SO_2 content of the gas is of the order of 0.1% and, in the presence of relatively large proportions of electronegative O_2 and H_2O, causes no noticeable electrical effect in the gas. Conversely, in acid-mist precipitation, SO_2 content may be about 6%, and in this case the effect of SO_2 cannot be ignored even in the presence of appreciable O_2 (65). (*b*) Current–voltage relations in room air as a function of voltage waveform. The electrodes are a concentric tube and wire of respective diameters 15.2 cm and 2.8 mm. In agreement with field observations, sparking voltage generally increases with waveform in the order unvarying dc, full wave, and half wave. This phenomenon is presumably due to the presence of dust on the positive collecting surface, for under exceptionally clean conditions all waveforms result in the same sparkover voltage (65).

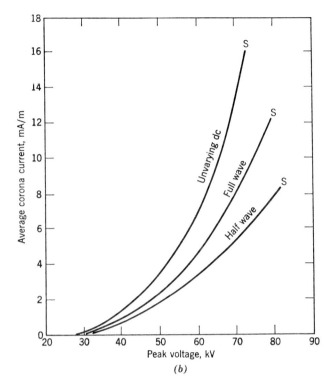

Fig. 3 (*continued*)

only common exception to the use of negative corona occurs in domestic air-cleaning applications. For while increased current and voltage alike make for superior efficiency of precipitation, negative corona often produces physiologically objectionable ozone in greater quantities.

Other electrochemical effects of the corona discharge in air include the formation of oxides of nitrogen, and in the presence of moisture, nitric acid. In addition, corona may cause insulating materials to deteriorate under the combined action of ion bombardment, ultraviolet radiation, and the chemically active compounds created in the discharge.

In the negative corona the chance that a secondary electron will be liberated from the wire by positive-ion bombardment depends on a number of factors: the work function of the wire, contaminants adhering to the wire, and the kinetic energy of the impacting positive ions. Electron ejection is therefore exceedingly capricious and likelihood of its occurrence varies from spot to spot, ever changing with time, over the cathode surface. It is this sporadic emission of secondary electrons coupled with avalanche formation that gives rise to the mutually repulsive corona tufts distributed along the negative wire.

Negative corona is also discontinuous in another sense. At the corona onset voltage, the discharge from a negative wire in air and other electron-attracting gases (Sec. II-B) are observable as small pulses of current, sometimes called Trichel pulses, which increase in repetition rate as the voltage is raised. Trichel pulses are easily detected on an oscilloscope and result when discrete clusters of positive space charge left behind by electron avalanches are swept into the cathode. With long wires having many discharging spots, or at higher voltages, the pulse repetition rate becomes very high, and it is no longer possible to distinguish individual discharges.

The positive corona discharge also produces characteristic current pulsations observable on an oscilloscope. These streamer pulses can occur individually or in bursts, and here too, at higher voltages, may become so frequent that separate pulses cannot be discerned. (Streamers are luminous conducting filaments of heavily ionized gas extending outward from the electrode surface.) As the voltage is still further increased, large pre-breakdown streamers appear and eventually bridge the interelectrode gap as a spark.

Appearance of corona pulses, whether positive or negative, on an oscilloscope, offers a convenient practical indicator of corona onset and a means of distinguishing corona from insulator leakage current in precipitator high-voltage circuits.

The foregoing description of the corona has been given with specific reference to the discharge in air at approximately room conditions. The same pattern remains essentially valid for a much broader range of temperatures and pressures, and for a variety of gases other than air. The effect of extreme temperatures and pressures is considered in Section VI-A. Certain physical attributes that serve to characterize the corona discharge in a given gas are considered below.

B. Electronegative Gases

When an electron collides with a neutral gas molecule it may attach itself and form a negative ion. The likelihood of doing so depends on the so-called electronegative nature of the gas and on the energy of the electron, slower moving electrons remaining for a longer time within the range of the atomic field of the gas molecule. In addition, the probability of electron attachment is markedly affected by the presence of certain gases and vapors as impurities. Relative ease of attachment in several gases is illustrated to order-of-magnitude accuracy in Table II. The average number of

TABLE II

Impacts Required for Electron Attachment[a]

Gas	β, avg. no. of collisions	Gas	β, avg. no. of collisions
Inert gases	∞	C_2H_6	2.5×10^6
N_2, H_2	∞	N_2O	6.1×10^5
CO	1.6×10^8	C_2H_5Cl	3.7×10^5
NH_3	9.9×10^7	Air	4.3×10^4
C_2H_4	4.7×10^7	H_2O	4.0×10^4
C_2H_2	7.8×10^6	O_2	8.7×10^3
		Cl_2	$<2.1 \times 10^3$

[a] From reference 35a. Used with permission of John Wiley and Sons.

collisions (β) is given which an electron need undergo before adhering to a neutral molecule. The inert gases and hydrogen, if *very* pure, do not form negative ions at all by electron attachment. Certain gases which themselves have no electron affinity are still capable of forming negative ions by an indirect process. The molecule is first dissociated by impact with an energetic electron and then attachment occurs to one of the fragments. This happens, for example, with CO_2, H_2S, and H_2O. Attachment to one atom of the molecule and subsequent dissociation occurs with the halide-acid gases, HCl, HBr, and HI. SO_2, a constituent of importance in many precipitator applications, attaches electrons directly.

A suitable electrode geometry alone is, in general, not the only condition that must be imposed to insure that a stable corona discharge will precede sparkover. Sparkover without antecedent corona will result, for example, for certain combinations of electrode size and high gas pressure (Sec. VI-A). But even at ordinary pressures, slow moving ions rather than electrons—the latter having drift velocities about three orders of magnitude greater—are needed to build up the interelectrode space charge needed for corona formation. In a gas not possessing the electronegative property, negative corona is impossible, sparkover alone occurring as the voltage is raised. It is for this reason that the electron-attachment process plays so central a role in the precipitation process.

Electrostatic precipitation, however, is feasible in nonattaching gases. We circumvent the difficulty merely by employing positive corona. In practice, nominally nonattaching gases of commercial grade are likely to contain sufficient electronegative impurities to obviate the problem or at least render it more tractable.

C. Ion Mobility

The motion of an ion acted upon by an electric field is impeded by repeated collisions with gas molecules in its path. The average drift velocity of the ion v_i (m/sec) is related to the field E (V/m) by the equation

$$v_i = bE \tag{1}$$

where the constant of proportionality b ($m^2/(sec\text{-}V)$) is called the mobility. The practical importance of this quantity in precipitation studies arises in calculating current–voltage characteristics and other electrical relationships involving the corona discharge. Theoretical expressions for mobility have been derived using the kinetic theory of gases, but mobilities calculated in this manner are of questionable reliability. Experimentally determined values are recommended whenever these are available. Table III gives a selection of mobilities for positive and negative ions in their parent gases. A more complete, but older, compilation of mobilities is given by Loeb (63). Even for the tabulated gases, the exact mobility to be used in actual situations remains subject to some uncertainty, for impurities have a pronounced effect in reducing the value. This is particularly true of pure nonattaching gases, the negative mobility of which is the electron mobility. The presence of impurities leading to negative ion formation have the apparent effect of

TABLE III

Mobility of Singly-Charged Gaseous Ions at 0°C and 1 atm[a]

Gas	Mobility ($m^2/(sec\text{-}V)$)		Gas	Mobility ($m^2/(sec\text{-}V)$)	
	$b_0\,(-)$	$b_0\,(+)$		$b_0\,(-)$	$b_0\,(+)$
He	$-^b$	10.4×10^{-4}	C_2H_2	0.83×10^{-4}	0.78×10^{-4}
Ne	$-^b$	4.2	C_2H_5Cl	0.38	0.36
A	$-^b$	1.6	C_2H_5OH	0.37	0.36
Kr	$-^b$	0.9	CO	1.14	1.10
Xe	$-^b$	0.6	CO_2 (dry)	0.98	0.84
Air (dry)	2.1×10^{-4}	1.36	HCl	0.62	0.53
Air (very dry)	2.5	1.8	H_2O (100°C)	0.95	1.1
N_2	$-^b$	1.8	H_2S	0.56	0.62
O_2	2.6	2.2	NH_3	0.66	0.56
H_2	$-^b$	$12.3(H_3^+)$	N_2O	0.90	0.82
Cl_2	0.74	0.74	SO_2	0.41	0.41
CCl_4	0.31	0.30	SF_6	0.57	

[a] From reference 64. Used with permission of McGraw-Hill Book Company.
[b] No electron attachment in the pure gas.

reducing the mobility by orders of magnitude. In other cases impurities seem to attach to ions and form clusters, thus increasing the mass and cross section of the ion, and lowering its mobility.

In accordance with the theoretical requirement that mobility be proportional to the mean free path, it is found experimentally that mobility is almost inversely proportional to the gas density over a wide temperature-pressure range. Thus

$$b = \frac{b_0}{\delta} \tag{2}$$

where b_0 is the mobility at standard conditions and δ is the gas density relative to standard conditions (i.e., $\delta = 1$ at $0°C$ and 1 atm) and is found from

$$\delta = \frac{273 p_a}{T + 273} \tag{3}$$

Here T ($°C$) is the temperature and p_a (atm) is the pressure.

There are situations in which the apparent ion mobility is much higher than given in Table III. This results because the ion is a high-mobility electron from the time it first appears until it attaches. For small interelectrode gaps, this phenomenon manifests itself in anomalously high corona currents. In addition, ion mobility, which is independent of E/p_a at lower values of this ratio, begins to increase when E/p_a exceeds a critical level, an effect that may be important in certain parts of the corona discharge. Abnormally heavy negative currents occurring at high temperatures are considered in Section VI-A-4.

The effect of adding a relatively low-mobility constituent to a higher-mobility gas (e.g., water vapor to dry air or SO_2 to N_2) is illustrated in Figure 3a.

III. CURRENT-VOLTAGE-FIELD RELATIONS

A. Introduction

The mathematical theory of the corona discharge relevant to the design and performance of electrostatic precipitators will be developed below. The rationalized mks system of units will generally be used. It is hoped that this will encourage its adoption in an industry in which the interchange between theory and practice has been traditionally burdened by the impracticalities of the electrostatic cgs or various mixed systems.

The following basic relations are required: the definition of electric field intensity E (V/m) in terms of the potential V (V)

$$\mathbf{E} = -\nabla V \tag{4}$$

and Poisson's equation

$$\nabla \cdot \mathbf{E} = \frac{\rho}{\varepsilon} \tag{5}$$

where ρ (C/m³) is the charge density and ε (F/m) is the permittivity. The permittivity may be expressed as

$$\varepsilon = \kappa \varepsilon_0 \tag{6}$$

where ε_0 is the permittivity of free space (8.85×10^{-12} F/m) and κ (dimensionless) is the relative dielectric constant of the medium. Since this quantity is essentially unity for gases under all realistic conditions of precipitation, ε_0 is hereafter used as the gas permittivity.

Although usually referred to in this text as a wire (its most common form), the discharge electrode may assume a number of shapes, some of which have been found, or are claimed, to be advantageous in specific applications. In addition to straight round wires, forms commonly encountered include barbed wires, multiple-twisted wires, twisted square rods, channels, and straight-edged or serrated ribbons. When a more complex discharge-electrode geometry is used, round-wire theory will often give a useful approximation if an equivalent round-wire size can be estimated.

B. Wire-Pipe Geometry

1. Particle-Free Condition

The simplest precipitator geometry for purposes of analysis is the coaxial wire–cylinder combination. For this arrangement, eqs. 4 and 5, respectively, become

$$E = -\frac{\partial V}{\partial r} \tag{7}$$

and

$$\frac{1}{r} \frac{d}{dr} (rE) = \frac{\rho_i}{\varepsilon_0} \tag{8}$$

where r (m) is the radial distance from the tube axis and ρ_i is the ion space-charge density. Prior to the onset of corona $\rho_i = 0$. Taking V_0 as the potential at the wire surface r_0 and grounding the tube ($V = 0$ at $r = r_1$), integration yields the electrostatic field

$$E = \frac{V_0}{r \ln (r_1/r_0)} \tag{9}$$

It is assumed throughout these calculations that the passive electrode is at ground potential. This procedure entails no loss of generality, corresponds to usual practice, and permits the potential of the discharge electrode to be equated to the voltage across the system.

Above the corona threshold the foregoing expression for the field remains a valid approximation only as long as the interelectrode space charge, whether due to ions or charged dust particles, is insignificant compared to the surface charge on the electrodes. In other words, the requirement must be satisfied that

$$|\bar{\rho} \cdot \pi r_1^2 L| \ll |V_0 C| \tag{10}$$

where $\bar{\rho}$ is the average space-charge density, $L(\text{m})$ the length of the system, and $C(\text{F})$ the capacitance. The capacitance is given by (66)

$$C = \frac{2\pi\varepsilon_0 L}{\ln (r_1/r_0)} \tag{11}$$

whence it follows that

$$|\bar{\rho}| \ll \left| \frac{2\varepsilon_0 V}{r_1^2 \ln (r_1/r_0)} \right| \tag{12}$$

Inserting the typical values $V_0 = 5 \times 10^4$ V, $r_0 = 1.4 \times 10^{-3}$ m, and $r_1 = 0.1$ m, we find $|\bar{\rho}| \ll 2 \times 10^{-5}$ C/m³. Since the average ion charge density $\bar{\rho}_i$ alone is characteristically of the order 10^{-6} C/m³ $\sim 10^{13}$ singly charged ions/m³ or more, it is clear that inequality 12, hence, eq. 9, fails under actual precipitating conditions.

It is possible to express the current per unit length of conductor j_l (m/sec)

$$j_l = 2\pi r \rho_i b E \tag{13}$$

whence, eliminating ρ_i from eq. 8 and integrating,

$$E = \left[\frac{j_l}{2\pi\varepsilon_0 b} + \left(\frac{r_0}{r}\right)^2 \left(E_c^2 - \frac{j_l}{2\pi\varepsilon_0 b} \right) \right]^{1/2} \tag{14}$$

E_c represents the minimal field intensity required to effect local breakdown of the gas close to the wire surface ($r \cong r_0$). This corona-starting field E_c is characteristic of the gas and wire radius, but is independent of the size or shape of the outer electrode (Sec. III-D). For sufficiently large r and j_l, eq. 14 simplifies to the convenient approximation

$$E = \left(\frac{j_l}{2\pi\varepsilon_0 b}\right)^{1/2} \tag{15}$$

which is independent of r. Although this formula is widely quoted in the literature, it should be noted that it may seriously underestimate the field at the relatively low discharge currents commonly encountered in industrial usage (see Fig. 5). At the higher current levels which are generally the rule in laboratory work, eq. 15 may give tolerable accuracy.

The corona current–voltage relation is found by integrating eq. 14. Following Townsend (67), and neglecting second-order terms,

$$\frac{V_0 - V_c}{V_c} \ln\left(\frac{r_1}{r_0}\right) = (1 + \phi)^{1/2} - 1 - \ln\frac{1 + (1 + \phi)^{1/2}}{2} \qquad (16)$$

where V_c(V) is the corona-starting potential, related to the starting field E_c by eq. 9. The quantity ϕ is defined as

$$\phi = \left(\frac{r_1}{E_c r_0}\right)^2 \frac{j_l}{2\pi\varepsilon_0 b} \qquad (17)$$

The most direct way of solving eq. 16 is to plot the dimensionless variables $((V_0 - V_c)/V_c) \ln (r_1/r_0)$ and ϕ against one another, as is done in Figure 4, and read off the appropriate quantities.

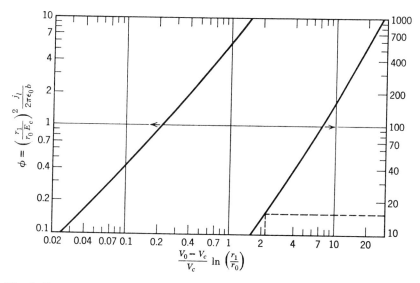

Fig. 4. Townsend's current–voltage relation (eq. 16) in terms of dimensionless variables. *Example:* Estimate the negative corona current to be expected in clean dry air at 25° C and 1 atm at a voltage of 50 kV using a 12-ft long wire-tube precipitator of respective diameters 109 mils and 9 in.

From the known quantities $r_0 = 1.38 \times 10^{-3}$ m, $r_1 = 0.114$ m, $V_0 = 5.0 \times 10^4$ V, and $V_c = 3.36 \times 10^4$ V (example of Sec. III-D), calculate the abscissa $[(V_0 - V_c)/V_c] \ln (r_1/r_0) = 2.16$. As shown by the dashed lines, this value corresponds to the ordinate $\phi = 15.6$. Setting $\varepsilon_0 = 8.86 \times 10^{-12}$ F/m, $b = 2.3 \times 10^{-4}$ m^2/(V-sec) (Table III and eq. 2), and $E_c = 54.9 \times 10^5$ V/m (example of Sec. III-D), we then have $j_l = 2\pi\varepsilon_0 b\theta(r_0 E_c/r_1)^2 = 0.88 \times 10^{-3}$ A/m. The current in a 12-ft (3.67 m) system is, therefore, 3.2 mA.

For low currents near the corona threshold (i.e., $\phi \ll 1$), eq. 16 may be represented by the simple approximation (68)

$$j_l = \frac{8\pi\varepsilon_0 b}{r_1{}^2 \ln{(r_1/r_0)}} V_0(V_0 - V_c) \tag{18}$$

Townsend's equations 16 and 18 are applicable to air and other ion-forming gases of atmospheric density and higher, but not for low pressures or small interelectrode spacings where the free-electron component of the current cannot be neglected.

Other current–voltage approximations for wire-pipe corona have been compiled by White (1b), Prinz (69), Winkel and Schuetz (70), and Simpson and Morse (71).

2. Particle Space Charge

The cylindrical corona field described by eq. 14 makes no allowance for the presence of suspended charged particles in the inter-electrode space. Such particles contribute to the total space charge, and owing to their low drift velocity relative to that of the gas ions, the space-charge effect due to the particles is often much greater than that of the ions, at least near the precipitator inlet.

The influence of particle space charge on the field may be approximated by assuming (cf. Sec. V-D-2) that the particle concentration per unit volume of gas N_p (particles/m^3) is constant over a given cross section of the precipitator, although decreasing with downstream distance. The aerosol's specific surface, that is, the particle surface area per unit volume of gas S (m^2/m^3) is

$$S = 4\pi a^2 N_p \tag{19}$$

where a is the radius (m) of assumed spherical particles. It is shown in Section IV-A that a particle charged by the ion-bombardment process (the dominant particle-charging mechanism in most applications) acquires a charge of $4\pi\varepsilon_0 pEa^2$ C where p (dimensionless) is given by eq. 51. The charged aerosol thus increases the initially present ion space charge density ρ_i by an amount ρ_p (C/m^3)

$$\rho_p = \varepsilon_0 pES \tag{20}$$

Including ρ_p in eq. 8 and solving (72),

$$E = \left\{ \left[\left(\frac{r_0}{r}\right)^2 \left(E_c{}^2 - \frac{j_l}{2\pi\varepsilon_0 b}\right) + \frac{j_l}{4\pi\varepsilon_0 b(pSr)^2} \right] \right.$$
$$\left. \times e^{2pSr} - \frac{j_l}{4\pi\varepsilon_0 b}\left[\frac{2}{pSr} + \frac{1}{(pSr)^2}\right] \right\}^{1/2} \tag{21}$$

If consideration is restricted to relatively large currents, to the region $r \gg r_0$ and to common situations in which the dimensionless term pSr is of the order of a few tenths or less,

$$E = \left[\frac{j_l}{2\pi\varepsilon_0 b}\left(1 + \frac{2pSr}{3}\right)\right]^{1/2} \cong \left(\frac{j_l}{2\pi\varepsilon_0 b}\right)^{1/2}\left(1 + \frac{pSr}{3}\right) \qquad (22)$$

The electric field as a function of radial distance, with and without space-charge effects, is shown in Figure 5. It is seen that space charge works to decrease the field near the wire and raise it near the tube. Indeed, with

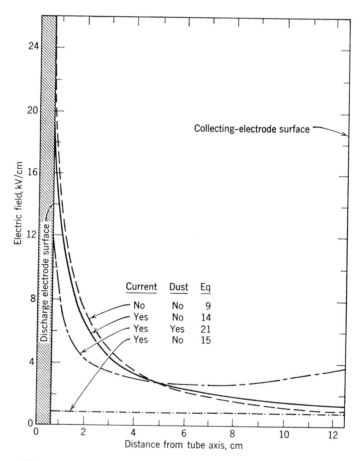

Fig. 5. Electric-field intensity in a typical wire-pipe precipitator. The current density is $j_l = 82\ \mu\text{A/m}$ (corresponding to a potential difference of 40 kV) and the specific surface is $S = 8.3\ \text{m}^2/\text{m}^3$ (corresponding to a dust concentration of 18 g/m³ for particles of diameter 6.6 μm and specific gravity 2) (10). (Used with permission of H. J. Lowe, Institute of Physics and Physical Society.)

adequate space charge, the minimum field no longer occurs at the collecting electrode surface. The reduction in field at the wire comes about because the space charge partly shields the discharge wires from the oppositely charged tube. Initiation of corona on a wire entails a critical corona-starting field at the wire surface (Sec. III-D). In order to attain this field in the presence of space charge, it is necessary to impress a higher voltage across the electrodes than would otherwise be necessary.

Space-charge effects may be considered from an alternative point of view which also yields useful results (73–75). To a first approximation, both particle and ion space-charge densities, ρ_p and ρ_i, respectively, are—unlike the case of eq. 21—assumed independent of position in the tube cross section. Poisson's equation, upon integration, then yields

$$E = \frac{V_c}{r \ln (r_1/r_0)} + \frac{(\rho_i + \rho_p)}{2\varepsilon_0} \frac{(r^2 - r_0{}^2)}{r} \qquad (23)$$

The field at any point r in the tube is the sum of two components: the electrostatic field prevailing in the absence of space charge, and the supplementary space-charge field. The earlier remark that space charge depresses the field near the wire is not inconsistent with eq. 23 which states that at $r = r_0$ the field is independent of the space-charge density. The former assertion assumes constant applied voltage, and the latter, conditions under which the voltage is raised to "absorb" the additional space-charge field.

Integration of eq. 23 gives the potential of the wire

$$V_0 = V_c + \left(\frac{\rho_i + \rho_p}{4\varepsilon_0}\right) r_1{}^2 \qquad (24)$$

where the potential of the tube is zero and r_0 is neglected relative to r_1. Expressing ρ_i in terms of j_l (eq. 13), the preceding equation may be rewritten

$$V_0 = V_c + \frac{\rho_p}{4\varepsilon_0} r_1{}^2 + \frac{j_l \ln (r_1/r)}{8\pi\varepsilon_0 bV_0} \qquad (25)$$

This expression, except for the particle space-charge term, is identical to Townsend's low-current approximation (eq. 18). Evidently, the presence of particle space charge raises the apparent corona-onset potential from the particle-free level V_c to a higher effective value

$$V'_c = V_c + \frac{\rho_p}{4\varepsilon_0} r_1{}^2 \qquad (26)$$

The particle space-charge density may be calculated using the relations of Section IV.

Equation 25 giving the potential of the discharge wire is a special case of the more general low-current relation

$$V(r) = V_0 - V_c \frac{\ln (r/r_0)}{\ln (r_1/r_0)} - \left[\frac{j_l \ln (r/r_0)}{8\pi\varepsilon_0 b(V_0 - V(r))} + \frac{\rho_p}{4\varepsilon_0} \right]$$

$$\times \left(r^2 - r_0^2 - 2r_0^2 \ln \frac{r}{r_0} \right) \quad (27)$$

giving the potential at any point r in the cylinder.

In most applications, the contribution of ρ_p to the current is inconsequential (but see Sec. VI-C).

C. Wire-Plate Geometry

1. Particle-Free Condition

A solution to Poisson's equation for a wire-plate electrode system (Fig. 1b) along the lines of Townsend's treatment of the wire-tube configuration is beset by formidable mathematical difficulties. Considerable simplification results, however, if we assume that the current is low, and that the resultant alteration of the potential by space charge can be represented by an additive correction analogous to that obtaining in eq. 25. The static potential at any point (x, y) in the duct (Fig. 6) is given by Cooperman (76) in terms of the rapidly converging infinite series

$$V(x, y) = V_0 \frac{\displaystyle\sum_{m=-\infty}^{\infty} \ln \left[\frac{\cosh [\pi(y - 2mc)/2s] - \cos (\pi x/2s)}{\cosh [\pi(y - 2mc)/2s] + \cos (\pi x/2s)} \right]}{\displaystyle\sum_{m=-\infty}^{\infty} \ln \left[\frac{\cosh (\pi mc/s) - \cos (\pi r_0/2s)}{\cosh (\pi mc/s) + \cos (\pi r_0/2s)} \right]} \quad (28)$$

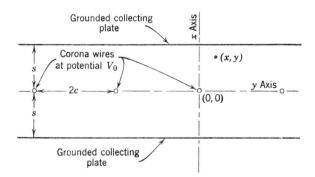

Fig. 6. Wire-plate (duct-type) electrode arrangement. All linear dimensions are in meters. The coordinates (0, 0) mark the center of a wire.

If to this we add the small-current perturbing potential (77)

$$\frac{j_l \ln (d/r_0)}{4\pi\varepsilon_0 bV_0} (s^2 - x^2) \tag{29}$$

it may be shown that

$$j_l = 4cj_s = \frac{4\pi\varepsilon_0 b}{s^2 \ln (d/r_0)} V_0(V_0 - V_c) \tag{30}$$

where j_s (A/m²) is the average current density at the plate. The parameter d(m) is represented closely by

$$d = \frac{4s}{\pi} \quad \text{for} \quad \frac{s}{c} \leqslant 0.6 \tag{31}$$

$$d = \frac{c}{\pi} e^{\pi s/2c} \quad \text{for} \quad \frac{s}{c} \geqslant 2.0 \tag{32}$$

$$\text{See Figure 7} \quad \text{for} \quad 0.6 < \frac{s}{c} < 2.0$$

The corona-starting voltage V_c in eq. 30 is (77)

$$V_c = r_0 E_c \ln \frac{d}{r_0} \tag{33}$$

where E_c is given by eq. 49.

Consideration of the foregoing relations leads to the following conclusions:

1. As the wire-to-wire spacing increases, j_l tends to become constant.

2. There is a wire-to-wire spacing, depending on the other dimensions, for which j_s is a maximum. This comes about because reduced spacing raises

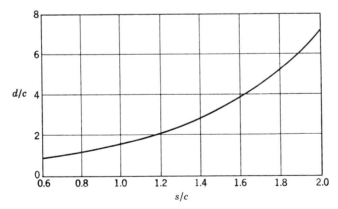

Fig. 7. Curve for determining the parameter d in terms of the dimensions of a wire-plate precipitator (eqs. 29 ff.) (77). (Used with permission of the Institute of Electrical and Electronic Engineers.)

the starting voltage and so tends to lower the current, but at the same time more wires are introduced between the plates tending to raise j_s. The current maximum is broad and the associated electrode dimensions are not critical.

3. j_l depends on wire-to-plate spacing inversely as a power lying between two and three.

Experimental data bearing on the foregoing considerations have been reported by Lagarias (78).

Dupuy (79) gives the field at a point $(x, 0)$ by the expression

$$E(x, 0) = \coth\left(\frac{\pi x}{2c}\right)\left\{\frac{4cj_{sx}}{\pi\varepsilon_0 b}\ln\left[\frac{\cosh\,(\pi x/2c)}{\cosh\,(\pi r_0/2c)}\right] + E_c{}^2\left(\frac{\pi r_0}{2c}\right)^2\right\}^{1/2} \qquad \left(s > \frac{3c}{2}\right)$$

(34)

which for a broad region near the plates simplifies to

$$E(x, 0) = \left[\frac{4cj_{sx}}{\pi\varepsilon_0 b}\left(\frac{\pi x}{2c} - \ln 2\right)\right]^{1/2}$$

(35)

The term j_{sx} represents the current density at a point on a plate directly opposite a wire. Troost's approximation (80)

$$E = \left(\frac{j_l s}{\pi\varepsilon_0 bc}\right)^{1/2} = \left(\frac{4j_s s}{\pi\varepsilon_0 b}\right)^{1/2}$$

(36)

is sometimes quoted; it gives fair agreement with eq. 35 when its use is restricted to the field at the plate ($x = s$) and $\pi s/2c \sim 2.5$, a value fortuitously representative of many precipitators. At a point $(x, 0)$ well removed from a wire, the potential is found by integrating eq. 35 (79)

$$V(x, 0) = \frac{4}{3\pi}\left(\frac{j_{sx}}{2\pi\varepsilon_0 b}\right)^{1/2}(\pi s - 2c\ln 2)^{3/2}\left\{1 - \left[1 - \frac{s - x}{s - (2c/\pi)\ln 2}\right]^{3/2}\right\} \quad (37$$

Less common precipitator electrode configurations have also been treated by Dupuy. Coplanar parallel wires opposite a *single* plate (Fig. 6 with one plate removed) yield the relations (79)

$$E(x, 0) = \frac{2}{1 - e^{-\pi x/c}}\left\{\frac{cj_{sx}}{2\pi\varepsilon_0 b}\left[e^{-\pi x/c} - e^{-\pi r_0/c} + \frac{\pi}{c}(x - r_0)\right] + E_c{}^2\left(\frac{\pi r_0}{2c}\right)^2\right\}^{1/2}$$

(38)

and

$$V(x, 0) = \frac{2}{3}\left(\frac{2j_{sx}}{\varepsilon_0 b}\right)^{1/2}\left(s - \frac{c}{\pi}\right)^{3/2}\left[1 - \left(\frac{\pi x - c}{\pi s - c}\right)^{3/2}\right]$$

(39)

In the special case $s = 2c$, the associated low-current approximation is (81

$$j_l = \frac{0.86\pi\varepsilon_0 b}{c^2[2\pi + \ln(c/\pi r_0)]} V_0(V_0 - V_c) \tag{40}$$

The above three relations are of interest in connection with certain novel designs in which the direction of gas flow is parallel to that of ion movement (82–85).

The field of a single wire parallel to, and midway between two plates (cf. two-stage precipitator, Fig. 2b) is described by (79)

$$E(x, 0) = \frac{2}{\sin(\pi x/2s)}\left\{\frac{j_{sx}}{\pi\varepsilon_0 b}\left[\cos\left(\frac{\pi r_0}{2s}\right) - \cos\left(\frac{x}{2s}\right)\right] + E_c^2\left(\frac{\pi r_0}{4s}\right)^2\right\}^{1/2} \tag{41}$$

Integration of this expression provides the current–voltage relation

$$\frac{V_0 - V_c}{V_c}\ln\left(\frac{4s}{\pi r_0}\right)$$

$$= \frac{1}{2}\left\{\sqrt{2\gamma + 1}\ln\left[\left(\frac{\sqrt{2\gamma + 1} - 1}{\sqrt{2\gamma + 1} + 1}\right)\left(\frac{\sqrt{2\gamma + 1} + \sqrt{\gamma + 1}}{\sqrt{2\gamma + 1} - \sqrt{\gamma + 1}}\right)\right]\right.$$

$$\left. - \ln\left[\left(\frac{\gamma}{2}\right)\left(\frac{\sqrt{\gamma + 1} + 1}{\sqrt{\gamma + 1} - 1}\right)\right]\right\} \tag{42}$$

where

$$\gamma = \frac{16s^3 j_{sx}}{\varepsilon_0\pi^3 b E_c^2 r_0^2} \quad \text{(dimensionless)} \tag{43}$$

This equation may be solved graphically, in the same manner as eq. 16.

2. Particle Space Charge

We return now to the usual wire-plate geometry of Figure 6 to consider the effect of dust space charge. Drawing an analogy between eqs. 18 and 30 governing wire-tube and wire-plate systems, respectively, Cooperman (75) has shown that in the latter case the ion space charge is given by the expression

$$\rho_i = \frac{j_l \ln(d/r_0)}{2\pi b V_0} \tag{44}$$

Assuming uniformly distributed particle space charge, the following relation, corresponding to eq. 25, may be written

$$V_0 = V_c + \frac{\rho_p}{2\varepsilon_0}s^2 + \frac{j_l \ln(d/r_0)}{4\pi\varepsilon_0 b V_0} \tag{45}$$

where the apparent corona-starting voltage has been raised to

$$V'_c = V_c + \frac{\rho_p}{2\varepsilon_0} s^2 \tag{46}$$

Comparison of eqs. 26 and 46 reveals that for equal tube diameter and plate-to-plate spacing, particle space charge elevates the effective duct starting voltage twice as much as for the tube, and this increase is independent of wire-to-wire spacing.

D. Corona Onset and Sparkover

In the absence of an adequate theory of breakdown in nonuniform fields it is not possible to calculate either the corona starting field and voltage, E_c and V_c, or the sparkover field and voltage, E_s and V_s, from atomic data. Furthermore, accuracy of measurement in the laboratory, and even more so in the field, is marred by surface asperities and dust deposits on the electrodes. Surface irregularities in combination with misaligned electrodes inevitably lower both corona-onset and sparkover levels in practical installations. Maximum sparkover voltages to be anticipated are conveniently determined in the laboratory; practical values, on the other hand, are best established through observation of similar installations. One indication of the uncertainties encountered is given by the empirical relation (86)

$$V_{sn} = V_{s1} - C_1 \ln n \tag{47}$$

where V_{s1} and V_{sn} are the respective sparkover voltages for one and n wires energized by a single power supply and C_1 is a constant. The physical basis of eq. 47 becomes clear when it is remembered that:

1. The instantaneous potential of all parallel-connected wires is set by the reduced potential of whatever wire in the system is then experiencing sparkover.

2. The greater the total wire length of the precipitator, the higher is the probability of occurrence, somewhere in the system, of those conditions tending to lower the sparkover potential.

Attempts to compensate for poor performance by adding wires—the new wires also being fed by the original electrical set—are apt to be self-defeating. For effective precipitation the length of corona wire energized by a single power supply must be limited. It is this consideration that leads to the general practice of dividing large precipitators into sections, each section energized by an independent supply (1c,87).

TABLE IV

Electrical Characteristics for Common Precipitator Applications[a]

Application	Tube diameter or duct width	Sparkover voltage, kV	Avg. corona current, mA/100 m²	Avg. corona energy density (W-sec)/m³
Fly ash	20–25 cm duct	40–70	10–40	100–250
Cement	20–25 cm duct	40–70	7–30	150–550
Paper mill	25 cm duct	70–80	7–30	100–550
Blast furnace	20 cm pipe	35–45	10–60	150–400
	30 cm pipe	65–75	10–30	70–400
Sulfuric acid	25 cm pipe	70–80	10–40	150–900
Copper and zinc smelters, converter gas	15 cm duct	~30	~10	~350
Roaster and reverberatory gas	30 cm duct	~75	~10	~300

[a] From reference 1d. Used with permission of Addison-Wesley Publishing Company.

Table IV gives rule-of-thumb values of sparkover voltage and associated variables useful for orientation purposes. The data are representative of most modern industrial precipitators, but are not, in general, applicable to the special situations discussed in Section VI. The sparking potential coincides with the cyclic peak of the imperfectly filtered unidirectional waveform.

Peak voltage gradients in the tabulated examples are typically of the order of 4–6 kV/cm averaged across the interelectrode gap. Both corona-starting and sparkover voltages are temperature–pressure dependent. Electrical breakdown, whether partial (corona) or complete (sparkover), depends on the likelihood of electrons accelerating to ionizing energies in the space of a mean free path, this distance being inversely proportional to relative gas density δ.

Texts on gaseous electronics show that a process occurring at a rate dependent essentially on the ratio E/δ (e.g., ionization by collision) conforms to the similarity principle (38a). An expression of this principle useful for present purposes is

$$\frac{E}{\delta} = f(l\delta) \tag{48}$$

where $E(V/m)$ is the breakdown field at some point and $l(m)$ is a characteristic dimension of the system. Breakdown in this context designates corona onset or sparkover, whichever occurs first as the voltage is raised.

It has been demonstrated empirically in numerous pure gases and gas mixtures that combinations of round wires and outer electrodes of arbitrary shape yield eq. 48 in the explicit form (88–90).

$$\frac{E_c}{\delta'} = A_g + \frac{B_g}{(r_0\delta')^{1/2}} \tag{49}$$

where A_g(V/m) and B_g(V/m$^{1/2}$) are constants. Values of A_g and B_g as reported by Thornton (91) are given in Table V. The relative gas density δ'

TABLE V

The Parameter A_g for Calculating Corona-Starting Fields in Various Gases.
$(E_c/\delta' = A_g + B_g/(r_0\delta')^{1/2}$ where $B_g = 1.23 \times 10^5$ V/m$^{1/2})^a$

Gas	A_g, $10^5 \dfrac{V}{m}$	Gas	A_g, $10^5 \dfrac{V}{m}$	Gas	A_g, $10^5 \dfrac{V}{m}$	Gas	A_g, $10^5 \dfrac{V}{m}$
Air	35.5[b]	NH_3	56.7	C_2H_2	75.2	$C_5H_{11}Cl$	264.0
H_2	15.5	N_2O	55.3	C_2H_4	21.3	CH_3Br	97.0
He	4.0	H_2S	52.1	C_3H_6	87.2	C_2H_5Br	98.0
Ne	4.5	SO_2	67.2	C_6H_6	86.7	C_3H_7Br	155.0
A	7.2	CS_2	64.2	CH_2Cl_2	126.0	CH_3I	75.0
Kr	9.5	CH_4	22.3	CH_3Cl_3	162.0	C_2H_5I	101.8
O_2	29.1	C_2H_6	26.2	CCl_4	204.0	CH_3OH	62.5
N_2	38.0	C_3H_8	37.2	CH_3Cl	45.6	C_2H_5OH	97.0
Cl_2	85.0[c]	C_4H_{10}	47.7	C_2H_5Cl	109.0	$(C_2H_5)_2O$	15.3
CO	45.5	C_5H_{12}	63.1	C_3H_7Cl	160.9	$(CH_3)_2CO$	5.4
CO_2	26.2	C_6H_{14}	72.0	C_4H_9Cl	200.0	$C_2H_4Cl_2$	240.0

[a] From reference 91. Used with permission of Taylor and Francis, Ltd.

[b] See text below

[c] This value seems too high (92).

is conventionally taken with respect to 1 atm and 25°C. This table is useful for obtaining approximate estimates of E_c; it is, however, not applicable to negative corona in pure nonattaching gases, and it may grossly overestimate the negative corona-starting field at elevated pressures [e.g., air: Sec. VI-A-3; methane: (93)], or temperatures (Sec. VI-A-4). For negative corona in air, the constants $A_g = 32.2 \times 10^5$ V/m and $B_g = 8.46 \times 10^4$ V/m$^{1/2}$, derived from the cumulative data of several investigators, are recommended (88,89).

Experimental corona-starting and spark-breakdown data for a variety of gases and electrodes are assembled in Landolt-Boernstein (94). Sparkover mechanisms with special reference to electrostatic precipitation is the subject of a number of studies by Penney et al. (95–98).

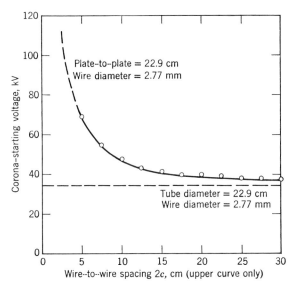

Fig. 8. Comparison of corona-starting voltages in wire-plate and wire-tube precipitators of corresponding dimensions. As the wire-to-wire spacing decreases below 5 cm the voltage required for corona onset becomes so high that only sparkover results. The curves are theoretical (eq. 33) and the points experimental (99). (Used with permission of Institute of Electrical and Electronic Engineers.)

Example. Determine the corona-starting voltage in room air for a duct precipitator (Fig. 6) of plate-to-plate spacing 9 in. ($2s = 0.229$ m), wire-to-wire spacing 4 in. ($2c = 0.102$ m), and wire diameter 109 mil ($2r_0 = 2.77 \times 10^{-3}$ m). Compare with a 109-mil diameter wire in a 9-in. diameter pipe.

The duct corona-starting voltage V_c is given by eq. (33) for which E_c, the corona-starting field at the wire, and the parameter d are required. The following quantities are, therefore, calculated: $\delta' = 1$ (p. 254); $E_c = 54.9 \times 10^5$ V/m (eq. 49); $s/c = 2.24$; $d = 0.548$ (eq. 32); and finally, $V_c = 45.7$ kV (eq. 33). The pipe starting voltage is given by eq. 9: $V_c = r_0 E_c \ln(r_1/r_0) = 33.6$ kV. Figure 8 shows the duct starting voltage as a function of wire-to-wire spacing. For equal duct width and pipe diameter and identical wire sizes, duct starting voltage will always exceed wire-pipe starting voltage. Starting voltages measured in industrial precipitators are invariably found to be lower than the calculated estimates. This effect is due to irregular electrode spacing, to extraneous discharges from electrode asperities and, in ducts, to the lower starting voltage of the end wires (99).

E. Power Supplies

Optimum precipitator performance requires, as a rule, the highest level of electrical energization attainable in a given set of circumstances. The physical principles and technical means for accomplishing this goal are today fairly well understood as a result of many years of research and practical observation. It was discovered by Cottrell in the early days of precipitation, and confirmed by long experience, that the pulsating half- or full-wave voltages obtained from imperfectly filtered rectifier sets yield collection efficiencies superior to those resulting from unvarying direct voltage. The relatively long decay periods for pulsating waveforms allow time for sparks to extinguish between current pulses and so permit operation at voltage and power levels not attainable in the absence of sparking. But although sparking is generally desirable in industrial practice, too frequent sparking will detrimentally lower the power input. It is primarily this consideration, the need to operate at some optimum spark rate (commonly of the order of 100 sparks per minute per electrical set), that determines the nature of the transformer-rectifier and associated automatic control equipment used to energize the precipitator. However, when spark rate provides the only feedback signal to the control system, transformers and rectifiers are likely to be vulnerable to damage from excess current. Consequently, a double-feedback system (monitoring current in addition to spark rate) is usually recommended to assure the most favorable average spark rate despite erratic variations in line and load conditions.

Air-cleaning or sampling precipitators are run below sparkover and present no control problems of consequence. An exception is the high-altitude air sampler for which current control is essential over wide variations in gas density (100).

The recent state of the art of industrial precipitator power supplies and control apparatus has been reviewed by White (1e), Rose and Wood (2a), and Coe (101). Other studies, mostly emphasizing solid-state devices (silicon and selenium rectifiers to replace vacuum tubes, thyristor controls to replace magnetic amplifiers) having advantages of improved response, lower power losses, and reduced equipment size, are listed in the references (102–113).

Pulse energization of precipitators was introduced by White (114) in 1952 and has been occasionally investigated since (115–118). Advantages offered by this technique include higher peak voltages, inherent current-limiting capability following sparking, and higher collection efficiency in certain applications. Under suitable circumstances, the total average current can be of the same order as the normal dc current. Commercial application of pulse precipitation must, however, await further technical and economic development of high-power electronic-switch components.

IV. PARTICLE CHARGING

When gases laden with suspended particulate matter are passed through an electrostatic precipitator, the great bulk of the particles acquire an electric charge of the same polarity as that of the discharge electrodes. This preferential charging occurs because the region of corona, that is, the region of intensive ion-pair generation, is limited to the immediate vicinity of a discharge wire, and so occupies only a small fraction of the total cross section of the precipitator. The remaining cross-sectional area, for reasons already given, contains a concentration of unipolar ions of the same sign as the wires.

Two distinct particle-charging mechanisms are generally considered to be active in electrostatic precipitation: (a) bombardment of the particles by ions moving under the influence of the applied electric field, and (b) attachment of ionic charges to the particles by ion diffusion in accordance with the laws of kinetic theory.

A. Ion Bombardment

The charge a particle acquires in this manner (also called field charging) has been calculated by Pauthenier and Moreau-Hanot (119). The following is assumed:

1. The particles are spherical.
2. Particle spacing is sufficiently great to render negligible particle-to-particle interactions.
3. The mean free path of the ions is small compared to the diameter of the particles.
4. The electric-field intensity, except in the immediate neighborhood of a particle, is uniform throughout the volume of the precipitator.
5. The ion concentration is also uniform throughout the precipitator.

Regarding the first assumption, Smith and Penney (120) have shown that departures from sphericity encountered in practice are not a serious source of error; the second assumption is true for most industrial and natural aerosols; the third is generally valid except for the finest particles or at much reduced pressures; while the other assumptions are approached in various degrees of approximation.

If a spherical particle bearing a uniformly distributed free surface charge Q(C) is placed in a uniform electric field E_0 in a gas, induced and free charge on the particle distort the original field E_0 and impart to it a radial component

$$E_g = -\frac{\partial V}{\partial r} = E_0 \cos \theta \left[2\left(\frac{\kappa_p - 1}{\kappa_p + 2}\right)\frac{a^3}{r^3} + 1 \right] + \frac{Q}{4\pi\varepsilon_0 r^2} \qquad (r \geqslant a) \quad (50)$$

Here r(m) is the radius vector from the center of the particle, θ is the polar angle between r and the undistorted field E_0, κ_p is the relative dielectric constant of the particle, and a(m) is its radius. An ion of charge q(C) is attracted to the particle and imparts its charge by attachment if the ion approaches from an angle θ for which the radial force $F = qE_g(\mathrm{N})$ is negative. Particle charging ceases at $F = 0$.

Figure 9a shows the lines of force in an initially uncharged sphere. As charging proceeds, the charge already present on the particle creates a repulsive force which modifies the configuration of the electric field and thereby reduces the rate of charging. Figure 9b shows the field configuration when the sphere is partly charged. The lines of force entering the sphere on the side facing the oncoming ions have been reduced. Eventually the lines of force will bypass the sphere completely and charging will halt. Setting $\theta = \pi$, $r = a$ and

$$p = 2\left(\frac{\kappa_p - 1}{\kappa_p + 2}\right) + 1 \tag{51}$$

the maximum free charge on the surface of the spherical particle is, from eq. 50,

$$Q_{\max} = 4\pi\varepsilon_0 p a^2 E_0 \tag{52}$$

Consideration of the gas-ion current to the particle shows that the particle charge as a function of time is

$$Q = Q_{\max}\frac{t}{t + \tau} \tag{53}$$

where τ (sec), the particle-charging time constant, is given by

$$\tau = \frac{4\varepsilon_0}{N_0 q b} \tag{54}$$

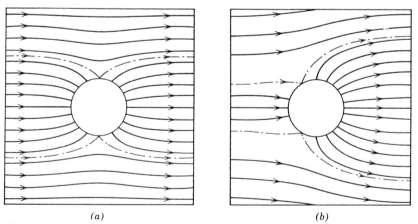

(a)   (b)

Fig. 9. The electric fields near (a) an uncharged and (b) a partly charged dielectric sphere. The dashed lines represent the limits of the fields passing through the spheres. The fields inside the spheres are not shown.

and $N_0(\text{m}^{-3})$ is the ion density. The time constant represents the period in which half the limiting charge is attained.

The preceding considerations remain valid for a conducting particle. In this instance, allowing the dielectric constant κ_p to increase without limit, $p = 3$.

Ion-bombardment charging in the presence of the bi-ionized field of the back corona is examined in Section VI-B-2.

Experimental studies of ion-bombardment charging have been made by a number of investigators. These include the measurements of White (121), Penney and Lynch (122), Hewitt (123), Drozin and LaMer (124), Edmondson (125), Langer (126), and Hignett (127) at atmospheric pressure, and the work of Liu et al. (128–133) at reduced pressure. Observations were more or less in agreement with theory. Hignett (127) points out that industrial precipitators commonly operate at space-charge densities as low as 5×10^{12} ions/m³, one or two orders of magnitude lower than the figure frequently quoted for laboratory precipitators (1f,2b,10). Charging-time constants as large as tenths of a second may result, which is of particular importance in two-stage applications. Elsewhere (134,135) Hignett examines particle-charge magnitude as a function of particle trajectory and dust burden.

B. Diffusion Charging

In addition to ion-bombardment charge, a suspended aerosol particle in an ionized gas acquires a charge by virtue of the random thermal motion of the ions and their consequent collision with and attachment to the particle. This process is called ion diffusion and its effect may be calculated in the approximate manner due to White (121).

It is shown in kinetic theory that the density of a gas (in this case an ion cloud) in a potential field varies in accordance with the expression

$$N = N_0 e^{-U(r)/kT_A} \tag{55}$$

where $N(\text{m}^{-3})$ is the number of ions per unit volume in the presence of a particle's field, $N_0(\text{m}^{-3})$ is the undisturbed uniform ion density, $U(r)$ (J) is the potential energy of an ion due to its position r in the field, k (J/°K) is Boltzmann's constant, and T_A(°K) is the absolute temperature. Neglecting the effect of the applied field, the potential energy of an ion of charge q(C) in the vicinity of a uniformly charged spherical particle is

$$U = \frac{qQ}{4\pi\varepsilon_0 r} \tag{56}$$

where Q(C) is the charge on the particle and r(m) is the distance from the center of the particle to the ion. The time interval t (sec) associated with the ion-particle collisions is given by

$$t = \frac{1}{\pi a^2 N \bar{v}_i} \tag{57}$$

where $\bar{v}_i$(m/sec) is the kinetic-theory rms velocity of the ions. If it is assumed that all ions impinging on a particle are attached by image forces, then charging will proceed at the rate

$$\frac{dQ}{dt} = \pi a^2 N q \bar{v}_i \tag{58}$$

Whence, for an initially uncharged particle,

$$Q = \frac{4\pi\varepsilon_0 akT_A}{q} \ln \left(\frac{aN_0q^2\bar{v}_i}{4\varepsilon_0 kT_A} t + 1 \right) \tag{59}$$

The foregoing derivation ignores the contribution of the externally impressed field E_0 to the potential U. Taking this additional quantity into account for a conducting particle, Pauthenier (136) finds

$$Q = \frac{4\pi\varepsilon_0 akT_A}{q} \ln \left[\frac{(8\pi)^{1/2}}{3} \frac{aN_0q^2\bar{v}_i}{4\pi\varepsilon_0 kT_A} \frac{\sinh(E_0qa/kT_A)}{E_0qa/kT_A} t + 1 \right] \tag{60}$$

The quotient $\sinh(E_0qa/kT_A)/(E_0qa/kT)$ approaches unity for very fine particles. It should be noted that although the electric field is now included in the calculation, the treatment still neglects the charge acquired by ion bombardment.

The kinetic-theory approach to the fine-particle charging problem is not the only means of solution that has been employed. Some investigators regard the charging mechanism as a diffusion process in which ions move continuously toward the particle under the action of a concentration gradient. The rate of capture of ions by the particle is taken to equal the ionic flux found by solving the steady-state diffusion equation with appropriate boundary conditions. Depending upon the conditions chosen, various solutions are obtained. Examples are given by the equations of Arendt and Kallmann (137), Bricard (138), Gunn (139), and Natanson (140). However, as has been pointed out by Liu, Whitby, and Yu (141,142), for these results to be valid, the ion concentration must be high enough for the charging process to be considered essentially continuous. At the usual maximum concentration of about 10^{15} ions/m³ encountered in precipitators, this requirement is not met: ion capture by micron and submicron particles is a discontinuous process and conclusions based on the solution of the continuous macroscopic diffusion equation cannot be relied upon. On the basis of this argument, Liu et al. (141) attack the problem of diffusion charging (in the absence of an applied electric field) from the standpoint of kinetic theory. Their calculations, although following a more rigorous procedure than that used earlier by White, nevertheless lead to an identical solution, that is, eq. 59.

Murphy (143,144), using a method of analysis similar to that of Liu et al., predicts higher charging rates than indicated by eq. 59, but as the latter authors show, this is a consequence of ignoring the effect of accrued particle charge on the trajectories of the incoming ions.

Diffusion charging at much reduced pressures, a phenomenon of importance in the operation of electrostatic aerosol samplers for stratospheric use, has been studied experimentally and theoretically by Liu et al. (128–133).

C. Combined Ion-Bombardment and Diffusion Charging

Both ion-bombardment and diffusion charging are simultaneously operative. For the range of particle sizes met in most industrial precipitator applications, charging by ion bombardment is dominant by far. A

TABLE VI

Number of Electronic Charges Acquired by Conducting Spherical Particles According to Various Theories[a]

Charging process		Ion bombardment, eq. 53				Ion diffusion (field neglected), eq 59.			
Treatment time, sec		0.01	0.1	1	∞	0.01	0.1	1	10
Particle radius, μm	0.1	1	3	4	4	3	7	11	15
	1	120	340	410	420	69	110	150	190
	10	12,000	34,000	41,000	42,000	1,100	1,500	1,900	2,300

Charging process		Bombardment and diffusion, eq. 61				Ion diffusion (field included), eq. 60			
Treatment time, sec		0.01	0.1	1	∞	0.01	0.1	1	10
Particle radius, μm	0.1	2	6	7	7	2	5	10	14
	1	120	340	410	420	150	190	230	270
	10	12,000	34,000	41,000	42,000	14,000	14,400	14,800	15,200

[a] The conditions on which the charge calculations are based are $E_0 = 2 \times 10^5$ V/m, $\kappa_p = \infty$, $T_A = 300°$K, $N_0 = 5 \times 10^{13}$ ion/m^3, $b = 1.8 \times 10^{-4}$ m^2/(V-sec), $\lambda_i = 1.1 \times 10^{-7}$ m, $k = 1.38 \times 10^{-23}$ J/(°K-molecule), and $\bar{v}_i = 510$ m/sec. The rms ion velocity is given by $\bar{v}_i = (3p_g/\rho_g)^{1/2}$ where p_g is the gas pressure (N/m^2) and ρ_g is the gas density (kg/m^3). These values approximately represent a wire-in-tube assembly of respective inner and outer radii 3.8 mm and 12.7 cm in room air at 40 kV with a negative discharge current of 0.13 mA/m. The ion-diffusion eqs. 59 and 60 are properly applicable only as long as the particles are only slightly charged and surrounded by a radially symmetrical atmosphere of ions, that is, as long as a significant repulsive zone (cf. Fig. 9) has not developed (146,151). Consequently, the large diffusion charges calculated for long times and large particles cannot be regarded as correct.

convenient, but not inviolate, rule is that ion bombardment predominates for particles of radius greater than a few tenths to 1 μm and ion diffusion for less than 0.1 μm.

Viewing diffusion charging in terms of ion-current flow to a particle and thereby generalizing the ion-bombardment mechanism to include the effects of diffusion, Cochet (145,146) derives the following equation that agrees with his experimental data

$$Q = \left[\left(1 + \frac{\lambda_i}{a} \right)^2 + \frac{2}{1 + (\lambda_i/a)} \frac{k_p - 1}{k_p + 2} \right] 4\pi\varepsilon_0 E_0 a^2 \frac{t}{t + \tau} \qquad (61)$$

Here λ_i(m) is the mean free path of the ions taken to be 10^{-7} m in room air.

Liu and Yeh (147) raise a twofold objection to Cochet's theory. These authors argue that the *ionic* mean free path is, in fact, an order of magnitude smaller than assumed by Cochet, and that irregular thermal motion which becomes increasingly dominant as particle size decreases cannot be expressed in terms of ordered movement along lines of force. Liu and Yeh have developed a new theory taking account of both the random motion of ions and the motion of ions as induced by an applied field. Good agreement is claimed with experimental data (132,147). Only numerical solutions are available.

Earlier attempts to deal simultaneously with the two interacting charging processes have been made by Murphy (144) and Armington (148). Foster (149) and Kraemer and Johnstone (150) report that simple addition of the ion-bombardment and diffusion charges given by eqs. 52 and 59 yields reasonable agreement with experiment.

Table VI compares the results of the charging theories of Pauthenier, White, and Cochet under typical discharge conditions in room air.

V. PARTICLE COLLECTION AND EFFICIENCY CALCULATION

A. Background of the Problem

Particle collection in an electrostatic precipitator is essentially a process of mass transfer through a moving gas, in a net direction that is normal to the collecting surface. It is necessary to distinguish at least three forms of particulate mass transfer from the main body of gas to the passive electrode: (a) electrostatic convection under the action of Coulomb forces and the electric wind, (b) turbulent diffusion of aerodynamic and electrodynamic origin, and (c) inertial drift. The Coulomb-force drift mechanism has been quantitatively investigated and its main features are quite clear. In contrast, the physical nature of mass transfer by turbulent diffusion is much more complicated and, in relation to electrostatic precipitators, it has so far not proved possible to set up general and rigorous quantitative definitions. Noteworthy first efforts in this direction have, however, been made by Williams and Jackson (152) and Cooperman (153–156). Inertial drift, another complex phenomenon, is a consequence of a particle's tendency, by virtue of its momentum, to continue moving in a straight line in the face of opposing or deflecting forces. Thus, the more massive a particle, the less likely it is to closely follow the motion of an eddy in which it is entrained (135,157).

Purely electrostatic mass transfer to the walls occurs in stationary gases and under the influence of the laminar-flow regime. It is a characteristic feature of laminar flow that the direction of gas motion as a whole coincides with the direction in which any separate part of the gas moves. Therefore, there is no macroscopic motion of the gas in the transverse direction (i.e., normal to the collecting electrode) under fully laminar conditions. In such circumstances, mass transfer to the precipitator collecting surface can occur only by electrostatic drift.

The laminar-flow precipitator is, however, a laboratory novelty; all, or virtually all, commercial precipitators and all single-stage precipitators operate with various degrees of turbulence. Various attempts have been made to develop a general and comprehensive theory of turbulent precipitation, the lack of which constitutes the major obstacle in the design of precipitators today. Before considering these attempts, however, it will be necessary to calculate a fundamental quantity for all theories of electrostatic precipitation, the particle migration velocity. This is the velocity exhibited by a charged suspended particle moving toward the collecting surface under the influence of an external electric field.

B. Theoretical Particle-Migration Velocity

The drag $F(\mathrm{N})$ on a spherical particle moving at velocity w through a gas is

$$F = \tfrac{1}{2} C_D \pi a^2 \rho_g w^2 \tag{62}$$

where $a(\mathrm{m})$ is the particle radius, $\rho_g(\mathrm{kg/m^3})$ the gas density and C_D the dimensionless drag coefficient. The drag coefficient is a function of the *particle* Reynold's number Re (dimensionless)

$$\mathrm{Re} = \frac{\rho_g w (2a)}{\mu} \tag{63}$$

where μ (decapoise) is the gas viscosity. In laminar flow (Re less than about unity), $C_D = 24/\mathrm{Re}$ and eq. 62 reduces to Stokes' law

$$F = 6\pi\mu a w \tag{64}$$

The electrostatic precipitating force on a charged particle is E_pQ, where E_p

is the precipitating field. For particles charged by ion bombardment, the limiting charge Q is given by eq. 52. Equating electrostatic and drag forces in this case, we have

$$w = \frac{2p\varepsilon_0 E'_c E_p a}{3\mu} \tag{65}$$

where E'_c designates the charging field. Corresponding relations are easily written for charging by the other mechanisms described in Section IV. See Figure 10. Appropriate combinations of large particle size, high gas

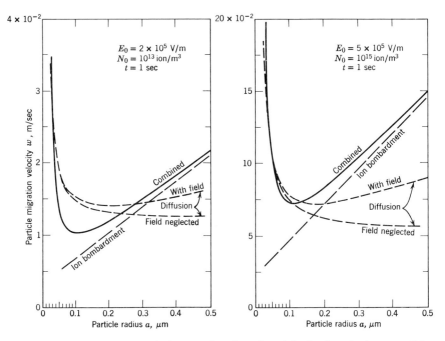

Fig. 10. Particle migration velocity as a function of particle size for submicron particles according to the following theories:

Mechanism		Author	Equation
Ion bombardment		Pauthenier et al. (119)	52, 53
Ion diffusion	Field neglected	Pauthenier (136), $E_0 = 0$	60
	Field considered	Pauthenier (136), $E_0 > 0$	60
Combined bomb't-diff'n		Cochet (145,146)	61

The curves include the Cunningham correction and assume that the charging and precipitating fields, E'_c and E_p, respectively, equal E_0 (145). Note the very high migration velocities possible for ultrafine particles ($a < 0.1\ \mu$m). This explains the observation that collected material at the mouth of a precipitator is often a mixture of course and ultrafine particles (159). (Used with permission of Gauthier-Villas éditeur.)

density (occurring at elevated pressure), and high fields can violate the criterion Re < 1; in such an event, eq. 65 will overestimate the drift velocity (10). If the particle size is comparable to the mean free path λ(m) of the gas molecules ($\sim 10^{-7}$ m in atmospheric air), there is a greater tendency for particles to slip between molecules, and w increases by a factor of about $1 + A_c(\lambda/a)$ (Cunningham correction) where A_c (dimensionless) is roughly unity (158).

Various practical difficulties, which may be insurmountable if rigor is demanded, face the investigator attempting to calculate particle-migration velocities under realistic precipitating conditions. No generally satisfactory means exist for selecting appropriate "effective" values of field intensity or space-charge density, both of which vary transversely and longitudinally throughout the precipitator, depend on variable dust concentrations, and vary in time unless pure dc voltage is used. In calculating ion-bombardment w it is sometimes argued that the maximum field near the wires be used for E_c since turbulence may be supposed to carry most particles into the high field region at some point before collection. Choice of E_p depends on viewpoint: According to the Deutsch-White theory (Sec. V-C), E_p is active only in the laminar sublayer—turbulence governing particle motion elsewhere—and therefore E_p must be assigned that value of electric intensity prevailing at the wall. In line with the diffusion theories of precipitation, however (Sec. V-D-2), particle convection is considered active throughout the cross section, whence it seems reasonable to set E_p equal to the average field.

Hignett (127) and White (160), on the basis of their own experimental studies and earlier work by others, conclude that particle-charging theory, whether for submicron or larger particles, may be relied upon to give at least order-of-magnitude agreement with observation.

C. The Deutsch Efficiency Equation

1. Derivation

An exponential efficiency equation for electrostatic precipitators was discovered by Anderson (161) in 1919 and given in the empirical form

$$\eta = 1 - k_A t \tag{66}$$

where η (dimensionless) is the fractional efficiency, t (sec) is the treatment time, and k_A is a precipitator constant. Three years afterward, Deutsch (162) showed that the precipitator constant could be evaluated in terms of physically significant quantities. A derivation of Deutsch's efficiency equation in terms of the physical picture given by White (1g) follows.

The precipitator cross section is considered to consist of two zones: a laminar boundary layer very close to the collecting wall, and a turbulent core occupying virtually the entire cross-sectional area. Particle motion in the core is assumed to be completely dominated by turbulence, turbulent mixing yielding a uniform particle concentration $C_p(\text{kg/m}^3)$ throughout a given cross section. In the boundary layer, the particle has a component toward the wall of velocity $w^*(\text{m/sec})$ which as a reasonable first approximation is assumed constant over the length of the precipitator. Ion-bombardment charge reaches virtually its limiting value in a fraction of a second, and ion-diffusion charge, although it increases with treatment time, does so relatively slowly after about the first second (Table VI). Over time interval dt the particulate matter lying within a distance $w\,dt$ is precipitated on collecting area dA' where $A'(\text{m}^2)$ is the cumulative collecting surface measured downstream from the mouth of the precipitator. This action reduces the particle concentration in the gas opposite dA' by dC_p. Equating rates of particle loss from the gas and particle accumulation on the surface in time dt

$$V_g\,dC_p = -wC_p\,dA' \qquad (67)$$

where $V_g(\text{m}^3/\text{sec})$ is the volume flow rate of the gas. Integrating this expression, and defining precipitator efficiency (dimensionless) in terms of respective inlet and outlet particle concentrations C_{in} and C_{out}, we have

$$\eta = 1 - \frac{C_{\text{out}}}{C_{\text{in}}} = 1 - e^{-Aw/V_g} \qquad (68)$$

where A is the total collecting area.

White (163) has shown that Deutsch's theory is equivalent to the assumption that collection occurs only if a particle, as a matter of chance, enters the laminar boundary layer where the Coulomb force of attraction alone is effective. By means of probability calculations, eq. 68 is then rederived.

Uzhov (4) provides nomographs for rapidly determining η in terms of A/V_g and w.

Contrary to the assertion sometimes made, it is not true that a uniform cross-sectional particle concentration is prerequisite to the derivation of a Deutsch-type exponential efficiency equation. Taking $C_w(A')$ to be the particle concentration in the gas at the wall, and $\bar{C}_p(A')$ the average concentration over the cross section, if the ratio

$$\frac{C_w}{\bar{C}_p} = \chi \qquad (69)$$

is a constant independent of A', it may be shown (164) that the efficiency equation takes the form

$$\eta = 1 - e^{-A\chi w/V_g} \qquad (70)$$

---

* The symbol w for this purpose is almost universally accepted. It was first used by Deutsch (162) to represent *Wanderungsgeschwindigkeit* (migration velocity).

The Deutsch derivation assumes that complete particle charging occurs within a negligibly small downstream interval at the precipitator inlet. Errors arising from this assumption are demonstrably small in most applications. Robinson (165) has calculated a correction applicable in the case of brief treatment times, such as are found in high-volume aerosol samplers.

2. Polydisperse Aerosols

Since, regardless of the charging mechanism, w is a function of particle size (e.g., eq. 65), the overall efficiency of a precipitator treating a distribution of particle sizes should be calculated in terms of a distribution of efficiencies. This has been done by Allander and Matts (166) on the assumption that particle size is log-normally distributed with geometric standard deviation σ_g (dimensionless), and that w is proportional to particle size (ion-bombardment charging). The resulting modification of the Deutsch equation is given graphically in Figure 11. Moderate departures from the monodisperse condition are shown to have decided effects in lowering efficiency, particularly for a high-performance precipitator; for example, setting $Aw_g/V_g = 6$ and increasing $ln\ \sigma_g$ from zero to 0.4 (the latter a representative value for fly ash) has the effect of reducing the overall efficiency from 99.75 to 98.5%. This is equivalent to raising the concentration of the precipitator exhaust 600%.

3. Precipitator Statistics

It is common practice in precipitator design to employ a single effective migration velocity w_e in the Deutsch equation, even for those aerosols having broad particle-size distributions. This circumstance, coupled with various uncertainties in the derivation or use of the Deutsch equation (e.g., cross-sectional uniformity of particle concentration, appropriate value of charging field E_c, neglect of agglomeration and reentrainment) severely limits the equation's reliability. The effective migration velocity for a given aerosol is best determined empirically, and preferably at the same linear gas velocity at which it is intended later to use the data. Typical variations of w_e with particle size and gas velocity are shown in Figure 12. No reason is offered in the Deutsch-Stokes' relations to expect w_e to increase initially with gas velocity; possible explanations of this phenomenon are noted in Section V-D. The decline of w_e at still higher gas velocities is usually accounted for by reentrainment of precipitated dust; but this cannot be the whole explanation because the effect has also been observed with wet and presumably nonreentraining aerosols (Fig. 12c). Since sizeable variations in w_e can be found in a given application, values of w_e reported in the literature without full background details (e.g., (20)) must be accepted with caution and, in new situations, should be used only as very approximate guides.

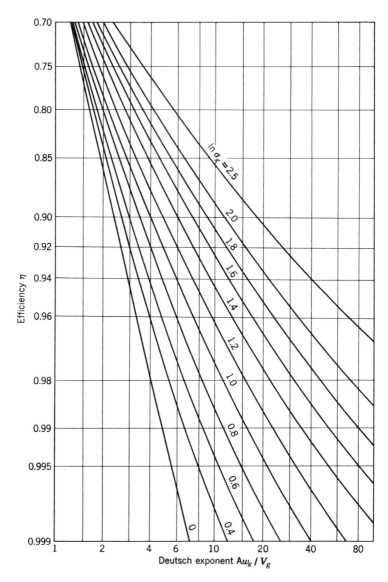

Fig. 11. The effect of a log-normal particle-size distribution on efficiency according to the Deutsch equation. The migration velocity for an individual particle is assumed proportional to particle size. w_g is the migration velocity calculated on the basis of the geometric mean particle radius and σ_g (dimensionless) is the geometric standard deviation. The curve for $\ln \sigma_g = 0$ is for a monodisperse aerosol (eq. 68) (166). (Used with permission of Staubforschungsinstitut des Hauptverbandes der gewerblichen Berufsgenossenschaften e.V.)

Designers of industrial electrostatic precipitators are called upon to harmonize two conflicting requirements: they must minimize relatively high capital costs and, simultaneously, in prescribing precipitator dimensions, introduce safety factors to reasonably insure meeting efficiency guarantees. But, in practice, performance is determined by a variety of unpredictable factors. Clearly, if performance data are available from existing precipitators of a similar type, the design of a new precipitator should not only be based on this information, but should logically be carried out by statistical methods. A systematic way of doing this has been described by Masuda (167).

In this method it is assumed that the effective migration velocities w_e are known for a sample group of about 30 or more nominally identical precipitators operated under comparable conditions. It is further assumed that the Deutsch equation or an equivalent exponential relation applies, that the effective migration velocities of the individual precipitators are normally distributed with standard deviation σ(m/sec) about a mean value $\bar{w}$, and that the condition $\sigma \leqslant \bar{w}/3$ is satisfied. It is required to design a new precipitator of the same kind as that of the sample units, but having an efficiency $\eta \geqslant \eta_0$ where η_0 is the minimum acceptable level. Because the migration velocity varies in a random manner from one installation to the next in the sample, and even from time to time in the same installation, it is impossible to fix the size of the new precipitator with certainty. Instead, we specify the probability $P(0 \leqslant P \leqslant 1)$ that the condition $\eta \geqslant \eta_0$ be fulfilled. The development of the theory will not be given here. The results, however, are summarized in Figure 13 in terms of the dimensionless variables P, B, and K, the last two of which are defined

$$B = (1 - \eta_0)e^{(A/V_g)\bar{w}} \tag{71}$$

$$K = \frac{A}{V_g} \sigma \tag{72}$$

If the sample size is less than about 30, the graphical approach of Figure 13 cannot be used; one must resort directly to equations given in the reference. The graphical method is best shown by means of an example:

Example. By measuring the efficiencies of many fly-ash precipitators having closely similar designs and operating conditions, it is found that the average migration velocity is $\bar{w} = 0.06$ m/sec and the standard deviation is $\sigma = 0.016$ m/sec (i.e., 68% of the w's fall within 0.06 ± 0.016 m/sec). The problem is to design a new precipitator of the same kind with 85% assurance that the guaranteed efficiency $\eta = 0.99$ will be met. The value of the required specific surface A/V_g is found by first making a reasonable estimate of its value based on experience, say 100 m²/(m³/sec). Provisional values of

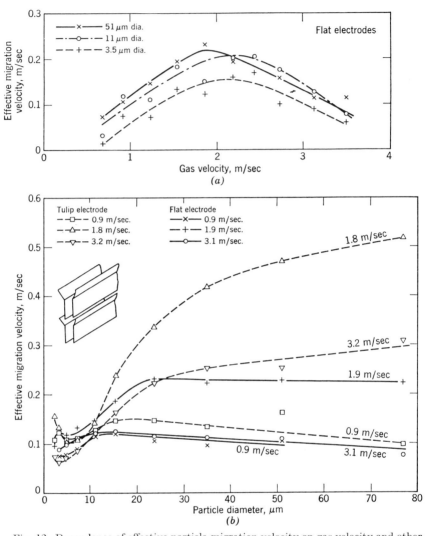

Fig. 12. Dependence of effective particle-migration velocity on gas velocity and other variables. (a) and (b) (168) are for fly ash and (c) (169) is for oil mist (mean diameter ~1 μm). The variation of w_e with precipitator length in (c) indicates that here w_e is high at the precipitator inlet and decreases with downstream distance. In addition to (b), numerous tests in actual installations show that w_e does not increase in proportion to particle size in accordance with eq. 65. For particle diameters greater than about 10 or 20 μm, w_e is very often approximately constant (170). This behavior is variously attributed to (1) the weaker surface forces of adhesion for larger particles (10); (2) the tendency of larger fly-ash particles to be carbonaceous and, therefore, conductive and more readily subject to electrostatic repulsion; (3) the ability of larger particles to more effectively erode precipitated dust (171–173); (4) the existence of a component of w_e due to eddy diffusion, which component decreases with increasing particle size and thus offsets the increase in w_e required by eq. 65 (174); (5) the influence of the electric wind (Sec. V-D-3). (Figs. (a) and (b) used with Permission of Éditions de Centre National de la Recherche Scientifique.)

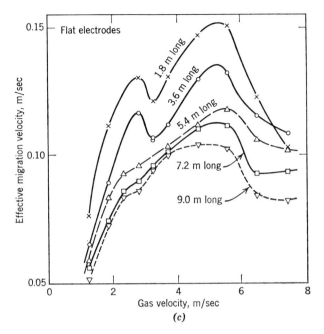

Figure 12 (*continued*).

B and K are next determined from eqs. 71 and 72

$$B = (1 - 0.99)e^{(100)(0.06)} = 5 \tag{73}$$

$$K = (100)(0.016) = 1.6 \tag{74}$$

Corresponding to these values of B and K, Figure 13 gives $P = 0.85$ as desired. If, however, the graphical value of P did not coincide with the required value of 0.85, a more favorable value of A/V_g would have to be assumed and the process repeated. In calculating the specific surface directly from the Deutsch equation for $\eta = 0.99$ and $w = \bar{w} = 0.06$ m/sec, it is found that A/V_g has the significantly smaller value of 77 m²/(m³/sec). It should be noted, however, that the probability of successfully meeting the guarantee has now fallen to 50%.

D. Quality of Gas Flow and Role of Turbulence

1. Introduction

The dominant role played by conditions of gas flow in electrostatic precipitation cannot be overemphasized. Disturbed flow in the form of uneven distribution, jets, or swirls, not only increases reentrainment losses from

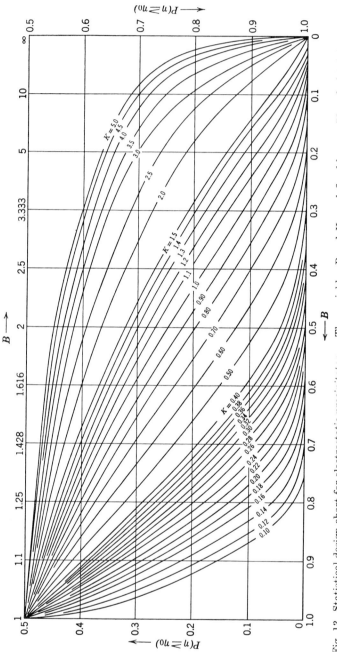

Fig. 13. Statistical design chart for electrostatic precipitators. The variables B and K are defined by eqs. 71 and 72 (167). (Used with permission of Staubforschungsinstitut des Hauptverbandes der geverblichen Berufsgenossenschaften e.V.)

electrodes and hoppers, but is responsible for poor collection initially. If longitudinal gas velocity varies from duct to duct across a precipitator, the application of eq. 68 to each duct individually will reveal a decay in overall efficiency. It is a common experience to improve efficiency from 60 or 70% to 95% or better by corrections in gas flow (1h). An efficiency equation applicable to unbalanced gas flow is given by Troost (80).

Characteristic solutions to the problems presented by large-scale gas-flow disturbances involve the development of appropriate equipment, e.g., turning vanes, diffusion screens, transitions, and plenum chambers. But in addition it must be recognized that poor gas flow is a system problem in total plant design. These very important practical considerations have been dealt with by White (1i) and Gottschlich (17).

The trend of future work in precipitation gas dynamics lies not in a further refinement of the criteria for good large-scale gas flow but rather in the following:

1. Understanding the small-scale process of particle diffusion as it affects precipitation.

2. Assessing the importance of turbulence and convection due to the electric wind.

3. Attempting to harness, as far as possible, these various particle-transfer mechanisms to reenforce electrostatic collecting forces.

The idea of utilizing diffusive forces to aid the precipitation process seems to have occurred independently to several investigators over the last few years. Recent research in the United States, the United Kingdom, and the Soviet Union has directed attention to the role of turbulence (and with it, particle diffusion) in modifying the effective particle migration velocity, if not the widely employed Deutsch precipitation equation itself.

2. Recent Theories of Precipitation

Friedlander's (175) analysis of the precipitator efficiency problem in 1959 seems to be the first treatment taking into account simultaneous eddy diffusion and movement under an external force field. Under the assumptions made, it is shown that the electrostatic migration velocity w in the Deutsch equation is to be replaced by an effective value w_e. The effective velocity w_e is dependent on gas velocity v and duct friction factor in such a manner that an increase in gas velocity is accompanied by a rise in w_e. A significant increase in w_e with v is actually observed in practice, at least for values of v which are sufficiently low so that serious reentrainment of the precipitate is avoided (Fig. 12). This apparent w_e-v dependence may, however, be explained otherwise. Robinson (164), for example, has shown that it may follow as a consequence of assuming the inlet dust to contain a nonprecipitable, reentraining fraction. Against this, though, is the observation that the w_e-v dependence occurs even with supposedly non-reentraining liquid particles (169). Heinrich (176) has suggested that increased turbulence associated with higher velocities causes more particles to be carried into the region of the discharge electrodes where they acquire a greater charge. This hypothesis is discounted by Brandt (177) on the grounds that any anticipated level of mainstream turbulence is probably overshadowed by the electric wind. Brandt offers his own explanation in terms of precipitator aspect ratio. Cooperman (178) raises the possibility that downstream particulate mass transfer by longitudinal diffusion is greater, relative to longitudinal convection, at lower gas velocities than at higher because then the longitudinal concentration gradient is higher. Effective treatment time may, thus, be regarded as less than proportional to reciprocal velocity, particularly at lower velocities.

The cross-sectional particle-concentration profile in Friedlander's theory is given in Figure 14b. Reentrainment of precipitate is not included in the theory.

A rather different approach is that of Williams and Jackson (152) who actually found a numerical solution of the diffusion equation with an added term to account for electrostatic particle convection. Unlike Friedlander's case, a laminar sublayer is assumed to exist at the collecting wall, all dust entering the sublayer being rapidly transferred to the wall by electric forces. The resulting particle concentration over the duct cross section is minimal at

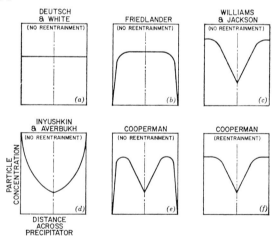

Fig. 14. Particle-concentration profiles over a precipitator cross section in accordance with various theories. The center line represents the discharge wire. The curves are intended for qualitative comparison only. (Used with permission of Air Pollution Control Association.)

the discharge wires, rising and leveling off as the walls are approached. A typical concentration profile appears in Figure 14c. The midstream dip in particle concentration (neglected by Friedlander) is, in fact, to be expected, since particles are swept toward the wall in both directions from the wires. In a purely laminar-flow precipitator, not included in this theory, there is a sharply defined clear zone, the width of which broadens progressively with downstream distance (2c).

Increased turbulent diffusion yields a more uniform particle distribution over the cross section. Net particle transport due to diffusion is therefore "negative," that is, unlike diffusive action in Friedlander's model, diffusion here opposes particle transport due to the electric field. Variations of w_e with v must be introduced empirically into the theory. Particle reentrainment is neglected.

In the studies of Inyushkin and Averbukh (179,180) experimental evidence obtained with a tubular laboratory precipitator is cited, showing that turbulent action exercises both a positive and negative effect on particle collection. The efficiency of a wetted-wall precipitator (presumably no reentrainment) producing a negligible electric wind was measured as a function of treatment time and pipe Reynolds number Re. For equal treatment times, efficiency η was observed to increase with increasing Reynolds number in passing from laminar flow (Re = 2000) through turbulent (Re up to 20,000) up to about $\eta = 99.8\%$. Above this value of η, laminar-flow efficiency exceeded turbulent, the former actually attaining 100%. The positive contribution of turbulence to particle collection is exemplified by an eightfold increase in "mechanical" (electrically un-energized) efficiency in going from Re = 2000 to Re = 20,000. The mechanism invoked to explain these mechanical efficiencies is particle penetration, under the action of inertial forces, of an assumed laminar boundary layer adjacent to the collecting wall. The negative contribution manifests itself in the incapacity of a finite turbulent precipitator to achieve 100% efficiency. Here turbulent eddies continuously redistribute over the cross section particles which the electric field tends to concentrate in the neighborhood of the walls (Fig. 14d).

On the basis of the foregoing considerations Inyushkin and Averbukh derive a Deutsch-type efficiency equation having a modified exponent.

By far the most thorough and sophisticated attack on the precipitator efficiency problem has been made by Cooperman (153–156). His model of the precipitation process is based on the coextensive mechanisms of eddy diffusion, electrostatic migration, and reentrainment, the last effect being measured in terms of a probability of reentrainment. For complete reentrainment the particle flux away from the wall is just balanced by the flux to the wall. Zero reentrainment, on the other hand, means the wall is a perfect collector, whence the particle concentration at the surface is zero. This condition is in accord with Friedlander's approach but constitutes the most important physical difference between the approaches of Cooperman and Williams and Jackson.

Both positive and negative turbulent particle transfer is postulated. But here, since the boundary layer is assumed turbulent, inertial ejection of particles through a laminar wall layer cannot be invoked to explain the observed increase of η with v at constant residence time. Instead, as indicated by Friedlander, positive diffusion is considered to sweep particles from the region of higher concentration away from the wall into the region of lower concentration near the wall, when under conditions of low reentrainment the prescribed concentration gradients exist. For heavy reentrainment, however, a dense cloud of particles is formed near the wall, and net particle transport by diffusion in this region is negative, that is, in a direction away from the wall. Whatever the gradient at the wall, diffusive transport in the vicinity of the wires is always negative, since the particle concentration in the gas near the wires is always less than in regions nearer the center of the interelectrode space (see Fig. 14e and f).

In qualitative agreement with the experimental data of Inyushkin and Averbukh (179), Cooperman's theory predicts laminar efficiencies less than turbulent (except when $\eta = 100\%$ is approached). The reason given is that although neither positive nor negative diffusion now exists, the effect of the former's absence is dominant.

When the diffusivity is low, gas flow is nearly laminar and the efficiency equation is not exponential. For larger diffusivities, the efficiency obeys an exponential law after the gas has traversed a short entrance section. However, the exponent is different from that given by Deutsch.

From a practical viewpoint, the most important parameter in this theory is the reentrainment probability. Reentrainment is responsible for a twofold action in lowering efficiency:

1. Removal of collected particles from the plate.
2. Removal of particles from the gas near the plate by enhanced negative diffusion.

Cooperman does not provide numerical solutions to the problem because estimates of the reentrainment probability are untrustworthy and because, by selection of appropriate values of the parameters, any desired collection efficiency can be produced. The theory is in the process of further development (156).

The attempts of Williams, Jackson, and Cooperman to obviate serious fallacies in the Deutsch equation have led to rather complex analyses. An approach to the problem by Robinson (164) represents an effort to retain the simplicity of the original exponential relation yet at the same time to meet, in some degree, some of the more serious objections to the old theory.

Similar assumptions are made to those of Inyushkin and Averbukh regarding inertial penetration of a laminar boundary layer. As with Williams and Jackson, diffusive particle transport is zero near the wall and negative elsewhere, a consequence of the assumption that maximum particle concentration and zero particle-concentration gradient are found at the wall. The concentration profile is not derivable from the theory.

Particle reentrainment is accounted for in terms of two particle fractions, one having a given nonzero probability of permanent capture, and the other having zero probability. Because of the latter reentrainment, the particle loss $(1 - \eta)$ is not an exponentially decreasing function of precipitator size A as in the case of the Deutsch relation. Nevertheless, the net flux of precipitate to the walls does remain exponential in A.

Hignett (134,135) has calculated the ion-bombardment charge acquired by particles as they move in the variable electric field of a tubular electrostatic precipitator. Variations in the field due to both geometry and ion and dust space charge are considered. The hypothetical case is assumed in which the turbulent force acting on a particle is constant and directed radially inward in opposition to the electrostatic precipitating force. Numerical solution of the resultant equations of charging and motion show that the trajectories of particles of diameter less than 20 μm are dominated by turbulence rather than the electric attractive force, and are the most likely to escape capture. Since about 80% by weight of the dust discharged from many modern power-station precipitators consists of particles smaller than 20 μm, Hignett suggests that improvement in precipitator performance may be achieved by redesigning the outlet zone to reduce the level of turbulence and increase the electric field strength. This might be done by a nondischarging section of relatively closely spaced parallel plates.

The efficiency of a parallel-plate, laminar-flow precipitator in collecting a monodisperse, uniformly charged aerosol is

$$\eta = \frac{wL}{2sv} = \frac{wA}{V} \qquad (0 \leqslant \eta \leqslant 1) \tag{75}$$

where L(m) is the length of the precipitator and v is its cross-sectional average gas velocity. Equation 75 is valid for any laminar velocity distribution; plug flow need not be assumed (181). In principle, 100% efficiency is attainable in finite length.

The conflicting assumptions made by various authors in the foregoing presentation reflect the lack of experimental data besetting the investigator who attempts to develop a theory of electrostatic precipitation. Observed results are particularly scanty regarding (*a*) the nature of the cross-sectional particle-distribution profile and (*b*) the effect of the electric wind of the corona discharge in modifying "normal" aerodynamic turbulence.

Photographs of the dust-concentration profile in a small duct precipitator have been published by Rose and Wood (2d). These pictures clearly reveal a relatively dust-free zone in the region of the discharge wires, the width of the zone progressively and markedly broadening as the gas moves downstream. Similar low-concentration zones have been observed by Robinson (182) and Hughes et al. (183) in single- and two-stage duct precipitators, respectively, and by Hignett (135) in a tubular system. The extent of the central clear zone and its tendency to widen with downstream travel appear to depend on the level of turbulence, particle size, and in duct precipitators, on the wire-to-wire spacing. Knowledge of particle concentration very close to the precipitator wall is also of prime importance in establishing the boundary conditions in any theory of precipitation, but this concentration is very difficult to determine experimentally. Measurements within a few millimeters of the wall have been made with glass beads (183) and nonreentraining oil droplets (184); the former give no indication that particle concentration decays very close to the collecting wall, and the latter are inconclusive.

The vertical maldistribution of dust in duct precipitators sometimes seriously affects precipitator performance. Particle concentrations in the gas near the bottom may be as much as twenty times as great as near the roof. The condition may be corrected by improving flow conditions at the inlet (185), or by inserting specially designed outlet baffles (186).

3. The Electric Wind

This phenomenon designating the movement of gas induced by the repulsion of ions from the neighborhood of a corona-discharge electrode has been known for a long time; a general historical review of the subject has been made by Robinson (187). The specific observation that smoke and dust particles could be borne along in the electric wind was reported as early as 1771 by Becarria (55), and again by Obermayer and Pichler (188) in 1886. Cottrell (189) also referred to the movement of the electric wind in his first precipitation patent of 1908. It was Haber (190), however, in 1921 who first suggested that the electric wind might significantly contribute to particulate motive power in electrostatic precipitation.

Fundamental quantitative electric-wind relations are derived as follows. Each infinitesimal volume of gas between the electrodes contains an ionic charge of density ρ_i and is acted upon by an electric field of intensity $\mathbf{E}$. Particle space-charge effects are assumed to be of secondary importance. The product $\rho_i \mathbf{E}$ therefore gives the electric force exerted on a unit volume of charged gas. Force per unit volume is the force per unit area per unit distance perpendicular to the area in question, and can be represented as a pressure gradient

$$\nabla p_g = \rho_i \mathbf{E} \tag{76}$$

where p_g is the pressure in N/m^2. The electric-wind velocity is of the order of 1 m/sec and negligible relative to the ion velocity. Thus, the current density is

$$\mathbf{j} = \rho_i b \mathbf{E} \tag{77}$$

whence

$$\nabla p_g = \frac{\mathbf{j}}{b} \tag{78}$$

The electric wind blows at a velocity w_{ew}(m/sec) for which the forward electrical pressure is matched by the aerodynamic back pressure p_g

$$p_g = \tfrac{1}{2} K_0 \rho_g w_{ew}^2 \tag{79}$$

where K_0, a dimensionless loss coefficient, is a function of the geometry and is, to a first approximation, independent of the velocity. Equations 78 and 79 give

$$w_{ew} = \left(\frac{2}{K_0 \rho_g b} \int_{r_0}^{r_1} j \, dr \right)^{1/2} \tag{80}$$

Since current density j is proportional to the current i(A),

$$w_{ew} = K_1 \left(\frac{i}{\rho_g b} \right)^{1/2} \tag{81}$$

where the system constant $K_1(m^{-1/2})$ is a function of geometry. The essential validity of eq. 81 has been demonstrated for a wire-pipe precipitator (190) as well as for nonprecipitator geometries (191).

Except for a flurry of interest in the early 1930's, the role of the electric wind has been only occasionally considered by precipitation workers. The one extensive study of the subject was by Ladenburg and Tietze (190), who observed electric-wind flow patterns produced by corona tufts on a negative wire in a tube. Wind velocities were measured by Schlieren photographs of a stream of CO_2 tracer gas injected into the precipitator and deflected toward the tube by the wind. The electric wind was found to affect particle migration velocity relative to the electrodes by superposing an additional net velocity component on the velocity of the particles relative to the gas. This effect was, in fact, held to be primarily responsible for the transportation toward the wall of submicron particles, although it was also credited with assisting the movement of particles as large as 10 μm. Continuity of flow was accounted for by postulating a rapid, directed movement of gas away from the discharge electrode, balanced by a relatively slow and diffuse return flow, the former more effectively conveying particles to the wall than the latter from it.

Objections to these conclusions regarding the importance of the electric wind were made by Deutsch (192) and others (193,194), chiefly on the grounds that (a) it was not necessary to propose a major electric-wind contribution to the particle-transportation rate to explain observed particle migration velocities and (b) requirements of flow continuity were not clearly satisfied.

According to eq. 65 for greater-than-micron particles, migration velocity depends on electric-field intensity, and not on corona current as such. Full-scale and pilot-plant observations, however, suggest that migration velocity is, indeed, specifically current dependent (170,195). Qualitatively at least, this dependence is such as might be expected on the assumption that electric-wind convection actually assists in particle transport. In such an event a new effective migration velocity w_e should be considered

$$w_e = w + w_{\text{ew}} \tag{82}$$

where w and w_{ew} are given by eqs. 65 and 81, respectively.

Further support for a relation of the form of eq. 82 has been provided by recent investigations. Chubb et al. (196) have studied the motion of individual dust particles in a point-plane corona using an interrupted-light photographic technique to measure particle velocities. In addition, a pulsed corona was employed to separate particulate forces due to the electric wind and electrostatic attraction. During the application of a high-voltage pulse, particles would move as directed by the vector sum of both forces. At the end of a pulse, however, the electrostatic force would abruptly cease, but air-flow inertia would briefly sustain the electric-wind pattern. Analysis of particle trajectories under these conditions revealed a significant electric-wind component in comparison to direct electrostatic forces for particles 5–10 μm in diameter.

Robinson (182) injected a stream of helium tracer from a point source axially into the center of a duct precipitator with positive corona, and observed the dispersion toward the wall of the helium with downstream travel. On encountering the first discharge wire, the single stream of helium split in two, each new helium stream then drifting progressively closer to its nearer wall as it continued to move downstream with the main flow. A net wire-to-wall gas flow, that is, the electric wind, was plainly indicated. The ensuing problem of flow continuity remained unresolved.

Striking differences in cross-sectional particle-concentration profiles with corona polarity have been reported, and ascribed to the electric wind which is itself polarity sensitive (184,191),

All theories of electrostatic precipitation that have been considered ignore the effect of the electric wind of the corona discharge on mass-transfer patterns, whether gas or particulate. Nevertheless, under electrical and flow conditions similar to those that often prevail in practice, the electric wind appears to make an appreciable contribution to transverse gas flow, and so presumably should be considered in any comprehensive analysis of precipitator performance.

4. Particle Adhesion and Reentrainment

The behavior of dust particles on or near the collecting electrodes is the subject of a continuing series of studies by Penney and his associates (197). Observations of individual particle trajectories using periodically interrupted illumination [cf. Chubb (196)] show that contrary to assumptions usually made in theoretical treatments of the precipitation problem, impact phenomena at the collecting electrode or precipitate surface cannot be neglected, particularly for particles greater than 10 μm (171,172). These larger particles may rebound on impact without losing their (normally) negative charge, or they may erode agglomerates of previously precipitated dust. Dust on the positive collecting electrode tends to acquire a like charge by induction, and if dislodged under low-current conditions can be forcefully accelerated away from that electrode. The repulsive force acting on the dust layer is opposed by an attractive electrical force due to the ion current. The net electrical force per unit area of dust surface $F'(N/m^2)$ is (198)

$$F' = \tfrac{1}{2}\varepsilon_0[(K_p j\rho_d)^2 - E^2]$$ (83)

where $E(V/m)$ is the field in the gas adjacent to the dust layer and F' is positive when attractive [cf. a similar expression by Lowe and Lucas (10)]. In addition, various "mechanical" surface forces must be considered, for example, van der Waals forces, adhesion due to moisture films, and reentraining forces due to wind drag. The effects of such forces on a dust bed are more difficult to analyze quantitatively; much work, however, has been done by Corn et al. (199,200), Bagnold (173,201), and others (10,202), and is reviewed by Löffler in this volume.

The presence of back corona (Sec. VI-B-1) can still further complicate the pattern of incoming particle trajectories by creating pockets of reverse ionization in the gas over the dust surface. Such localized regions of charge can recharge individual particles in an unpredictable way and cause them to follow highly irregular paths (171,172).

Investigation of the factors influencing dust adhesion on collecting electrodes has led Penney et al. to conclude that significant differences in work function and contact potential are to be found over the surface of many kinds of dust particles (198,203). Contact potential differences across opposite faces of a particle produce a dipole moment capable of orienting the particle in an electric field. The particle thus deposits in such a way that attractive Coulomb forces are developed between adjacent layers. For fly ash, this condition results in interparticle adhesive forces many times greater than would exist if the dust had been mechanically deposited.

Subsequent work by Niedra and Penney (204) shows that particles do not necessarily require large enough dipole moments to effect particle alignment in the electric field of the corona. Instead, many particles seem to orient themselves in the field established by particles already deposited. Again, firm adhesion is thought to be achieved by electrostatic attraction between small areas of opposite polarity brought together at points of particle-to-particle contact.

The foregoing experimental studies of dust behavior at the collecting electrode are the subject of a recent review (205).

A problem of great importance in electrostatic precipitation is the effective removal of collected particles from the passive electrode and its transfer, with minimal reentrainment, to the hoppers. In modern precipitators dust may fall as much as 12 m through a transverse gas stream before reaching the hoppers; 15-m high plates are under consideration.

The mechanics of electrode rapping has not received the extensive attention the subject deserves. Brief reviews of the rapping art have been given by White (1j), Rose and Wood (2e), and Strauss (3a). Elsewhere, vibration studies of collection (206) and discharge (207) electrodes have been described. Recent progress has been reported by Sproull (208) whose own experimental findings are summarized as follows:

1. The maximum acceleration attained in a rapping blow determines the fraction of the dust layer dislodged by the blow.

2. As a general rule, for ordinary industrial dusts, the maximum acceleration in shear rapping should have an order of magnitude of 200 g. About half that acceleration is needed in perpendicular rapping. Some dusts, such as American pulverized-coal fly ash, can be satisfactorily rapped with much lower accelerations.

3. Excluding cases in which the temperature passes through the dew point, or the dust melts, an increase in temperature facilitates rapping.

4. The fraction of the dust layer dislodged by a rap of given intensity rapidly increases until a certain thickness is reached, after which a further increase in thickness has little effect.

American and European rapping practice appears to exhibit a fundamental difference in outlook. The tendency in the United States toward continuous rapping (i.e., every few minutes or less) aims at the elimination of visible rapping puffs in the stack discharge, psychologically so objectionable. Higher long-term collection efficiencies are, however, more likely when the rapping is intermittent (i.e., at intervals of as much as several hours). Ideally, the rapping interval should be adjusted to the needs of different parts of the precipitator (168,209,210).

A number of collecting-electrode configurations are described in the monographs (1k,2f,3b), the various designs purporting to increase efficiency by

(a) providing baffles to shield deposited dust from the reentraining forces of the gas stream, (b) providing catch pockets which convey precipitated dust into a quiescent gas zone behind the collecting plate, (c) minimizing protrusions from the plate surface in order to raise sparkover voltage, and (d) facilitating rapping and dust transfer to the hopper.

VI. PRECIPITATION UNDER EXTREME CONDITIONS

A. High-Pressure/High-Temperature

1. Applications

In recent years, electrostatic precipitators have been used in chemical-processing, power-generation and mass-transport applications involving temperatures and pressures well in excess of conventionally accepted limits. Successful pilot or full-scale trials have been run at pressures up to 55 atm and temperatures (not simultaneous) to 800°C. In some of these instances, the precipitator is used to combat, in a single operation, both the pollution of the atmosphere and the fouling or eroding of process equipment. But even other high-pressure/high-temperature cases which do not strictly constitute environmental pollution problems are of interest, and often provide application data directly pertinent to air-pollution situations. A number of actual and potential uses of high-pressure ($\delta > 1$)/high-temperature ($>300°C$) precipitation are listed in Table VII.

2. High-Pressure Electrical Characteristics (Positive Corona)

The increase in sparkover voltage with gas pressure, a well-known feature of parallel-plate electrodes, is not observed for all electrode geometries. In particular, the positive corona discharge from air and other electronegative gases exhibits a maximum in the sparkover voltage which, with increasing pressure, declines, eventually to coincide with the corona-starting voltage (37,234–243). The pressure at this point of intersection beyond which

TABLE VII
High-Pressure/High-Temperature Precipitator Applications

Application	Problem	Status	References
Coal-fired gas turbine (700–800°C, 7–8 atm)	Turbine-blade erosion, air pollution	Pilot plant	211–213
Solid-waste fired gas turbine (800–900°C, 7–12 atm)	Turbine-blade corrosion, air pollution	Pilot plant	214, Figure 15
Oxygen-injection steel production (precipitator preceding heat exchanger)[a]	Heat-exchanger erosion, air pollution	Proposal	215
Open-ended magneto-hydrodynamic generator (500°C, 1 atm)	"Seed" recovery, air pollution	Pilot plant	216,217, Figure 16
High-temperature, coal-fired fuel cell (850°C, 1–2 atm)	Plugging of cell passages, formation of hot spots	Proposal	218
Synthesis-gas production (160°C, 26 atm)	Catalyst-bed contamination	Pilot plant	219
High-temperature, gas-cooled nuclear reactor (350–800°C, 50–100 atm)	Particle buildup in heat exchanger	Proposal	218
Coal gasification (ambient temperature, 9 atm)	Tar deposition in gas pipeline	Full scale	220
Flue-gas desulfurization process (470°C, 1 atm)	Fouling of $SO_2 \rightarrow SO_3$ converter and air heater, air pollution	Commercial prototype	221,222, Figure 17 in Chap, 3, Section IV-C-I
Pulverized-coal fired boiler (precipitator preceding air heater) 320°C, 1 atm)	Erosion and fouling of preheater, air pollution	Full scale	221,223,224
Natural-gas pipelines (ambient temperature, 55 atm)	Reduction in pipeline capacity by entrained lubricating oil droplets	Commercial prototype	225–227
Cleaning of high-pressure carbon dioxide gas	CO_2 purification by distillation and adsorption hampered by oil and water droplets	Pilot plant	228
Manufacture of high-purity elemental phosphorous (370–540°C, 1 atm)	Contamination of P_4 vapor by furnace dust	Full scale	229,230
Turbine powered by high-pressure blast furnace gas (2 atm?)	Turbine-blade erosion, air pollution	Full scale (?)	6a
Dry-process cement kiln (400–500°C, 1 atm)	Air pollution	Full scale	231
Open-hearth furnace (600°C, 1 atm)	Air pollution	Full scale	232
Zinc–lead hearth roaster (400–650°C, 1 atm)	Air pollution	Full scale	233
Molybdenum–sulfide hearth roaster (400°C, 1 atm)	Air pollution	Full scale	233

[a] Low-temperature precipitators ($\sim$250°C) are currently employed downstream of the heat exchanger.

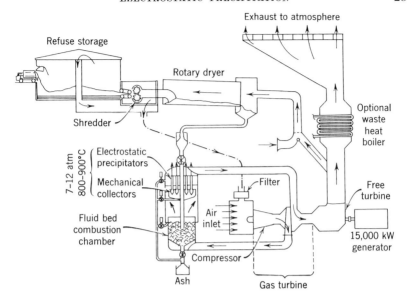

Fig. 15. Gas-turbine incinerator under development by the Combustion Power Company. Solid waste is shredded, dried, and fed at the rate of 400 ton/day into the refuse combustor. Air at 7–12 atm from the compressor produces burning at 800–900°C and yields more efficient recovery of heat energy than is possible in conventional incinerators functioning at atmospheric pressure. The ability to remove particulate matter from the combustor exhaust by a combination of inertial collection and electrostatic precipitation is a key to the successful operation of the system. Particles that are not removed quickly corrode and erode the turbine. Electric power is a salable byproduct of the incineration process and helps offset operating expenses (214). (Used with permission of Combustion Power Company.)

sparkover alone without antecedent corona obtains, is termed the critical pressure. Figure 17 illustrates the critical-pressure (or density) effect for the positive corona in air at room temperature. It is clear that knowledge of the maximum sparkover level and the critical pressure, limits beyond which conventional electrostatic precipitation is not known to be possible, is prerequisite to the sound design of a high-pressure precipitator.

If the wire is too large relative to the tube, sparkover will occur without corona even at ordinary pressures. Experience shows this to happen for values of the ratio r_1/r_0 less than 10 or 15 (Fig. 17c, (244)), contrary to the old rule (245) requiring a ratio $r_1/r_0 < e = 2.71$.

Subject to this restriction, Robinson (88) has shown that the corona-starting relation for round wires

$$\frac{E}{\delta'} = A_g + \frac{B_g}{\sqrt{r_0 \delta'}}$$

$(A_g = 32.2 \times 10^5 \text{ V/m}, B_g = 8.46 \times 10^4 \text{ V/m}^{1/2} \text{ in air})$ \hfill (49)

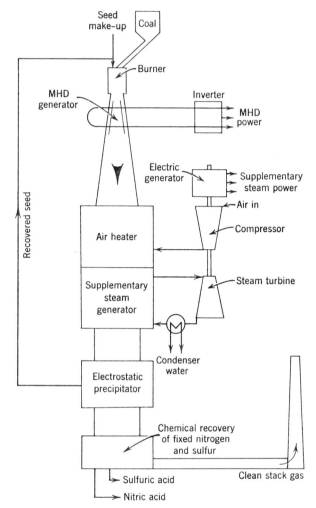

Fig. 16. Magnetohydrodynamic (MHD)—steam power plant. The efficiency of such an installation could reach 60% or more, considerably above the attainment of an overall plant efficiency of 41% of the best steam-turbine electrical plant. Furthermore, the use of fossil fuels in an MHD generator results in far lower levels of air pollution than conventional fossil-fuel electric power plants and substantially less thermal pollution than nuclear plants. In place of the solid conductors of the usual turbine-driven generator, MHD devices substitute a gas raised to 2000–2500°C. The gas is the combustion product of a fossil fuel "seeded" with an alkali metal (about 1%) to promote ionization. Ionized gas is passed through a magnetic field to generate a current. Gas-cleaning equipment satisfies both the need for air pollution control and the economic requirement for seed recovery (217). The success of electrostatic precipitation in this application has been demonstrated on a pilot scale (216). (Used with permission of Industrial Research, Inc.)

long established for the *positive and negative* discharge at atmospheric and much lower gas densities, remains valid for *positive* corona in air at least to $\delta' = 35$, and at least over the interval 1.8×10^{-4} m $< r_0 < 6.3 \times 10^{-3}$ m, as long as the critical density is not exceeded. But these results taken in conjunction with Table V go much further: they suggest that eq. 49 holds for gases generally, from the critical level downward, possibly to the lowest densities at which precipitation is practical. (Useful particle migration velocities have been measured at relative densities as low as $\delta \sim 1/300$ (130).) The positive relative critical density δ_{cr}^+ for wire-pipe electrodes in room-temperature air obeys the empirical relation (234)

$$\delta_{cr}^+ = k_p \left(\frac{1}{r_0} + c_0 \right) \tag{84}$$

Here $c_0 = 900$ m^{-1} and k_p(m) is a function of the pipe radius given by

$$k_p = 0.11(1 + 0.1r_1) \tag{85}$$

over the range 0.02 m $< r_1 < 0.08$ m and possibly for $r_1 > 0.08$ m. Clearly, fine wires and large interelectrode spacings elevate the critical density. Fine wires provide the additional advantages of minimizing corona-starting voltages and raising the peak sparkover voltage. Judicious choice of electrode dimensions can locate the sparkover voltage well above the precipitator's operating voltage, so that the breakage of fine wires by sparking need not be a problem.

Experimental positive current-voltage characteristics for various gas pressures are shown in Figure 18. The curves are in fair agreement with eq. 16.

The corona-current and corona-starting relations, 16 and 49, respectively, describe phenomena dependent not on pressure or temperature separately but rather on their combined effect as manifested in relative gas density. Observed departures from this rule at high temperature are considered in Section VII-A-3.

As the pressure is raised (at constant temperature), there is an interplay of two opposing effects: (a) shorter mean-free paths impede ionization by collision and so tend to raise the sparkover level, and (b) enhanced photoionization and reduced ion diffusion tend to facilitate streamer propagation from the anode across the gap. As the pressure increases, the initially dominant first effect gives way to the second, the streamer develops across the gap and, at the critical pressure, spark breakdown ensues (37a). Increasing the gas temperature raises the critical pressure, for the greater gas diffusivity now obtaining tends to more effectively quench the streamer before it bridges the interelectrode space (236,238). Similarly, the wider the gap, the better the

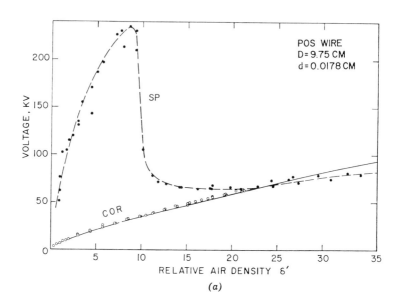

(a)

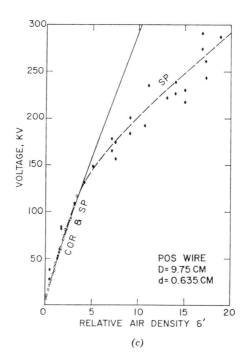

(c)

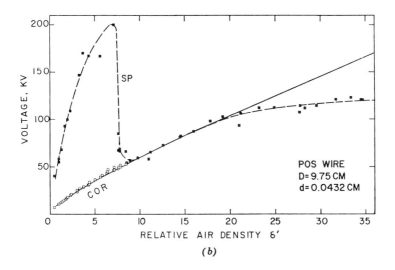

(b)

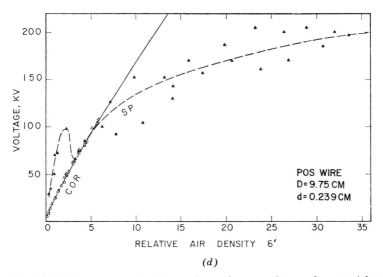

(d)

Fig. 17. Positive corona-starting and sparkover voltages for coaxial wire-pipe electrodes in air (25°C). D and d are the respective pipe and wire diameters. The voltage is unvarying dc. The solid lines are the corona-starting curves according to eqs. 9 and 49 are are in good agreement with the data below the critical density; extensions of these curves beyond the critical density are without physical significance (234). (Used with permission of American Institute of Physics.)

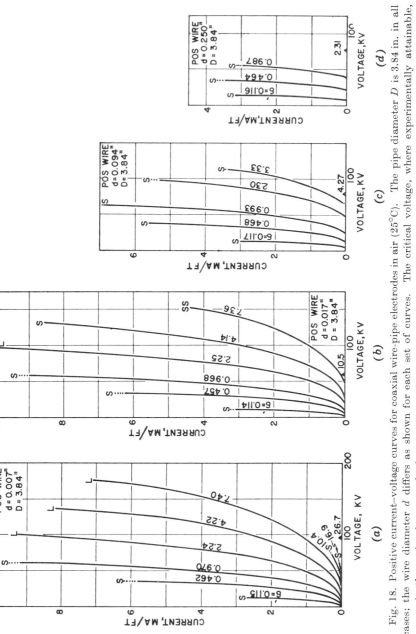

Fig. 18. Positive current–voltage curves for coaxial wire-pipe electrodes in air (25°C). The pipe diameter D is 3.84 in. in all cases; the wire diameter d differs as shown for each set of curves. The critical voltage, where experimentally attainable, occurs in the region between the last current–voltage curve and the symbol ▲. L designates the limit of the power supply, S sustained sparkover at the point shown, and SS sporadic sparkover at various points on the curve culminating, however, in sustained sparking at the point marked SS. The voltage is unvarying dc (246).

opportunity for the streamer to dissipate, hence the higher the critical pressure. Limited data suggest that the addition of a gas of high thermal conductivity to a gas of lower conductivity (e.g., helium to impure nitrogen) tends to elevate the critical pressure; presumably the advancing streamer is suppressed by the more rapid conduction of thermal energy from its tip (238,240,241).

3. High-Pressure Electrical Characteristics (Negative Corona)

Experimental negative corona-starting and sparkover voltages are shown as a function of δ' for room-temperature air in Figure 19. Unlike the positive corona-starting voltage which may be closely calculated from the field given by eq. 49, the negative starting voltage is observed to scatter over a band of voltages which grows increasingly wider with pressure, the *upper limit* of the band being given by the field of eq. 49 (88,247). Useful negative corona currents may, in fact, be drawn at voltages well below the level required by eq. 49. A striking practical example of this is found in an 800-psi natural gas (96% methane) precipitator: significant corona currents and high particle migration velocities result despite the fact that the operating voltage is a fraction of the value ostensibly needed to initiate corona (225, cf. 93). The effect, however, can be erratic and must not be relied upon indiscriminately. Anomalously low starting voltages often yield very slowly increasing initial currents (88).

The observed instability of the negative starting voltage is not altogether unexpected. In the negative discharge, the wire cathode is the source of electrons sustaining the flow of current. Electron emission from the cathode is, in turn, governed by the microgeometry of the cathode surface, the presence of oxide films and dust, and other factors not subject to experimental control. The effects of surface contamination and asperities are aggravated by the diminution of lateral diffusion resulting from increased gas density (38b). The absence of a precisely defined negative corona-starting voltage at high δ finds its counterpart in the vagaries of uniform-field breakdown data (248). In positive corona, the cathode is in a low-field region and consequently does not contribute appreciably to the maintenance of the discharge. Secondary electrons originate in the gas surrounding the anode wire. The positive corona-starting voltage is, therefore, relatively independent of the condition of the electrode surfaces and less likely to fluctuate in value.

Further contrasting Figures 17 and 19, we see that the negative discharge offers both higher maximum sparkover voltages and higher critical densities. (As a general rule, negative corona remains preferable at elevated pressures.) The negative critical density δ_{cr}^- for wire-pipe electrodes in room-temperature

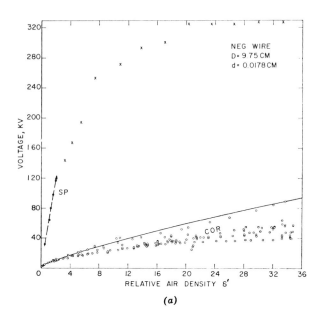

(a)

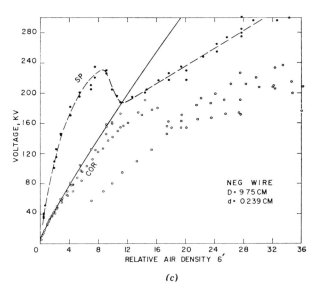

(c)

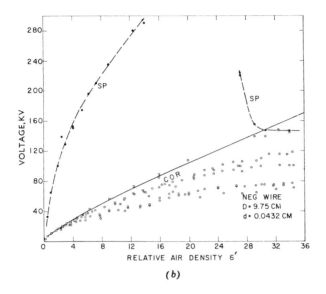

(b)

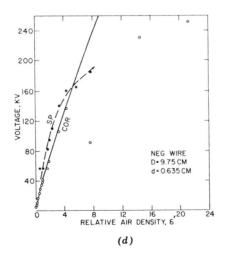

(d)

Fig. 19. Negative corona-starting and sparkover voltages for coaxial wire-pipe electrodes in air (25°C). D and d are the respective pipe and wire diameters. The voltage is unvarying dc. Solid curves represent corona-starting voltages corresponding to the fields given by eq. 49 and indicate an upper limit to the band of starting voltages observed experimentally (88,247). The points $\times$ in (a) indicate the maximum voltages attainable; sparkover occurred at higher values.

air conforms to the empirical relation (247)

$$\delta_{cr}^- = 0.12 r_1 \left(\frac{1}{r_0} + 700 \right) \tag{86}$$

at least over the range 0.023 m $< r_1 <$ 0.077 m.

The negative critical density, as used above, is defined as that value of δ' at which the theoretical corona-starting voltage (calculated from eqs. 49 and 9) and sparkover voltage intersect (Fig. 19).

Still another peculiarity lacking a positive counterpart is exhibited by the high-pressure negative corona: the existence of a semistable discharge above δ_{cr}^-. This "postcritical" discharge is shown by the hatched regions in the current–voltage curves of Figure 20. These curves are erratic and not accurately reproducible. The discharges are punctuated by sporadic, self-quenching sparks at various points. At sufficiently high voltages, sustained sparkover together with heavy currents and collapse of voltage occurs (246). There are no data available on precipitator performance in the postcritical region.

4. High-Temperature Precipitation

Owing to its possible application in several coal-utilization processes (e.g., gas-turbine driving, coal gasification), Shale et al. at the U.S. Bureau of Mines have engaged in a continuing study of the practicability of electrostatic precipitation at temperatures as high as 820°C. The most promising results to date have been obtained with a pilot precipitator consisting of a cluster of 16 tubes, each 1.83 m long and 15.2 cm in diameter, having negative discharge wires 0.21 cm in diameter (249,250). At a fixed relative density $\delta = 1.5$ in a gas mixture consisting of the combustion products of natural gas and excess air, current-voltage curves at various temperatures from 31 to 730°C were observed approximately to coincide, thus indicating the absence of an independent temperature effect (other than that manifesting itself in δ) on electrical variables. Between 730 and 820°C the curves became unstable with time, a pattern of behavior suggesting misalignment of electrodes due to uneven thermal expansion.

Fly ash was separately injected into the system and removal efficiencies of 90–98% were measured under the following conditions: Linear gas velocity, 1.5 m/sec; temperature, 800°C, $\delta = 1.5$; voltage, 38 kV; current, 6.6 mA/m. Rapping at such high temperature produced severe misalignment and eventually incapacitated the precipitator. Shale concludes that negative corona precipitation is feasible up to 730°C and probably at least to 800°C provided that the requirements of adequate relative density and structural rigidity are met.

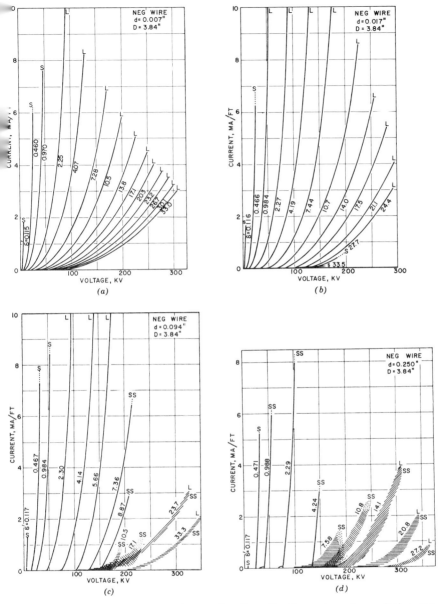

Fig. 20. Negative corona current as a function of voltage at various relative gas densities for wire-pipe electrodes in air (25°C). The pipe diameter D is 3.84 in. in all cases; the wire diameter d differs as shown for each set of curves. There is no sharply defined critical voltage as with the positive discharge. The postcritical discharge, where it occurs, is generally recognized by its shallower current-voltage slope and poor reproducibility. L designates the limit of the power supply (unvarying dc), S sustained sparkover at the point shown, and SS sporadic sparkover at various points on the current–voltage curve culminating, however, in sustained sparkover at the point marked SS. A broken line or dotted area marks the range over which sustained sparkover was observed in three trials (246).

Earlier studies by Shale et al. (212,251–253) with a 4.93-cm diameter collecting tube and a 0.051-cm diameter wire in air reveal significantly higher positive sparkover voltages than negative in the range from 190 to at least 820°C at pressures from 1 to 3.3 atm.   Below 190°C the negative sparkover voltage is found to exceed positive, an observation in accord with common experience.   The higher sparkover voltage of the negative corona at lower temperature is not, however, invariably observed.   This is shown in Figures 18 and 20, and the work of Cooperman (254).

Shale and Holden (255) have subsequently described a more general pattern into which their earlier data fit.   Using a 5.38-cm tube in air, corona onset and sparkover are shown to be dependent on wire size–in addition to gas temperature and density—in the manner illustrated in Figure 21.   Relatively large-diameter wires at 820°C lead to corona thresholds much below those predicted by eq. 49 and to sparkover levels that increase significantly with wire size.

Comparative positive corona data have not been published.   Shale does, however, report that negative corona is more effective than positive in removing entrained solids at 800°C and 6.4 atm in the multitube precipitator described above (256,257).   This is attributed to the much higher negative corona power input that could be provided (1ℓ,170).   For although the negative operating voltage as limited by sparkover was somewhat less than the maximum attainable positive voltage, the corresponding negative current was very much higher.

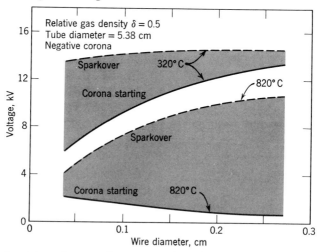

Fig. 21. Coaxial sparkover and corona-starting voltages in air for a 5.38-cm diameter tube as a function of wire size and gas temperature.   The wire is negative.   At 320°C, the corona-starting curve is in accord with eq. 49 (255).   (Used with permission of Institute of Electrical and Electronic Engineers.)

The higher positive sparkover levels for fine wires at elevated temperature had been reported earlier by Thomas and Wong (258) and Winkel and Schuetz (70). The latter workers examined the positive and negative corona discharge in air between a 0.03-cm wire and a 3.5-cm tube at atmospheric pressure between 100 and 600°C. Current at a given voltage was found to increase more rapidly with temperature in the case of the negative discharge. This effect is at least partly attributable to the heavier free-electron current associated with relatively large mean-free paths (low δ) and narrow interelectrode spacing. Shale (253), citing Frommhold (259) and Loeb (260), explains the relatively high currents and low negative sparkover voltages in his experiments in 800°C air as the consequence of an independent temperature effect. In this view, the important variable to be considered is the fraction of time an electron can remain detached from a molecule or atom, the average detachment interval increasing with temperature. The ion of interest in this reaction is probably O^-, the O_2^- ion which figures more prominently at lower temperatures having shed its electron at about 300°C. Thus in air at 800°C the equilibrium attachment–detachment conditions that appear to prevail yield heavy free-electron concentrations that, on the one hand, give rise to large currents and, on the other, to loss of corona stabilization otherwise resulting from negative-ion space charge.

The studies of Thomas and Wong (258) cover the respective temperature-pressure span of 30–830°C and 1–8 atm. Using a 5.1-cm diameter tube and 0.08-cm wire in air, these workers found that over the ranges in question positive current–voltage relations are a function of relative gas density δ only, whereas negative corona characteristics depend on gas temperature as well, the independent temperature effect becoming more pronounced the higher the current. For both polarities, pressure-dependent instabilities set in between 600 and 830°C. No temperature dependence of the negative corona in air was observed below 550°C. This result is in agreement with the earlier findings of Koller and Fremont (261) for air and methyl chloride at pressures of 1–5 atm and from room temperature to 500°C.

Currently a pilot precipitator that treats incinerator exhaust gases at 900°C and 7 atm is under test (233). Results are not yet available.

The existence of upper temperature limitations to practical electrostatic precipitation have been considered by Cooperman (262). His calculations —which remain to be verified experimentally—lead to the following conclusions:

1. At temperatures in the neighborhood of 800°C, thermal ionization rates may become high enough to play a significant role in the precipitation process; at 1400°C, the corona may be overwhelmed by thermal ionization with resultant currents so excessive that effective precipitation is impossible.

2. Ionization rates may usually be expected to depend on trace quantities of the alkali metals (elements of low ionization potential) rather than the principal gaseous constituents. Potassium, because of its relative abundance, is the most likely source of thermal ions in industrial processes.

3. The highly mobile electrons will be rapidly swept out of the system, except possibly in the presence of gases having very high probabilities of electron attachment (probability $= \beta^{-1}$, Table II). Thus, the space charge produced by thermal ionization will be positive and a corona discharge of the same polarity will generally be required.

The main difference between the explanations of Cooperman given above and Loeb, Frommhold, and Shale given earlier for high currents in high-temperature, oxygen-bearing gases is in the identification of the trace impurity in the gas. Loeb et al. contend that the impurity is the O^- ion with an electron affinity of 1.5 eV, well below the 4.3-eV ionization potential of potassium. However, if the potassium is present in sufficient concentration, its effect could dominate.

Comparisons of the anticipated high-temperature performance of electro-static precipitators with thermal precipitators, fiber filters, and pebble beds have been made by Strauss and Lancaster (218) and Thring and Strauss (215). Initial considerations suggest the relative superiority of electrostatic precipitation as a high-temperature collection mechanism, it being noted that limits of operation in electrostatic precipitation are still not defined with certainty and that the effects of materials of construction and electrode geometry may be sadly unpredictable.

B. High-Resistivity Precipitate

1. Back-Corona Formation

In the precipitation of high-resistivity dusts it is generally observed that after a brief initial period of operation, collection efficiency deteriorates, current increases, and the sparkover voltage—assuming negative corona—drops. The current increase is due to secondary emission, so-called back corona, that originates in the dust deposit on the collecting electrode, and assumes the form of a luminous sheet of tuftlike discharges of polarity opposite to that of the primary discharge. Back corona may be simulated in the laboratory by covering the normally passive electrode with a porous or perforated dielectric material of suitable resistivity (e.g., glass fabric, mica). Oscillographic observation of the back-corona current from a dielectric sheet having a single perforation reveals a relaxation oscillation the frequency of which increases with the impressed voltage (263). The dielectric in the neighborhood of the perforation thus appears to serve as a condenser which,

first charged by primary ions, discharges when the voltage attains the breakdown level of the gas in the pocket.   In most practical cases, the breakdown
strength E is of the order of 10^6 V/m (1m,264), and its value in atmospheric
air has been reported as high as 7×10^6 V/m (263).   It is, however, a
function of the gas, its relative density, and the interstitial geometry.
Lowering the relative gas density intensifies back corona; raising the density
tends to suppress it (265,266).   Industrial corona-current densities j are
generally less than about 10^{-3} A/m² (Table IV).   At first thought we should
not expect back-corona disturbances to appear unless the dust resistivity
ρ_d (ohm-m) exceeds

$$\rho_d = \frac{E}{j} = \frac{10^6}{10^{-3}} = 10^9 \text{ ohm-m} \tag{87}$$

a figure greater than the "critical" resistivity of about 10^8 ohm-m widely
quoted in industrial practice.   Definition of the back-corona threshold in
terms of particle resistivity is, however, not so simple a matter that it can be
fully accounted for in terms of eq. 87.   The problem is complicated by dust
resistivity generally being dependent on applied voltage for a given electrode
spacing, the resistivity falling with increased voltage (267,268).   Additionally, values of resistivity are, to some extent, dependent on the experimental
technique employed.   Various schemes have been described (1n,264,267,
269–277) some yielding results that differ by an order of magnitude or more
for ostensibly identical samples.

According to a theory proposed by Cooperman (278), back corona sets in
when the time constant of the dust τ_d(sec)

$$\tau_d = \varepsilon_0 \kappa_d \rho_d \tag{88}$$

is greater than that of the surrounding gas τ_g(sec)

$$\tau_g = \varepsilon_0 \kappa_g \frac{E}{j} \tag{89}$$

where κ_d (dimensionless) is the relative dielectric constant of the dust and E
is the field in the immediate vicinity of the back-corona electrode.   The
critical resistivity is then given by

$$\rho_d = \frac{\kappa_g}{\kappa_d} \frac{E}{j} \tag{90}$$

This somewhat lowers the estimate of ρ_d in eq. 87.   An apparently similar
approach to the question of critical resistivity has been taken by Boehm
279), but is not reported in detail.

Precipitator energization by high-voltage pulses of low-duty cycle, though
subject to other disabilities such as reduced average power input, may be used

to minimize back corona.   Since the precipitate serves as a leaky condenser, it tends to smooth out current pulses, the voltage across the dust layer then being determined more by the time-average current than by the peak.   The narrower the pulse width relative to $\tau_d = \varepsilon_0 \kappa_d \rho_d$, the more pronounced is this effect (1o,118).

A precipitator having a series of automatically cleaned disc plates mounted on a horizontal shaft and rotating in a vertical plane has been claimed to perform well with high-resistivity dusts (82) despite the fact that extremely thin dust deposits have elsewhere been observed to result in back corona. A subsequent report (83), however, indicates that despite laboratory successes, precipitator designs involving heavy moving parts in hot, dusty gases are unsuited to the rigors of industrial service.   An important earlier attempt (280) to clean plates by removing them from the gas stream also proved impractical and was abandoned.

Contrary to the frequent assertion that the presence of highly conductive particles, such as unburnt carbon, help combat the high-resistivity problem by providing conductive paths through the dust layer (281), Hesselbrock (267,282) claims that conductive particles exercise a deleterious effect because they give rise to localized high-field concentrations in the precipitate, and subsequent back corona.   Reduction in sparkover voltage by conducting particles in an otherwise high-resistivity dust has been observed by Douglass and Penney (283).   Considerations of this kind, incidentally, enter into determining whether a mechanical collector used in conjunction with a precipitator should precede or follow.   It is, for example, sometimes observed that a preceding mechanical collector that catches the larger, often conductive, particles, lowers precipitator efficiency.   On other occasions, the preceding mechanical collector clearly enhances precipitator performance, possibly for the reason given by Hesselbrock (267), or possibly for the more usual one that the heavier, more abrasive, particles are kept from scouring the precipitator's collecting plates.   Matters affecting series combinations of mechanical collectors and electrostatic precipitators are discussed by numerous authors (4,18,224,281,284–293).   Precipitators followed by wet scrubbers also find service in some applications (294–299).

Lowe, Dalmon, and Hignett (300) have noted that as long as the applied voltage is maintained constant, precipitator efficiency is unaffected to $\rho_d = 10^{11}$ ohm-m or even above.   The reason suggested is that positive ions leaving *points* of back ionization on the collecting electrode are confined to discrete narrow tubes occupying only a small fraction of the interelectrode volume and, therefore, do not significantly countercharge negatively charged particles.   This explanation is in direct conflict with the model of Pauthenier described in the next section.

Dzoanh's observation (301) that at low pressures a transverse magnetic field superimposed on a back corona discharge significantly increases the threshold and sparking potentials may be of practical interest in specialized applications.

2. Particle Charging in a Bi-Ionized Field

In the presence of back corona and the resultant bi-ionized inter-electrode field (the total corona current consisting of negative ions migrating in one direction and positive in the other), the equations given earlier governing particle charging are no longer applicable. Assuming that the oppositely charged ions are uniformly interspersed in the gas, it may be shown that eq. 53 is to be replaced by (302)

$$Q = Q_{\max} \frac{1 - e^{-\alpha t}}{1 - \left(\dfrac{1 - \xi}{1 + \xi}\right)e^{-\alpha t}} \tag{91}$$

where

$$\alpha = \frac{1}{\varepsilon_0}(b_f\rho_f b_b\rho_b)^{1/2} \tag{92}$$

$$\xi = \left(\frac{b_b\rho_b}{b_f\rho_f}\right)^{1/2} = \left(\frac{j_b}{j_f}\right)^{1/2} \qquad (0 \leqslant \xi \leqslant 1) \tag{93}$$

and the limiting charge $Q_{\max}$ is given by (302,303)

$$Q_{\max} = 4\pi\varepsilon_0 pa^2 E_0 \frac{1 - \xi}{1 + \xi} \tag{94}$$

Ion mobility b, space charge density ρ, and current density j due to "forward" and back corona are distinguished by the respective subscripts f and b. Unlike monopolar ion bombardment which effectively ceases after a few time constants τ (eq. 54), bipolar ion bombardment continues indefinitely. A maximum charge $Q_{\max}$ is approached because positive and negative ion currents to the particle tend to become equal and so neutralize each other. $Q_{\max}$ could even be zero in the special case $\xi = 1$; in such a bi-ionized field no electrostatic precipitation occurs. As ρ_b approaches zero (absence of back corona), charging eq. 91 reduces to the companion eq. 53 for the mono-ionized field. Equation 94 has been found to be in agreement with experiment (304).

The debilitating effect of back corona on an electrostatic precipitator is strikingly shown by considering a case in which the back corona current is only one-third of the total current $j_t = j_b + j_f$. One then has $\xi = (0.5)^{1/2} = 0.71$, and $Q_{\max}$, (whence the particle migration velocity w) is reduced by a factor $(1 - \xi)/(1 + \xi) = 0.17$. According to the Deutsch equation, precipitator length must be increased $1/0.17 = 5.8$ times to restore collection

efficiency to the level corresponding to zero back corona.   Other considera
tions, such as the reduction of applied voltage by sparking, and the lowering
of the interelectrode field by increased voltage drop across the high-resistivity
precipitate, are likely to reduce efficiency still further.

Photographs by Seman and Penney (171) of individual particle trajectories
in back corona suggest that the uniform charging and countercharging
assumed by Pauthenier is no more than statistically correct, even for identical
particles.   Pockets of positive and negative charge seem to exist in the gas
over discrete interelectrode regions.

Other aspects of the theory of the bi-ionized field, extensively studied by
Pauthenier and his school (81,263,265,302,304–309), may be briefly noted as
follows.   Poisson's equation for a wire-pipe system containing bipolar space
charge is

$$\frac{1}{r}\frac{d}{dr}(rE) = \frac{1}{\varepsilon_0}(\rho_f - \rho_b) \tag{95}$$

Experimental observations with a negative forward corona justify the
assumption that the component forward and back corona currents per unit
length, j_{fl} and j_{bl}(A/m), respectively,

$$j_{fl} = 2\pi r b_f E \rho_f \tag{96}$$

$$j_{bl} = 2\pi r b_b E \rho_b \tag{97}$$

are conserved between wire and cylinder.   That is, $j_{fl} + j_{bl}$ give the total
current per unit length j_{tl} as measured directly.   The parameter ξ of eq. 93
is thus independent of the radius vector r for a given set of operating conditions.   Eliminating ρ_f and ρ_b from eq. 95

$$rE\frac{dE}{dr} + E^2 + \frac{j_{tl}}{2\pi\varepsilon_0}\left[\frac{b_b - \xi^2 b_f}{(1 + \xi^2)b_f b_b}\right] = 0 \tag{98}$$

whence

$$E = \left\{\frac{j_{tl}}{2\pi\varepsilon_0}\left(\frac{b_b - \xi^2 b_f}{b_f b_b(1 + \xi^2)}\right) + \left(\frac{r_0}{r}\right)\left[E_c^2 - \frac{j_{tl}}{2\pi\varepsilon_0}\left(\frac{b_b - \xi^2 b_f}{b_f b_b(1 + \xi^2)}\right)\right]\right\}^{1/2} \tag{99}$$

For j_{tl} and r not too small (cf. eq. 15)

$$E = \left[\frac{j_{tl}}{2\pi\varepsilon_0}\left(\frac{b_b - \xi^2 b_f}{b_f b_b(1 + \xi^2)}\right)\right]^{1/2} \tag{100}$$

Analysis of experimental data shows that the negative primary current
in the presence of back corona is itself greater than it would be without back
corona.   This condition is due, in part, to the bombardment of the central
electrode by back-corona ions, and in part to photoionization originating
both at the wire and at the surface of the precipitate.

3. Conditioning; Sulfur Oxides in Flue Gas

The passage of an electric current through a layer of precipitate occurs both over the surface and through the volume of the individual particles. Surface conductivity is dependent on surface moisture and chemical films adsorbed on the particles, whereas volume conductivity is a property inherent in particle composition. For semiconducting materials usually of interest in electrostatic precipitation, volume resistivity ρ_v(ohm-m) is a decreasing function of temperature in reasonable agreement with the relation

$$\rho_v = \rho_\infty e^{E/kT_A} \tag{101}$$

where the quantity ρ_∞(ohm-m) and the activation energy E (J) are constants of the material (1p,310). Depending upon the nature of the dust, the validity of eq. 101 in a humid gas may not become apparent until the temperature exceeds 100–300°C or more and conducting surface films are driven off. Figure 22, illustrating the transition from surface to volume conduction, is qualitatively typical of curves for numerous substances (3c).

Fly-ash resistivity at temperatures up to 820°C have been measured by Shale et al. (271) who find, for their test conditions, that the resistivity remains in a satisfactory range for electrostatic precipitation (Fig. 23).

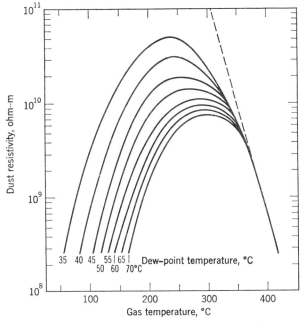

Fig. 22. Resistivity of a cement-kiln dust as a function of gas temperature and dew point. The broken line shows the effect of volume conduction in accordance with eq. 101. Fly-ash resistivities commonly peak between 100 and 200°C (269). (Used with permission of Staubforschungsinstitut des Hauptverbandes der geverblichen Berufsgenossenschaften e.V.)

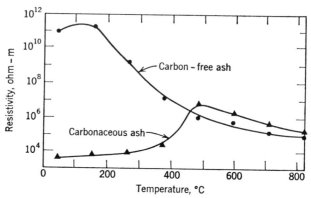

Fig. 23. Resistivity as a function of temperature for virtually carbon-free ash and for 30% carbon ash. The resistivity of the latter initially increases with temperature owing to the oxidation of conductive combustibles (271). (Used with permission of U.S. Bureau of Mines.)

Particles of very low resistivity present a problem of their own. Such particles rapidly discharge upon capture, assume the same polarity as the collecting surface, and are then repelled into the gas stream (pith-ball effect). The lower resistivity limit usually quoted for effective precipitation is 10–100 ohm-m. Carbon black, a material for which the resistivity may be as low as 10^2–10^3 ohm-m, can be collected by a small, high-velocity precipitator followed by an inertial collector. In this arrangement, the submicron carbon particles are first agglomerated, either at the collecting plates, or in the gas stream by the polarizing action of the electric field. [Soviet practice favors longer, low-velocity precipitators without a following mechanical collector (6b).] Paper flakes emitted from refuse incinerators are another material that can prove difficult to collect in precipitators, in this case because of large particle size as well as low resistivity. When it occurs, the problem is largely eliminated by cyclone after-collectors (288).

Precipitators, of course, can be used as agglomerators even if no low-resistivity condition exists: Oxygen-steelmaking converter processes may employ a precipitator-agglomerator followed by cyclones (289), and in the pulp and paper industry, wet scrubbers following precipitators yield scrubber efficiencies in the 90% range at very low pressure drops (1–2 in. WG), a result attributable to the agglomeration capabilities of the precipitators (221).

Control of particle resistivity by conditioning the carrier gas plays an important part in many applications of electrostatic precipitation. The availability of moisture alone in some cases, or moisture with small quantities of chemical conditioners in others, often suffices to lower the particle resistivity by one or more orders of magnitude. In this fashion, by eliminating back-corona disturbances, precipitator performance may be dramatically upgraded.

Inlet gas may most economically be humidified by water atomization within a spray chamber installed immediately ahead of the precipitator. Frequently, however, the temperature of the gas is too low ($<200°C$) to effect speedy evaporation of the mist and, in consequence, a slurry may build up in the chamber, especially if its size is small. Masuda et al. (311) claim to have developed a practical humidification system suited to such cases, with a new design of nozzle (312).

The physical adsorption of a moisture film on a chemically inert surface is characterized by relatively low binding energies. A water film alone consequently produces less effective conditioning that does a chemical binder that is strongly adsorbed on the particle surface and which in turn strongly adsorbs moisture. This activated adsorption effect is often accomplished for weakly basic particles by strong acids, for example, HCl, H_2SO_4, and for weakly acidic particles by strong bases, for example, NH_3 (1q).

Reverse conditioning to increase resistivity is also possible. The addition of $<1\%$ naphtha soap or soap stock to rotary cement-kiln gas has been found to raise particle resistivity by an order of magnitude under the temperature–humidity conditions examined (313).

Sulfate deriving entirely from H_2SO_4 in flue gas provides a natural conditioning substance in the combustion of coal, in which sulfur is normally present from a small fraction of a per cent to about 5%. Although the actual concentrations of sulfur oxides produced in combustion vary depending on the coal, the furnace design, and the operating conditions, the usual orders of magnitude are 0.1% for SO_2 and 0.001% for SO_3 (1r).

The conditioning of low-sulfur flue gas by SO_3 injection, viewed unfavorably by American electric utility companies partly because of anticipated adverse public reaction, has been successfully employed on a limited commercial scale in Great Britain, and extensively in New South Wales, Australia. The small amount of SO_3 supplied, generally less than 50 ppm, is completely adsorbed by the dust and does not lead to increased emission of sulfur compounds into the atmosphere (18 279,314).

The physical mechanism of fly-ash conditioning by SO_3 is convincingly demonstrated in Figure 24. Electron micrographs of unconditioned fly-ash particles reveal a smooth surface in contrast to the conducting surface film present after conditioning. On the basis of a compatible model, Masuda (276,315,316) has developed a theory expressing the apparent resistivity of spherical particles in terms of true volume and surface resistivities.

Conditioning has the advantage that if the high-resistivity problem is not recognized until after a plant is commissioned, injection equipment can be installed without shut-down. Moreover, there are older installations that

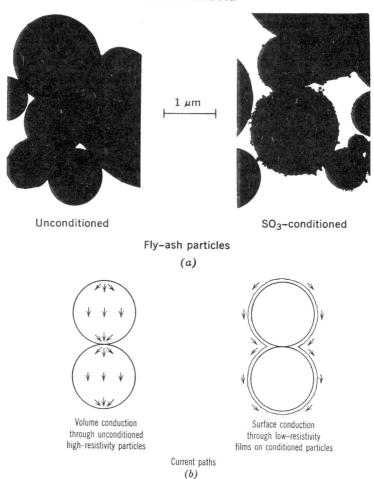

Fig. 24. (a) Electron micrographs of unconditioned and SO₃-conditioned dust. (b) Corresponding schematic representations of current flow (314). (Figure a, b used with permission of K. Darley and the Central Electricity Research Laboratories.)

cannot conform to the requirements of recent legislation. The expense of fitting in additional precipitation equipment, even if room for this is available, may be prohibitive.

The cost of SO₃ conditioning is significant (18,317). Darby (18), however, observes that if the effective migration velocity can be doubled through the use of 10 ppm of SO₃, total expenses (including injection equipment and conditioning agent) capitalized over 20 years shows an appreciable economy as compared with the alternative solution of installing a larger precipitator. Systems for SO₃ injection (318,319), and for the catalytic conversion of naturally present SO_2 to SO_3 for conditioning purposes (320), are described in the patent literature.

Experience in the ammonia conditioning of power-boiler flue gases is reported by Reese and Greco (321) and Baxter (327). The former consider the economics of ammonia relative to SO_3 conditioning.

The addition of 1–2 ppm by weight of yellow phosphorus is said to produce the same increase in efficiency as the injection of 15 ppm of SO_3, while resulting in less corrosion (323).

Other considerations being (apparently) equal, the sulfur content of coal may still be a misleading indicator of expected precipitator performance. Thus, American and United Kingdom coals generally give higher migration velocities than certain South African or Australian coals having the same sulfur content. The reason for the anomaly is not clear at present, but perhaps it arises from variations in catalytic substances serving to convert SO_2 to SO_3. New South Wales coals present a particularly perplexing problem (18,324–328).

Despite increasing pressures to minimize gaseous as well as particulate air pollution, the development of equipment and processes for controlling the emission of sulfur products lags far behind progress in particulate collection. The most direct approach to the reduction of SO_2 stack emission from utility boilers and other sources is legislation prohibiting the use of fossil fuel containing more than a prescribed proportion of sulfur. In New York City, for example, this limit, effective in 1969, is set at 1% by weight. Additionally, a 99.0% precipitator efficiency on all coal-burning power boilers was required by 1969. Pending legislation is still more stringent.

In complying with these laws, several alternatives present themselves:

1. Upgrade precipitator performance by increasing electrical sectionalization.

2. Add new precipitator sections in series or in parallel.

3. Change to low-sulfur oil or natural gas.

4. Adjust the precipitator inlet temperature to a point well removed from the peak resistivity level.

(Fly-ash resistivity–temperature curves, with sulfur content replacing moisture as the parameter, are qualitatively similar to Figure 22, the higher peaks corresponding to the lesser sulfur content.) Alternative *4* finds application in the world's largest precipitator, the Ravenswood Station of the Consolidated Edison Company of New York. Here the precipitator is installed upstream of the air heater to give a precipitator inlet temperature of 370°C. At this high temperature, the resistivity of the ash is at a satisfactory level and determined by volume conduction alone, that is, resistivity is now independent of the sulfur content of the fuel (221,223,224). Reducing the temperature to achieve a tolerable value of resistivity is also possible, but it then becomes necessary to drop below the acid dew point (140–150°C), inviting corrosion and lowering the sparkover voltage (224,329). For cases in which precipitators follow boilers designed to discharge below the acid dew point, collector performance may be restored by the addition of small quantities of ammonia to the flue gas (321). Caution must be observed in selecting the ammonia injection point: injection before the air heater causes plugging of that unit for temperatures below 200°C (321,322).

Methods have been proposed employing electrostatic precipitation in the removal of SO_2 from flue gas. A recent suggestion involves the introduction of boiler gas into an absorption tower where powdered, activated MnO diffuses through it. The resulting $MnSO_4$ absorbent is then collected by a multicyclone and precipitator (330,331). Other approaches call for the addition to the furnace of alkaline-earth carbonates or dolomite for reaction with the sulfur content of the gas. Dolomite injection, however, severely reduces precipitator efficiency. In tests, dust resistivity was observed to increase by several orders of magnitude, evidently a result of SO_3 absorption. Had sufficiently higher temperatures been used, though, effective precipitation probably would have resulted regardless of the SO_3 level of the combustion gas (332,333).

Pretorius and Mandersloot (334) raise a general objection to schemes employing suspended-solids techniques for SO_2 removal. In their opinion a limited mass-transfer rate from the gas to the absorbent-particle surface, and the limited concentration of particles that can be held aloft without deposition, mean that systems using suspended solids are not very attractive for practical situations where 90% pollutant removal is necessary.

A review of limestone sorption processes by the Tennessee Valley Authority (333) draws somewhat similar conclusions. It is pointed out that unless new technology for improving sorption is developed, 25–30% utilization of the injected limestone is probably the highest attainable. Excess limestone can be used to increase SO_2 removal, but adverse effect on the power plant operation and the higher cost of much larger precipitators then become major disadvantages.

Relations linking precipitator performance and sulfur content with electrical sectionalization, gas temperature, and other variables have been considered by Ramsdell (335), Katz (281), Engelbrecht (336), and Archbold (337). Gartrell (338) has reported a 75% reduction in ground-level SO_2 concentrations in an area exposed to coal-fired power-plant emissions following the installation of electrostatic precipitators. The reasons for this have not been determined. The effect of this on the emissions of sulfur dioxide are detailed by Strauss (p. 100).

Nuclear sources appear to offer the most satisfactory long-term solution to the elimination of sulfur oxide and fly-ash contamination arising in electric-power generation. But unless world reserves of uranium ores are much in excess of estimates, nuclear energy cannot, in the immediate future, satisfy increasing demands for power. Between 1980 and 1990, however, breeder reactors may provide a solution since their production of energy per unit mass of uranium is 50 times the present rate (339,340b). Although the construction of large fossil-fueled plants has become less frequent, particularly in large centers of population, increasing power requirements are expected to raise the rate of coal consumption in the United States to about 700 million tons per year by 1980, one and one-half times the rate of only a decade and a half earlier (340a).

C. Corona Quenching

1. Mild Quenching by Suspended Dust

One of the reasons for the failure in 1884 of the pioneering attempt by Lodge and Walker to precipitate lead oxide fume on a commercial scale (61) seems to have been the finely divided nature of the suspended particles. The heavy particle space charge often resulting in such cases tends to elevate the corona threshold and so depress the current at a given voltage. As long as it can be assumed that the current is carried essentially by free ions (unattached to aerosol particles), the treatment given in Sections III-B-2 and III-C-2 remains applicable. In other words, it is valid to assume

$$j_t = j_i + j_p = E(\rho_i b + \rho_p b_p) \cong E\rho_i b \tag{102}$$

where $j_t(\text{A/m}^2)$ is the current density at a point, the subscripts t, i, and p designating total, free ion, and particle, respectively. Particle mobility $b_p[(\text{m/sec})/(\text{V/m})]$ is commonly 2 or 3 orders of magnitude less than the ion mobility b; therefore, although ρ_p may be larger than ρ_i, j_p is ordinarily no greater than a few per cent of j_t.

A curious polarity dependence of corona suppression is reported by Winkel and Schuetz (70). Comparing eqs. 15 and 22 for identical fields at the tube surface r_1 of a coaxial cylindrical precipitator, they write

$$\frac{j_{lc}}{j_{ld}} = 1 + \frac{2pr_1}{3}S \tag{103}$$

where j_{lc} and $j_{ld}(\text{A/m})$ are, respectively, the linear current densities in clean and dirty gas. This relation is qualitatively supported by experiment (Fig. 25). With negative corona, quenching above room temperature is more severe than with positive, becoming still more so with increasing fume concentration and gas temperature—a phenomenon perhaps to be associated with the greater free-electron component of the current in the less dense, hotter gas.

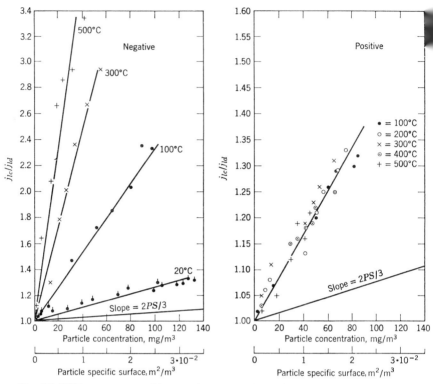

Fig. 25. Mild corona quenching by iron-oxide fume for positive and negative discharges with a 0.3-mm wire in a 3.5-cm diameter tube at atmospheric pressure. The bottom curve for each polarity is that expected on the basis of eq. 103 (70). (Used with permission of Staubforschungsinstitut des Hauptverbandes der geverblichen Berufsgenossenschaften e.V.)

2. Severe Quenching by Suspended Dust

This condition, first analyzed by Deutsch (74) in 1931 and later named by Sproull (341) to distinguish it from its milder counterpart, occurs if the particle concentration is so dense that all free ions are captured. Equation 102 then takes the form

$$j_t = E\rho_p b_p \tag{104}$$

In this event, the current consists solely of ions riding along on slowly migrating dust particles and j_t is characteristically reduced to 1% or less of its former value. The maximum charge Q that a particle can acquire is determined not by the relevant law of charging (Sec. IV) but by the limited availability of free ions. Specifically

$$Q = \frac{\rho_p}{N_p} = \frac{\rho_i'}{N_p} \tag{105}$$

where $\rho_i'(\text{C/m}^3)$ is the limited density of ions available for charging and N_p (particles/m³) is the particle number density. Thus, with particularly severe quenching, the charge per particle may be far below that attainable in ordinary circumstances and the associated particle migration velocity and collection efficiency, correspondingly low. Over that length of the precipitator in which severe quenching prevails

$$w = w_s \frac{\rho_i'}{N_p Q_s} \qquad (\rho_i' \leqslant N_p Q_s) \tag{106}$$

where Q_s is the charge that would be acquired were ρ_i' adequate, and w_s is the associated migration velocity. The Deutsch differential eq. 67 and Stokes' law then give (342,343)

$$\eta = \frac{E_p \rho_i' A}{6\pi\mu a V_g (N_p)_{\text{in}}} \tag{107}$$

where $(N_p)_{\text{in}}$ is the inlet particle number concentration. Despite severe quenching, particle collection still occurs, albeit at a reduced rate.

Sproull (341) reports that a two-stage precipitator using alternating current in the charging stage and direct current in the collector is less sensitive to severe quenching than a precipitator of conventional design. Since for equal effective voltages, ac corona dissipates more power than direct current —the power loss being proportional to the frequency—one must expect that the ion density ρ_i' will be greater in the ac case. According to Darby (344) the corona-suppression problem may be overcome to some extent by the use of strip discharge electrodes having sawtooth or other irregular edges.

3. Buildup on Discharge Electrodes

Lowe, Dalmon, and Hignett (300) have shown that the deposition of a film of highly resistive dust ($\rho_p > 10^8$ ohm-m) on the discharge electrodes could satisfactorily account for certain instances of "difficult" precipitation that have been observed from time to time in power-station precipitators. Fine particles ($<10 \mu$m), the motion of which is largely governed by turbulence (135), can be driven directly onto the negative discharge electrode, or sufficiently close to acquire a positive-ion charge in the corona region, and then be precipitated on the discharge electrode. As the resultant dust layer builds up, the potential drop across it increases, and, if the particle resistivity is high enough, the interelectrode field will drop sufficiently to materially reduce the current. Figure 26 illustrates the current suppression anticipated in a characteristic case. Since dust deposited on the corona electrodes is likely to consist of tenaciously adhering fine particles, it is unlikely that conventional electrode rapping or vibrating can offer an adequate solution to the problem. Conditioning of the gas to lower the particle resistivity appears to offer the most promising remedy in such cases (300). Dransfield and Lowe (292), observing the sensitivity of the corona to dust

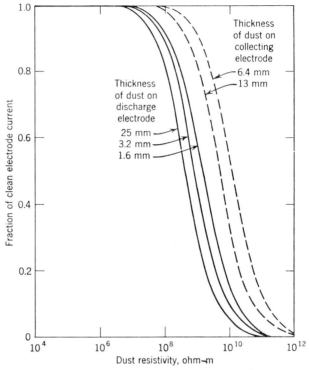

Fig. 26. Estimated reduction of corona current by a layer of resistive dust on either the discharge (solid curves) or collecting electrode (broken curves) of a tubular precipitator. The tube and wire diameters are 25.4 cm and 6.4 mm, respectively. At a negative wire potential of 50 kV the current is 0.13 mA/m when the electrodes are dust free. Since the resistance R (ohm) of a cylindrical layer of dust of length L (m) and inner and outer radii r_{in} and r_{out} (m) is given by

$$R = \frac{\rho_d}{2\pi L} \ln \left(\frac{r_{out}}{r_{in}} \right) \tag{108}$$

it is seen that the resistance, and hence the voltage drop, is greater at the wire than at the tube for a given dust thickness (300). (Used with permission of Institution of Electrical Engineers.)

buildup on the active electrodes, recommend that discharge wires be rapped (or vibrated) continuously. This is in contrast to their contention that collecting electrodes are best rapped intermittently.

VII. APPLICATIONS

A. Introduction

Adequate coverage of the numerous practical aspects of electrostatic precipitation presents a twofold difficulty, the resolution of which cannot be attempted in this chapter. First, a comprehensive treatment requires a compilation and critical evaluation of operational data. Much of this comes from sources of indeterminate reliability scattered throughout the literature. Second, much of the essential design data, particle migration velocities, for instance, is regarded as confidential proprietary information by manufacturers who have accumulated it at considerable cost in time and effort. It is for the latter reason that published descriptions of industrial precipitators, no matter how detailed in certain respects, are commonly deficient in revealing certain critical variables of the process.

The degree to which a full-scale precipitator performs in accordance with theory is highly variable; therefore, no substitute exists in industrial precipitation practice for diversified, long-term experience. Approaches to precipitator design, however, that are *governed* by empiricism are likely to lead to confusion. Observations in the field, which by their nature are often without adequate controls or other experimental safeguards may, when presented injudiciously, be cited in support of almost any conclusion. Furthermore, the subjective approach that unfortunately goes hand in hand with much field work provides a very limited base for new developments and fails to permit design and performance to be analyzed in terms of fundamental physical relationships.

B. Precipitator Utilization

Today, an estimated 4000 industrial precipitators in the United States treat some 200,000 Nm^3/sec of various polluted gases. Fly-ash collection in electric-power generating stations accounts for $\sim$60% of the total capacity. The balance is principally divided among steel and cement production ($\sim$10% each), and the processing of paper and non-ferrous metals ($\sim$7% each). Precipitation in the chemical industry, including the now obsolete applications of fuel-gas detarring and carbon-black collection, is responsible for most of the remainder ($\sim$6%) (12). Surveys of the practical state of the art, considering established and growing areas of application, have been made by a number of authors (221,229,345–349). Guepner (210) has considered various factors for optimizing performance that should be borne in mind by precipitator designers.

The diversity of precipitator applications as reflected in European practice of a decade or two ago is illustrated in the extensive listing—together with limited operational data—compiled by Heinrich and Anderson (13) and quoted, in part, by Strauss (3d). Uzhov (6) has likewise described numerous precipitator installations of the same period in the Soviet Union.

For an up-to-date view of current trends, reference should be made to the abstract journals. Especially helpful are:

1. Air Pollution Control Abstracts (Air Pollution Control Association, Pittsburgh) for the broadest overall coverage.

2. *Electrical and Electronics Abstracts, Science Abstracts, Series B*, (Institution of Electrical Engineers, London, and Institute of Electrical and Electronics Engineers, New York) highlighting electrical developments.

3. *Chemical Abstracts* (American Chemical Society, Washington, D.C.) emphasizing the chemical industry and reporting on obscure sources otherwise frequently neglected.

4. *Nuclear Science Abstracts* (United States Atomic Energy Commission, Oak Ridge, Tennessee) useful for its identification of unpublished United States Government reports and its emphasis on radioactive aerosols.

5. The abstract section of the *Journal of the Iron and Steel Institute.* (Iron and Steel Institute, London) for special reference to ferrous metallurgy.

6. *Fuel Abstracts and Current Titles* (Institute of Fuel, London).

7. *Engineering Index* (Engineering Index, Inc., New York).

8. *Applied Science and Technology Index* (H. W. Wilson Co., New York).

9. *British Technology Index* (Library Association, London), the last three for inclusion of the trade journals.

As evidenced by recent (1963–1968) contributions to the literature, areas of special interest to precipitator users and designers include:

Electric power: coal-fired boilers (18,221,222,224,279,281,287,322,336,347, 350–353); oil-fired boilers (224,354,355); flue-gas desulfurization (330,331); flue-gas conditioning systems (318–320).

Ferrous metals: open-hearth, blast, electric, and cupola furnaces, oxygen steelmaking, sinter plants (18,232,289,298,356–387).

Cement: (18,314,388–392).

Pulp and paper: (290,294–296,299,393–398).

Nonferrous metals: zinc, copper, lead (399–403); gold (404); aluminum (405,406); tungsten (407).

Chemicals: sulfuric acid (18,408,409); phosphates (410,411); phosphoric acid (412); phosphorus (299); petroleum refining (413); gypsum (414); carbon black (415,416); hydrocarbon pyrolysis gas (220,415,417–419); water gas (420); oleum (421); magnesite (422); Me esters (423); miscellaneous (424–427).

Refuse incineration (with and without subsidiary power generation): (288,428–440, (junk automobiles) 441).

Food processing: smokehouse exhausts, food-smoking acceleration, deep-fat fryers (442).

Radioactive materials: (443–447).

Miscellaneous processing: natural-gas pipelines (225); zinc galvanizing (448); roofing-felt saturators (449,450); coal preparation (451); bituminous pipe coating (452); carbon-electrode baking furnaces (453).

Air cleaners: (454–458).
Aerosol samplers: (26,100,447,459–467).
Aerosol classifiers: (181,468,469).

Among recent departures from conventional precipitators the following designs may be mentioned:

1. A two-stage sampler using a pulsed collecting voltage to achieve a uniformly dense precipitate over the full length of the collecting electrode (460).

2. A two-stage sampler for accomplishing the same purpose by maintaining the collecting surface in reciprocating motion (470).

3. An electrostatic bubble scrubber in which a corona-charged aerosol is bubbled through a conducting liquid (471).

4. A two-stage precipitator in which particles charged by corona are precipitated by their own space charge in the absence of an external field (472).

5. A modification of the preceding in which water mist is periodically injected and charged in order to maintain the space-charge collecting field (473).

6. A collector using electrostatically atomized and charged water droplets to supplement the usual precipitation process (474,475).

7. An electrostatic cyclone separator (476).

8. A device claimed to impart a positive particle charge from the thermal radiation of a positively charged mass (477–479).

9. Precipitators using perforated or other collecting plates with gas flow directed parallel to the collecting field (82–85).

10. A sectionalized tubular precipitator having two individually energized stiff discharge electrodes, one projecting inward from each end of the tube (480).

A new concept of commercial importance in the precipitation field is the modular or package unit, a development that has made the electrostatic precipitator, traditionally a high capital-cost item, available to some of the smaller producers of particulate pollutants. Modular units, typically handling 5–15 m^3/sec, can be installed in segments and expanded to meet increased gas flows or efficiency demands. The easily replaceable modules are particularly suitable for use with corrosive gases.

Two-stage precipitators [Fig. 2, (24)], mostly of importance in indoor air cleaning, find occasional service in aerosol sampling and other applications: food processing (442), asphalt saturating (449,450), radioactive particle collection, and (unusually) sulfuric acid manufacture (409).

C. Financial Considerations

The most economical choice of proper gas-cleaning equipment for a specific problem can be made only after considering a number of factors including capital costs, water availability (where scrubbing or wet precipitation provide acceptable solutions), interest charges, power and maintenance costs, tax credits, and the value of the collected material. A total view is essential. In certain cases the cumulative water and power costs alone of an initially relatively inexpensive high-energy scrubber can, in a few years, exceed the much greater capital cost of a precipitator designed for the same service. In this vein Smith and Snell (411) report an economic evaluation of a fabric baghouse, a wet scrubber, and an electrostatic precipitator, each achieving 99% efficiency, for use with a fluid-bed roaster for the calcination of phosphate rock. The precipitator, which was the most costly to purchase and install, was found to be the least expensive to depreciate and operate over a ten-year period. Cost-estimating generalities must be applied with caution. Depending primarily on sulfur and ash content, precipitator costs for treating pulverized coal-fired boiler gas may vary over a range of 3 to 1. Chamberlin et al. (481,482) have provided cost data with detailed examples and comparisons with other gas-cleaning equipment in the power, steel, and cement industries. Additional cost comparisons are cited by Stairmand (293) who also shows that, contrary to popularly held belief, electrostatic precipitators are no more expensive than should be expected on the basis of their relatively high performance. Precipitator economics with reference to refuse incineration—an application of importance in Europe but only in its infancy in the United States (221,434)—has been considered by Fife and Boyer (483). Incidental information on capital and operating costs with reference to specific processes is found sporadically throughout the applications literature. Problems peculiar to foreign precipitator sales, import barriers, currency restrictions, and the like, are briefly noted by Schneider (346).

D Pilot-Plant and Model Studies

The pitfalls inherent in pilot-plant extrapolation and the extension of pilot data to new gas velocities and electrical conditions must be emphasized. Pronounced differences that may arise in gas-flow distribution and sparkover levels (cf. eq. 47) can require the application of generous safety factors in carrying over pilot observations to full-scale designs. These precautions with respect to gas-flow and electrical characteristics have to be observed even though the pilot plant is —as it always should be—a section of the full-size unit and not a scaled-down replica. *Properly executed and interpreted,* however, pilot-plant tests are capable of yielding results that can be used with assurance, and that provide valuable guidelines for first designs.

Analog model studies for investigating the character of the gas flow in a projected full-scale precipitator and its connecting ductwork are well established in the industry despite conflicting estimates of their utility in many cases. White (1s) reports excellent quantitative agreement between model and field results in gas-flow patterns when the conditions of geometric and Reynolds similarity, especially the former, have been satisfied. According to Opfell and Sproull (484), however, meeting these requirements alone provides no assurance that gas-velocity distribution in the prototype will be accurately reproduced. This is especially a matter of concern in the case of a model built to the $\frac{1}{16}$ or smaller scale common today because of the large size of many industrial precipitators. Dependable results, it is pointed out, could be derived from models built to as large a scale as is economically reasonable, modeling only those parts of the installation where the gas-flow pattern is most critical, or most doubtful. For each part of the equipment so modeled the dominant dimensionless characteristic number (Reynolds, Froude, Euler, or other) should be matched as closely as practicable with the corresponding prototype value.

Without attempting to discredit gas-flow model studies generally, it must be remarked that such studies are frequently a gratuitous adjunct to precipitator engineering and serve merely to confirm a satisfactory design already selected, without benefit of models, by experienced designers. Nevertheless, it should not be overlooked that the relatively small investment required for a model study (usually less than 1 or 2% of the total price) is often capable of providing cheap insurance against possible costly modifications in the full scale.

In electrified model work Hughes et al. (183) recommend the use of electromechanical similarity relations. These include an electroviscous number, the ratio of electrostatic to viscous forces, and an electric Reynolds number, the ratio of inertial to electrostatic forces.

E. Operation and Maintenance; Explosion Hazards

Normal operational considerations of a general nature are given by Rose and Wood (2g), Gottschlich (17), Sickles (485), and White (1t). Recommendations for fly-ash precipitators are made by Dransfield and Lowe (292), and for kraft-mill precipitators by Rabkin and Crommelin (486); these are, however, much broader in application than indicated.

Chemical changes induced by electrical discharges are sometimes unexpectedly encountered in the gas, or the particles, or even in structural members of the precipitator. Polarity-dependent alterations in the chemical structure of aerosol particles have been observed (423), and green corrosion and cracking found in stressed brass parts have been attributed to the action of nitrogen oxides formed by sparking in moist air (487).

The incendiary nature of the corona discharge and electric spark naturally bars the application of electrostatic precipitation in explosive atmospheres. There are situations, however, where the particle-laden gas, although not normally explosive, may become so in the event of upstream process malfunction. Denisov et al. (402) discuss the destruction of a wet precipitator caused by the ignition of elemental sulfur during oxygen flash smelting of copper-zinc concentrates. According to Guepner (488), precipitator gas explosions may be prevented by adding to a partial stream of the gas mixture a component which increases the mixture's incendivity, and then continuously testing the stream with an ignition source. Kornev et al. (418) describe a methane-pyrolysis precipitator monitored by a photoelectronic system that disconnects the high-voltage primary when the flow rates of natural gas and oxygen change. Safety precautions to be observed with precipitators in the chemical industry are comprehensively treated in a monograph by Uzhov (5).

F. Industrial Standards

In order to facilitate the specification, guarantee, and purchase of electrostatic precipitators, pertinent standards and definitions have been developed in several countries. For example, in the United States, this has been done by the Industrial Gas Cleaning Institute (IGCI) and the Air Pollution Control Association (APCA), and in the German Federal Republic by the Verein Deutscher Ingenieure (VDI). The efforts of these organizations go far in achieving a measure of national standardization within the industry. Relevant publications are given in the references (273,485–494).

References

1. H. J. White, *Industrial Electrostatic Precipitation*, Addison-Wesley, Reading, Massachusetts, 1963; (a) pp. 1–30; (b) pp. 94–96; (c) pp. 41–42; (d) p. 235; (e) pp. 196–237; (f) p. 135; (g) pp. 162–163; (h) p. 238; (i) ch. 8; (j) p. 365; (k) pp. 349–354; (l) p. 204; (m) p. 321; (n) pp. 295–300; (o) p. 326; (p) p. 307; (q) pp. 306–313; (r) pp. 315–316; (s) p. 282; (t) ch. 11; (u) p. 35.
2. H. E. Rose and A. J. Wood, *An Introduction to Electrostatic Precipitation in Theory and Practice*, 2nd ed., Constable, London, 1966; (a) pp. 175–192; (b) p. 55; (c) pp. 106–108; (d) p. 109; (e) pp. 162–169; (f) pp. 169–173; (g) ch. 7.
3. W. Strauss, *Industrial Gas Cleaning*, Pergamon, Oxford, 1966; (a) pp. 370–375; (b) pp. 375–384; (c) pp. 364–365; (d) pp. 393–395.
4. V. N. Uzhov, *Ochistka Promyshlennykh Gazov Elektrofiltrami (Cleaning of Industrial Gases by Electrofilters)*, Goskhimizdat., Moscow, 1967.
5. V. N. Uzhov, *Tekhnika Bezopasnosti pri Ekspluatatsii Elektrofil'trov na Predpriyatiyakh Khimicheskei Promyshlennosti (Safety Technique During Electrofilter Operation in the Chemical Industry)*, Khimiya, Moscow, 1964.
6. V. N. Uzhov, *Sanitary Protection of Atmospheric Air*, 1955; B. S. Levine, English Transl., U.S. Dept. Comm. Publ. No. 59-21092, no date: (a) pp. 98–99; (b) p. 124

7. Schneerson, B. L. *Elektricheskaya Otchistka Gazov (Electric Gas Purification)*, Publisher and date wanting; through reference 6, p. 106.

8. Lutyński, J. *Elektrostatyczne Odpylanie Gazów (Electrostatic Dust Collectors for Gases)*, WNT, Warszaw, 1965.

9. O. J. Lodge, *Electrical Precipitation*, Oxford University Press, London, 1925.

10. H. J. Lowe and D. H. Lucas, *Brit. J. Appl. Phys., Suppl. No. 2*, S40-S47 (1953).

11. L. M. Roberts, in *Encyclopedia of Chemical Technology*, Vol. 5, R. E. Kirk and D. F. Othmer, Eds., Interscience, New York, 1950, pp. 646-662.

12. L. M. Roberts and A. B. Walker, in *Encyclopedia of Chemical Technology*, Vol. 8, R. E. Kirk and D. F. Othmer, Eds., Interscience, New York, 1965, pp. 75-92.

13. R. F. Heinrich and J. R. Anderson, in *Chemical Engineering Practice*, Vol. 3, H. W. Cremer and T. Davies, Eds., Academic Press, New York, 1957, pp. 484-534.

14. R. Meldau, *Handbuch der Staubtechnik*, Vol. 2, VDI Verlag, Duesseldorf, 1958, pp. 218-236.

15. G. Underwood, *Intern. J. Air Water Pollution*, **6**, 229 (1962).

16. C. F. Gottschlich, in *Air Pollution*, 2nd ed., Vol. 3, A. C. Stern, Ed., Academic Press, New York, 1968, pp. 437-456.

17. C. F. Gottschlich, *Electrostatic Precipitators*, American Petroleum Institute, New York, 1961.

18. D. W. Ertl, *South African Mech. Engr.*, **8**, 159 (1967).

19. D. W. Ertl, *South African Mech. Engr.*, **9**, 13 (1967).

20. H. Simon, in *Air Pollution Engineering Manual*, J. A. Danielson, Ed., U.S. Public Health Service Publ. No. 999-AP-40, 1967, pp. 135-156.

21. M. Pauthenier, *Météorol.*, **69**, 221 (1963).

22. Am. Ind. Hyg. Assoc., *Air Pollution Manual*, Part II, Detroit, 1968, pp. 85-94.

23. Proc. Eng. Seminar Electrostatic Precipitation, Penn. State University, University Park, Pennsylvania, 1954-1960.

24. R. C. Adrian, in *Air Pollution Engineering Manual*, J. A. Danielson, Ed., U.S. Public Health Service Publ. No. 999-AP-40, 1967, pp. 156-166.

25. H. H. Jones, in *Air Sampling Instruments*, 2nd ed., Am. Conf. Govt. Ind. Hygienists, Cincinnati, 1962, pp. B-4-1 to B-4-22.

26. M. Lippmann, in *Air Sampling Instruments*, 3rd ed., Am. Conf. Govt. Ind. Hygienists, Cincinnati, 1966, pp. B-4-1 to B-4-22.

27. Predicasts, *Industrial Air Pollution*, Cleveland, Ohio, 1968.

28. National Center for Air Pollution Control, Request for Proposal PH 22-68-Neg. 50, Cincinnati, Ohio, 1968.

29. R. A. Strehlow, Rept. No. SO-1003, CFSTI, 1951.

30. W. J. Scheffy, Publ. No. COO-1016, CFSTI, 1956.

31. L. B. Loeb, *Electrical Coronas*, University of California Press, Berkeley, 1965.

32. J. D. Meek and J. M. Craggs, *Electrical Breakdown in Gases*, Oxford, 1953.

33. J. D. Cobine, *Gaseous Conductors*, McGraw-Hill, New York, 1941.

34. A. von Engel, *Ionized Gases*, Oxford, 1965.

35. L. B. Loeb, *Fundamental Processes of Electrical Discharge in Gases*, Wiley, New York, 1939; (a) p. 267.

36. L. B. Loeb, *Basic Processes of Gaseous Electronics*, University of California Press, Berkeley, 1955.

37. L. B. Loeb and J. M. Meek, *The Mechanism of the Electric Spark*, Stanford University Press, 1941; (a) pp. 94-95.

38. F. Llewellyn Jones, *Ionization and Breakdown in Gases*, Wiley, New York, 1957; (a) pp. 71-73; (b) pp. 98f, 113f.

39. F. G. Cottrell, Publ. 2307, U.S. Govt. Printing Office, Washington, 1914.
40. F. G. Cottrell, U.S. Patent 1,067,974, July 22, 1913.
41. M. Robinson, *Elec. Eng.*, **82**, 559 (1963).
42. M. Robinson, *J. Electrochem. Soc.*, **115**, 131C (1968).
43. W. A. Schmidt and J. R. Anderson, *Elec. Eng.*, **57**, 332 (1938).
44. H. J. White, Paper No. 57-35, Air Pollution Control Assoc., 1957.
45. W. Gilbert, *De Magnete*, English Transl. by S. P. Thompson, London, 1900, pp. 24–25.
46. Academie del Cimento, *Essayes of Natural Experiments*, English Transl. by R. Waller, London, 1684, pp. 128–132.
47. O. von Guericke, *Experimenta nova magdeburgica*, Amsterdam, 1672, pp. 147–150.
48. R. Boyle, *Experiments and Notes About the Mechanical Origine of Electricity*, London, 1675, pp. 13–14.
49. J. Priestly, *History and Present State of Electricity*, 5th ed., London, 1794, p. 6.
50. F. Hauksbee, *Physico-Mechanical Measurements*, London, 1709, pp. 46–47.
51. I. Newton, *Opticks*, 2nd ed., London, 1718, pp. 315–316.
52. J. A. Nollet, *Essai sur l'électricité des corps*, Paris, 1746.
53. A. H. Smith, *Writings of Benjamin Franklin*, Vol. 2, New York, 1905, p. 303.
54. W. Watson, *Trans. Roy. Soc. (London)*, **10**, 296 (1746).
55. G. Becarria, *Elettricismo artificiale*, Turin, 1772, par. 781, 783.
56. T. Cavallo, *A Complete Treatise of Electricity*, London, 1777, pp. 309–310.
57. C. A. de Coulomb, *Memoires de l'Académie des Sciences, Paris*, 1785, p. 612.
58. M. Hohlfeld, *Archiv gesamte Naturlehre*, **2**, 205 (1824).
59. M. Faraday, *Experimental Researches in Electricity*, Vol. I, London, 1839; (a) par. 1569; (b) par. 1595.
60. C. F. Guitard, *Mechanics Magazine*, **53**, 346 (1850).
61. O. J. Lodge, *Nature*, **28**, 297 (1883).
62. A. O. Walker, *Engineering*, **39**, 627 (1885).
63. L. B. Loeb, in *International Critical Tables*, Vol. 6, E. W. Washburn, Ed., McGraw-Hill, New York, 1929, pp. 110–114.
64. S. C. Brown and J. C. Ingraham, in *Handbook of Physics*, E. U. Condon and H. Odishaw, Eds., McGraw-Hill, New York, 1967, p. 4—181.
65. P. Cooperman, Rept. No. 54, Research-Cottrell, Inc., Bound Brook, N.J. 1953.
66.·S. S. Attwood, *Electric and Magnetic Fields*, Wiley, New York, 1949, p. 78.
67. J. S. Townsend, *Phil. Mag.*, [6] **28**, 83 (1914).
68. J. S. Townsend, *Electricity in Gases*, Oxford, 1915, pp. 375–376.
69. H. Prinz, *Arch. Elektrotech.*, **31**, 756 (1937).
70. A. Winkel and A. Schuetz, *Staub*, **22**, 343 (1962).
71. J. H. Simpson and A. R. Morse, *Natl. Res. Council Can.*, *Radio Elec. Eng. Div. Bull.*, **14** (1), 18 (1964).
72. M. Pauthenier and M. Moreau-Hanot, *J. Phys. Radium*, **6**, 257 (1935).
73. G. Mierdel and R. Seeliger, *Arch. Elektrotech.*, **29**, 149 (1935).
74. W. Deutsch, *Ann. Physik*, **10**, 847 (1931).
75. P. Cooperman, Paper No. CP 62-253, Am. Inst. Elec. Engrs., 1962.
76. P. Cooperman, Rept. No. 46, Research-Cottrell, Inc., Bound Brook, N.J., 1952.
77. P. Cooperman, *Trans. Am. Inst. Elec. Engrs.*, **79 I**, 47(1960).
78. J. S. Lagarius, *Trans. Am. Inst. Elec. Engrs. (Commun. Electron.)*, **78**, 427 (1959).
79. J. Dupuy, *Compt. Rend.*, **242**, 1140 (1956).
80. N. Troost, *Proc. Inst. Elec. Engrs. (London)*, **101**, 369 (1954).
81. J. Dupuy, *Compt. Rend.*, **242**, 2309 (1956).

82. W. Seidel and W. Kaufmann, *Staub*, **24**, 405 (1964).

83. H. Schnitzler, *Staub* (*English Transl.*), **25**, (3), 43 (1965).

84. H. Schnitzler, *Staub*, **23**, 78 (1963).

85. S. L. Soo and W. Rodgers, Paper No. 68-104, Air Pollution Control Assoc., 1968.

86. G. H. Horne, Unpublished rept., Western Precipitation Co., Los Angeles, 1921; through H. J. White, *Industrial Electrostatic Precipitation*, Addison-Wesley, Reading, Massachusetts, 1963, p. 220.

87. D. O. Heinrich, *Staub*, **24**, 131 (1964).

88. M. Robinson, *Power App. Systems*, **86**, 185 (1967).

89. D. W. Ver Planck, *Trans. Am. Inst. Elec. Engrs.*, **60**, 99 (1941).

90. J. B. Whitehead, in *International Critical Tables*, Vol. 6, E. W. Washburn, Ed., McGraw-Hill, New York, 1929, pp. 107–108.

91. W. M. Thornton, *Phil. Mag.*, **28**, 666 (1939).

92. E. E. Charlton and F. S. Cooper, *Gen. Elec. Rev.*, **40**, 438 (1937).

93. W. T. Sproull, Paper No. CP 57-46, Am. Inst. Elec. Engrs., 1957.

94. H. H. Landolt and R. Boernstein, *Zahlenwerte und Funktionen*, 6th ed., Vol. 4, Part 3, Springer, Berlin, 1957, pp. 105–203.

95. G. W. Penney and S. F. Craig, *Trans. Am. Inst. Elec. Engrs.*, **79 I**, 112 (1960).

96. T. I. Agnew and G. W. Penney, Paper No. CP 62-303, Am. Inst. Elec. Engrs., 1962.

97. G. W. Penney, Paper No. CP 63-526, Inst. Electric. Electron. Engrs., 1963.

98. L. S. Schmitz and G. W. Penney, *Inst. Elec. Electron. Engr. Trans. Power App. Systems*, **86**, 360 (1967).

99. M. Robinson, *Inst. Elec. Electron. Engr. Intern. Conv. Record*, Part 7, 73 (1965).

100. L. R. Solon, P. Lilienfeld, and H. J. DiGiovanni, Rept. No. TID-23860, CFSTI, 1966.

101. E. L. Coe, *J. Air Pollution Control Assoc.*, **17**, 724 (1967).

102. J. Weber and G. F. Goetz, *Siemens Rev.*, **30**, 20 (1963).

103. G. F. Goetz, E. Schwarz, and R. Schlitt, *Siemens Rev.*, **29**, 406 (1962).

104. S. Bartos, *Electrotech. Obz.*, **54**, 278 (1965); through *Chem. Abstr.*, **66**, 67159 (1967).

105. H. Schummer and W. Steinbauer, *Siemens Rev.*, **34**, 458 (1967).

106. L. L. Nagel, *3rd Conf. Rectifiers in Industry*, Inst. Electric. Electron. Engrs., 161 (1963).

107. E. Schwarz, R. Schlitt, and G. F. Goetz, *Siemens Rev.*, **29**, 313 (1962).

108. R. Schlitt and A. Goller, *Siemens-Z.*, **39**, 229 (1965).

109. A. Hoffmann and R. Zabel, *Siemens-Z.*, **39**, 1022 (1965).

110. S. Machat, *Power*, **111** (2), 64 (1967).

111. R. A. Edwards, J. B. Howarth, and F. C. Cole, *Colloq. Electrostatic Precipitators*, Inst. Elec. Engrs., London, 1965.

112. S. A. Stevens and C. F. Amor, *Colloq. Electrostatic Precipitators*, Inst. Elec. Engrs., London, 1965.

113. D. Walker and J. Vermeulen, *AEI Eng.* (*Assoc. Elec. Ind.*), **3**, 123 (1963).

114. H. J. White, *Trans. Am. Inst. Elec. Engrs. Commun. Electron.*, **71 I**, 326 (1952).

115. J. B. Thomas, T. R. Williams, and T. Suzuki, *Trans. Am. Inst. Elec. Engrs.*, **79 I**, 1 (1960).

116. J. B. Thomas and T. R. Williams, *Trans Am. Inst. Elec. Engrs.*, **79 I**, 136 (1960).

117. J. B. Thomas and T. R. Williams, in *Gas Discharges and the Electrical Supply Industry*, J. S. Forrest, P. R. Howard, and D. J. Littler, Eds., Butterworth, London, 1962, pp. 209–218.

118. E.-M. Koschany, *Staub* (*English Transl.*), **27** (4), 5 (1967).

119. M. Pauthenier, and M. Moreau-Hanot, *J. Phys. Rad.*, **3**, 590 (1932).

120. P. L. Smith and G. W. Penney, *Trans. Am. Inst. Elec. Engrs.*, *(Commun. Electron.)* **80 I,** 340 (1961).

121. H. J. White, *Trans. Am. Inst. Elec. Engrs.*, **70,** 1186 (1951).

122. G. W. Penney and R. D. Lynch, *Trans. Am. Inst. Elec. Engrs.*, **76 I,** 294 (1957).

123. G. W. Hewitt, *Trans. Am. Inst. Elec. Engrs.*, **76 II,** 300 (1957).

124. V. G. Drozin and V. K. LaMer, *J. Colloid Sci.*, **14,** 74 (1959).

125. Edmondson, H., Ph.D. thesis, University of Leeds, 1961.

126. G. Langer, Publ. No. ARF-3187-5, CFSTI, 1961.

127. E. T. Hignett, *Proc. Inst. Elec. Engrs.*, **114,** 1325 (1967).

128. B. Y. H. Liu and K. T. Whitby, Publ. No. TID-20020, CFSTI, 1963.

129. B. Y. H. Liu and K. T. Whitby, *2nd Progr. Rept. to U.S. At. Energy Comm.*, Contract No. AT(11-1)-1248, CFSTI, 1964.

130. B. Y. H. Liu and K. T. Whitby, Publ. No. COO-1248-4, CFSTI, 1965.

131. B. Y. H. Liu, K. T. Whitby, and H. H. S. Yu, Publ. No. COO-1248-9, 1966.

132. B. Y. H. Liu and K. T. Whitby, Rept. No. COO-1248-14, CFSTI, 1967.

133. B. Y. H. Liu and K. T. Whitby, Rept. No. COO-1248-17, CFSTI, 1968.

134. E. T. Hignett, Lab. Note RD/L/N 86/66, Cent. Elec. Res. Lab., Leatherhead, Surrey, England, 1966.

135. E. T. Hignett, *Colloq. Electrostatic Precipitators*, Inst. Elec. Engrs., London, 1965.

136. M. Pauthenier, *Compt. Rend.*, **240,** 1610 (1955).

137. P. Arendt and H. Kallmann, *Z. Physik*, **35,** 421 (1925).

138. J. Bricard, in *Problems of Atmospheric and Space Electricity*, S. C. Coroniti, Ed., Elsevier, Amsterdam, 1965.

139. R. Gunn, *J. Météorol.*, **11,** 339 (1954).

140. G. L. Natanson, *Soviet Phys.-Tech. Phys. (English Transl.)*, **5,** 538 (1960).

141. B. Y. H. Liu, K. T. Whitby, and H. H. S. Yu, *J. Colloid Interface Sci.*, **23,** 367 (1967).

142. B. Y. H. Liu, K. T. Whitby, and H. H. S. Yu, *J. Appl. Phys.*, **38,** 1592 (1967).

143. A. T. Murphy, Ph.D. thesis, Carnegie Institute of Technology, Pittsburgh, Pennsylvania, 1956.

144. A. T. Murphy, F. T. Adler, and G. W. Penney, *Trans. Am. Inst. Elec. Engr.*, **78,** 318 (1959).

145. R. Cochet, *Compt. Rend.*, **243,** 243 (1956).

146. R. Cochet, *Colloq. Intern. Centre Natl. Rech. Sci. (Paris)*, **102,** 331 (1961).

147. B. Y. H. Liu and H. C. Yeh, *J. Appl. Phys.*, **39,** 1396 (1968).

148. R. E. Armington, Ph.D. thesis, University of Pittsburgh, 1957.

149. W. W. Foster, *Brit. J. Appl. Phys.*, **10,** 206 (1959).

150. H. F. Kraemer and H. F. Johnstone, *Ind. Eng. Chem.*, **47,** 2426 (1955).

151. M. Pauthenier, Centre de Recherches sur la Pollution Atmosphérique, Paris, private communication, 1968.

152. J. C. Williams and R. Jackson, *Proc. Symp. Interaction Fluids Particles*, Inst. Chem. Engrs. (London), pp. 282–288, Discussion pp. 291–293, 297–298 (1962).

153. P. Cooperman, Paper No. CP 60-300, Am. Inst. Elec. Engrs., 1960.

154. P. Cooperman, Paper 65-132, Air Pollution Control Assoc., 1965.

155. P. Cooperman, Paper 66-124, Air Pollution Control Assoc. 1966.

156. P. Cooperman, *Atmos. Environ.*, in press.

157. S. L. Soo, *Fluid Dynamics of Multiphase Systems*, Blaisdell, Waltham, Massachusetts, 1967, p. 51ff.

158. N. A. Fuchs, *Mechanics of Aerosols*, Macmillan, New York, 1964, pp. 25–29.

159. M. Pauthenier, *Compt. Rend.*, **240,** 1761 (1955).

160. H. J. White, *Colloq. Intern. Centre Natl. Rech. Sci. (Paris)*, **102**, 37 (1961).
161. E. Anderson, unpublished rept., Western Precipitation Co., Los Angeles, 1919.
162. W. Deutsch, *Ann. Physik*, **68**, 335 (1922).
163. H. J. White, *Ind. Eng. Chem.*, **47**, 932 (1955).
164. M. Robinson, *Atmos. Environ.*, **1**, 193 (1967).
165. M. Robinson, *Anal. Chem.*, **33**, 109 (1961).
166. C. Allander and S. Matts, *Staub*, **52**, 738 (1957).
167. S. Masuda, *Staub (English Transl.)*, **26** (11), 6 (1966).
168. J. Dalmon and H. J. Lowe, *Colloq. Intern. Centre Natl. Rech. Sci. (Paris)*, **102**, 363 (1961).
169. W. H. Cole, Rept. No. PRJ 63-5, Research-Cottrell, Inc., Bound Brook, N. J., 1963.
170. R. F. Heinrich, *Trans. Inst. Chem. Engrs.*, **39**, 145 (1961).
171. G. W. Seman and G. W. Penney, *Trans. Inst. Elec. Electron. Engr. Power App. Systems*, **86**, 365 (1967).
172. G. W. Seman and G. W. Penney, *Inst. Elec. Electron. Engr. Intern. Conv. Record*, Part 7, **13**, 69 (1965).
173. R. A. Bagnold, *Intern J. Air Pollution*, **2**, 357 (1960).
174. C. J. Stairmand and J. C. Williams, *Trans. Inst. Chem. Engrs.*, **39**, 145 (1961).
175. S. K. Friedlander, *Chem. Eng. Progr. Symp. Ser.*, **55**, 135 (1959).
176. D. O. Heinrich, *Staub*, **23**, 83 (1963).
177. H. Brandt, *Staub*, **23**, 378 (1963).
178. P. Cooperman, Fairleigh-Dickenson Univ., Teaneck, N.J., private communication, 1968.
179. N. V. Inyushkin and Ya. D. Averbukh, *Soviet J. Non-Ferrous Metals (English Transl.)*, **1962**, 35.
180. N. V. Inyushkin and Ya. D. Averbukh, *Izv. Vyssihikh Uchebn. Zavedenii, Khim. i Khim. Tekhnol.*, **6**, 1031 (1963).
181. J. W. Thomas and D. Rimberg, *Staub (English Transl.)*, **27**, 18 (1967).
182. M. Robinson, *J. Air Pollution Control Assoc.*, **18**, 235 (1968).
183. J. M. Hughes, J. J. Stukel, and S. L. Soo, Paper 68-103, Air Pollution Control Assoc., 1968.
184. M. Robinson, *J. Air Pollution Control Assoc.*, **18**, 688 (1968).
185. R. Pistor, *Staub (English Transl.)*, **26** (4), 41 (1966).
186. M. Robinson, Rept. No. PRJ 68-2, Research-Cottrell, Inc., Bound Brook, N.J., 1968.
187. M. Robinson, *Am. J. Phys.*, **30**, 366 (1962).
188. A. V. Obermayer and M. von Pichler, *Sitzber. Wien. Akad. Wiss.*, **93 II**, 408, 925 (1886); through R. Ladenburg and W. Tietze, *Ann. Physik*, **6**, 581 (1930).
189. F. G. Cottrell, U.S. Patent 895,729 (August 11, 1908).
190. R. Ladenburg and W. Tietze, *Ann. Physik*, **6**, 581 (1930).
191. M. Robinson, *Trans. Am. Inst. Elec. Engrs., Commun. Electron.*, **80 I**, 143 (1961).
192. W. Deutsch, *Ann. Physik*, **9**, 249 (1931).
193. E. Anderson, *Physics*, **3**, 23 (1932).
194. G. Mierdel and R. Seeliger, *Naturwiss.* **19**, 753, (1931).
195. C. Scheidel and H. G. Eishold, *Elektrotech. Z.*, **A79**, 953 (1958).
196. J. N. Chubb, W. D. Bamford, and J. B. Higham, *Colloq. Electrostatic Precipitators*, *Inst. Elec. Engrs.*, London, 1965.
197. G. W. Penney, *J. Air Pollution Control Assoc.*, **17**, 588 (1967).
198. G. W. Penney, and E. H. Klinger, *Trans. Am. Inst. Elec. Engrs., Commum. Electron.*, **81**, 200 (1962).

199. M. Corn and F. Stein, *Am. Ind. Hyg. Assoc. J.*, **26**, 325 (1966).

200. M. Corn, in *Aerosol Science*, C. N. Davies, Ed., Academic Press, New York, 1966, pp. 359–392.

201. R. A. Bagnold, *Physics of Blown Sand and Desert Dunes*, Methuen, London, 1941.

202. G. W. Penney and R. E. Probst, *Inst. Electric. Electron. Engrs. Intern. Conv. Record*, Part 4, 164 (1964).

203. G. W. Penney, *Arch. Environ. Health*, **4**, 301 (1962).

204. J. M. Niedra and G. W. Penney, *Intern. Conv. Record, Inst. Electric Electron. Engrs.*, Part 7, 88 (1965).

205. G. W. Penney, Paper No. 67-WA/APC-1, Am. Soc. Mech. Engrs., 1967.

206. M. Sarna, *Energetica*, **22** (4) 124 (1968); through *Elec. Electron. Abstr.*, **71**, 22998 (1968).

207. T. Hiramatsu and Y. Kobori, *Hitachi Rev.*, **45**, 130 (1963); through *Air Pollution Control Assoc. Abstr.*, **10**, 6538 (1965).

208. W. T. Sproull, *J. Air Pollution Control Assoc.*, **15**, 50 (1965).

209. D. H. Lucas, *J. Inst. Fuel*. **36**, 203 (1963).

210. O. Guepner, *Staub*, **23**, 478 (1963).

211. F. W. Schmitz, Rept. No. EDR 63-3, Research-Cottrell, Inc., Bound Brook, N.J. 1963.

212. C. C. Shale, W. S. Bowie, J. H. Holden, and G. R. Strimbeck, Rept Invest. No. 6325, U.S. Bur. Mines, 1963.

213. C. C. Shale, W. S. Bowie, J. H. Holden, and G. R. Strimbeck, Rept. Invest. No. 6397, U.S. Bur. Mines, 1964.

214. Combustion Power Co., Inc., Palo Alto, Calif. CPU-400 Tech. Proposal to U.S. Dept. HEW, 1967.

215. M. W. Thring and W. Strauss, *Trans. Inst. Chem. Engrs.*, **41**, 248 (1963).

216. M. Robinson, Rept. No. PRJ 64-4, Research-Cottrell, Inc., Bound Brook, N.J., 1964.

217. R. J. Rosa and F. A. Hals, *Ind. Res.*, **10** (6), 68 (1968).

218. W. Strauss and B. W. Lancaster, *Atmos. Environ.*, **2**, 135 (1968).

219. M. Robinson and R. F. Brown, Rept. No. PRJ 61-8, Research-Cottrell, Inc., Bound Brook, N.J., 1961.

220. L. M. Roberts and R. L. Chamberlin, *1st Europ. Symp. Cleaning Coke Oven Gas*, Europ. Fed. Chem. Eng., 1963.

221. A. B. Walker, *Symp. Proc. Metropolitan Engrs. Council Air Resources*, New York, 12 (1967).

222. R. F. Bovier, *Proc. Am. Power Conf.*, **26**, 138 (1964).

223. W. G. Henke, *Air Pollution Control News*, No. 3, Research-Cottrell, Inc., Bound Brook, N.J., 1968.

224. W. J. Cahill, *Proc. Metropolitan Engrs. Counc. Air Pollution*, New York, 74 (1967).

225. H. J. Hall, R. F. Brown, J. B. Eaton, and C. W. Brown, *Oil Gas J.*, **66** (37), 109 (1968).

226. E. Anderson, *Pacific Coast Gas Assoc. Proc.*, **30**, 142 (1939); *Gas Age*, **83** (13), 23 (1939).

227. F. E. Vandeveer, in *Gas Engineer's Handbook*, C. G. Segeler, Ed., Industrial Press, New York, 1965, pp. 9/129–9/130.

228. A. Hersiezky and A. Bendes, *Az Erjedésipari Kutató Intézet Közlemenyei*, 53 (1959); through *Fuel Abstr. Current Titles*, **1**, 48 (1960).

229. H. J. Hall, *Chem. Eng. Progr.*, **59**, 67 (1963).

230. P. Elwood, *Chem. Eng.*, **72** (2), 54 (1965).

231. W. A. Schmidt, *J. Ind. Eng. Chem.*, **4**, 719 (1912); *Orig. Commun. 8th Intern. Congr. Appl. Chem.*, **5**, 117 (1912).
232. Z. Iwasaki and S. Komelji, *Fuji Elec. J.*, **35**, 1013 (1962); through *Air Pollution Control Assoc. Abstr.*, **9**, 5768 (1964).
233. Research-Cottrell, Inc., Bound Brook, N.J., private communication, 1968.
234. M. Robinson, "Critical Pressures of the Positive Corona Discharge Between Concentric Cylinders in Air," *J. Appl. Phys.*, in press.
235. I. Goldman and B. Wul, *Tech. Phys. USSR*, **1**, 497 (1934).
236. I. Goldman and B. Wul, *Tech. Phys. USSR*, **3**, 16 (1936).
237. I. Goldman and B. Wul, *Tech. Phys. USSR.* **5**, 355 (1938).
238. A. H. Howell, *Trans. Am. Inst. Elec. Engr.*, **58**, 193 (1939).
239. D. Berg and C. N. Works, *Trans. Am. Inst. Elec. Engrs.*, **77 III**, 820 (1958).
240. H. C. Pollack and F. S. Cooper, *Phys. Rev.*, **56**, 170 (1939).
241. T. R. Foord, *Proc. Inst. Elec. Engr. (London)*, **100 II**, 585 (1953).
242. A. Kusco, *D.Sc. thesis*, Massachusetts Institute of Technology, Cambridge, Massachusetts, 1951.
243. A. Boulloud, *Compt. Rend.*, **242**, 2542 (1956).
244. E. Uhlmann, *Arch. Elektrotech.*, **33**, 323 (1929).
245. F. W. Peek, *Dielectric Phenomena in High Voltage Engineering*, 3rd ed., McGraw-Hill, New York, 1929, p. 29.
246. M. Robinson, "Positive and Negative Coaxial Cylindrical Corona in Air at High Pressure," to be published.
247. M. Robinson, "Critical Pressures of the Negative Corona Discharge Between Concentric Cylinders in Air," to be published.
248. J. G. Trump, R. W. Cloud, J. G. Mann, and E. P. Hanson, *Elec. Eng.*, **61**, 961 (1950).
249. C. C. Shale, Paper No. 66-125, Air Pollution Control Assoc., 1966; abridged in *J. Air Pollution Control Assoc.*, **17**, 159 (1967).
250. C. C. Shale, *Combustion*, **35** (10), 42 (1964).
251. C. C. Shale and A. S. Moore, *Combustion*, **32** (2), 42 (1960).
252. C. C. Shale, Paper No. 64-8, Air Pollution Control Assoc., 1964.
253. C. C. Shale, U.S. Bur. Mines Inform. Circ. No. 8353, 1967.
254. P. Cooperman, Paper No. CP 63-172, Am. Inst. Elec. Engr., 1963.
255. C. C. Shale and J. H. Holden, Paper No. ES-THU-5, *Inst. Electric. Electron. Engrs. Ind. Gen. Appl. Conf. Record*, 1968.
256. C. C. Shale, U.S. Bur. Mines Coal Res. Center, Morgantown, W.Va., private communication, 1968.
257. C. C. Shale and G. E. Fasching, *U.S. Bur. Mines Rept. Invest.*, to be published.
258. J. B. Thomas and E. Wong, *J. Appl. Phys.*, **29**, 1226 (1958).
259. L. Frommhold, *Fortschr. Physik*, **12**, 597 (1964).
260. L. B. Loeb, University of California, Berkeley, private communication, 1967.
261. L. R. Koller and H. A. Fremont, *J. Appl. Phys.*, **21**, 741 (1950).
262. P. Cooperman, *Commun. Electron.*, **75**, 792 (1964).
263. M. Pauthenier, *Compt. Rend.*, **222**, 1219 (1946).
264. W. Simm, *Chem.-Ing. Tech.*, **31**, 43 (1959).
265. N. T. Dzoanh, *Compt. Rend.* **250**, 1001 (1960).
266. M. Robinson, "Suppression of Back Corona in Compressed Air," to be published.
267. H. Hesselbrock, *Staub (Engl. Transl.)*, **25**, 32 (1965).
268. S. Franck, *Z. Physik*, **87**, 323 (1934).
269. H. G. Eishold, *Staub (Engl. Transl.)*, **26** (1), 14 (1966).
270. W. Loszek, *Proc. Intern. Clean Air Congr., Part I*, London, 105 (1966).

271. C. C. Shale, J. H. Holden, and G. E. Fasching, U.S. Bur. Mines Rept. Invest. No. 7041, 1968.
272. H. Loquenz, Staub (English Transl.), 27 (5), 41 (1967).
273. S. W. Randolph, J. Air Pollution Control Assoc., 8, 249 (1958).
274. J. Tůma, Staub (English Transl.), 26 (11), 1 (1966).
275. K. Bielanski, Energetyka, 22 (4), 136 (1968); through Electric. Electron. Abstr., 71, 22996 (1968).
276. S. Masuda, Electrotech. J. Japan, 7 (3), 108 (1962).
277. L. Cohen and R. W. Dickinson, J. Sci. Instrum., 40, 72 (1963).
278. P. Cooperman, Fairleigh-Dickenson Univ., Teaneck, N.J., private communication, 1965.
279. J. Boehm, in M. Tomaides J. Air Pollution Control. Assoc., 18, 681 (1968).
280. H. Klemperer and J. E. Sayers, Trans. Am. Soc. Mech. Engrs., 78, 317 (1956).
281. J. Katz, J. Air Pollution Control Assoc., 15, 525 (1965).
282. H. Hesselbrock, Energie, 16, 497 (1964).
283. D. A. Douglass and G. W. Penney, Inst. Electric. Electron. Engrs. Ind. Gen. Appl. Conf. Record, 585 (1967).
284. R. F. Heinrich, Mitt. Ver. Grosskesselbesitzer, 68, 322 (1960).
285. W. Barth, Staub, 24, 441 (1964).
286. R. L. Chamberlin and E. J. Malarkey, Ind. Coal Conf., Purdue Univ., 1963.
287. H. J. Ochs, Wasser Luft Betrieb, 11, 471 (1967).
288. C. Cederholm, Proc. Clean Air Congr., Part 1, London, 122 (1966).
289. R. Flossmann and A. Schuetz, Staub, 23, 433 (1963).
290. T. E. Kreichelt, D. A. Kemnitz, and S. T. Cuffe, U.S. Public Health Serv. Publ. No. 999-AP-17 (1967).
291. J. E. Sayers, J. Inst. Fuel, 33, 542 (1960).
292. F. Dransfield and H. J. Lowe, in Gas Purification Processes, G. Nonhebel, Ed., George Newnes, London, 1964, pp. 536–549.
293. C. J. Stairmand, in Gas Purification Processes, G. Nonhebel, Ed., Newnes, London, 1964, ch. 12.
294. R. O. Blosser and H. B. H. Cooper, Paper Trade J., 151 (11), 46 (1967).
295. E. J. Malarkey and C. Rudosky, Paper Trade J., 152 (40), 57 (1968).
296. R. O. Blosser and H. B. H. Cooper, Atmos. Pollution Tech. Bull. 32, Natl. Council Stream Improvement, New York, 1967.
297. G. A. Hansen, J. Air Pollution Control Assoc., 12, 409 (1962).
298. M. G. Teplitskii, I. Z. Gordon, G. A. Aleinikov, N. A. Kudryavaya, and A. N. Pyatigorskii, Stal', 28, 756 (1968); through Chem. Abstr., 69, 88971 (1968).
299. R. O. Blosser and H. B. H. Cooper, Tappi, 51 (5), 73A (1968).
300. H. J. Lowe, J. Dalmon, and E. T. Hignett, Colloq. Electrostatic Precipitators, Inst. Elec. Engrs., London, 1965.
301. N. T. Dzoanh, Abstract B-8, 21st Gaseous Electronics Conf., 1968; to be published.
302. M. Pauthenier, Colloq. Intern. Centre Natl. Rech. Sci., 102, 279 (1961).
303. R. Reffay, Colloq. Intern. Centre Natl. Rech. Sci., 102, 255 (1961).
304. T. Nhan, D.Sc. thesis, University of Paris, 1958.
305. J. Dupuy, Compt. Rend., 244, 1737 (1957).
306. M. Pauthenier, Compt. Rend, 247, 187 (1958).
307. T. Nhan, Compt. Rend., 246, 3028 (1958).
308. M. Pauthenier, Colloq. Intern. Centre Natl. Rech. Sci., 102, 263 (1961).
309. M. Pauthenier, Proc. 1st Natl. Conf. Aerosols, Liblice near Prague, 53 (1962).
310. F. Seitz, Modern Theory of Solids, 1st ed., McGraw-Hill, New York, 1940, p. 191.

311. S. Masuda, T. Onishi, and H. Saito, *Ind. Eng. Chem., Proc. Design Develop.*, **5**, 135 (1966).

312. S. Masuda, U.S. Patent 3,137,445, June 16, 1964.

313. N. A. Olesov, *Tsement*, **34** (3), 7 (1968); through *Chem. Abstr.*, **69**, 88324 (1968).

314. K. Darby and D. O. Heinrich, *Staub (English Transl.)* **26** (11), 12 (1966).

315. S. Masuda, *Staub (English Transl.)*, **25** (5), 1 (1965).

316. S. Masuda, *Bull. Dept. Elec. Eng. Univ. Tokyo*, **11**, 1 (1963).

317. J. Coutaller and C. Richard, *Pollution Atm.*, **9** (33), 9 (1967).

318. K. Darby and K. R. Parker, Brit. Patent 933,286, August 8, 1963.

319. Lodge Cottrell, Ltd., French Patent 1,445,982, July 15, 1966.

320. A.-G. Metallgesellschaft, French Patent 1,448,415, August 5, 1966.

321. J. T. Reese and J. Greco, *J. Air Pollution Control Assoc.*, **18**, 523 (1968).

322. W. A. Baxter, *J. Air Pollution Control Assoc.*, **18**, 817 (1968).

323. N. Orne, U.S. Patent 3,284,990, November 15, 1966.

324. H. G. T. Busby and K. Darby, *J. Inst. Fuel*, **36**, 184 (1963).

325. J. F. Pottinger, *Australian Chem. Proc. Eng.*, **20**, (2), 17 (1967).

326. J. B. Kirkwood, *Proc. Clean Air Conf.*, *(Sydney)*, Paper No. 14 (1962).

327. K. S. Watson and K. J. Blecher, *Intern. J. Air Water Pollution*, **10**, 573 (1966).

328. K. S. Watson, B. P. Flanagan, and J. Blecher, *Colloq. Electrostatic Precipitators*, Inst. Elec. Engrs. (London), 1965.

329. J. P. Carey, R. G. Ramsdell, C. F. Soutar, and W. B. White, *Proc. Am. Power Conf.*, **29**, 495 (1967).

330. S. Fukuma and K. Kamei, *Jap. Chem. Quart.*, **4** (3), 12 (1968).

331. S. Ludwig, *Chem. Eng.*, **75** (3), 70 (1968).

332. B. D. Pfoutz, Final Rept. Proj. 812–33, Research-Cottrell, Inc., Bound Brook, N.J., 1967.

333. Tennessee Valley Authority, Publ. No. PB 178972, CFSTI, 1968.

334. S. T. Pretorius and W. G. B. Mandersloot, *Powder Technol.*, **1**, 129 (1967).

335. R. G. Ramsdell, *Proc. Am. Power Conf.*, **30** (1968).

336. H. L. Engelbrecht, *Proc. Am. Power Conf.*, **28**, 516 (1966); *Air Eng.*, **8** (8), 20 (1966).

337. M. J. Archbold, *Proc. Am. Power Conf.*, **23** 371 (1961).

338. F. E Gartrell, *Proc. Am. Power Conf.*, **27**, 117 (1965).

339. A. M. Squires, *Acqua Ind.*, **9**, 13, 17 (1967); through *Chem. Abstr.*, **69**, 2807 (1968).

340. Federal Power Commission, *Air Pollution and the Regulated Electric Power and Natural Gas Industries*, Washington, 1968; (a) p. 3; (b) p. 24.

341. W. T. Sproull, Paper No. 63-57, Air Pollution Control Assoc., 1963.

342. M. Robinson, Rept. No. TM 68-7, Research-Cottrell, Inc., Bound Brook, N.J., 1968.

343. Ye. M. Bulabanov, *Elektrichestvo*, No. 2, 57 (1965); *Elec. Technol. USSR (English Transl.)* **1**, 109 (1965).

344. K. Darby, *Colloq. Electrostatic Precipitators*, Inst. Elec. Engrs., London, 1965.

345. H. J. Hall, *Power Eng.*, **70** (1), 44 (1966).

346. G. G. Schneider, *Proc. Intern. Clean Air Congr., Part I*, London, 149 (1966).

347. W. W. Moore, *Natl. Conf. Air Pollution*, Washington, 1966.

348. H. F. Lund, *U.S. Public Health Serv. Publ. No. 1649*, 207 (1966).

349. H. J. Ochs, *Wasser Luft Betrieb*, **10**, 473 (1966).

350. G. S. Chekanov and L. P. Yanovskii, *Elek. Sta.*, **39** (5), 13 (1968); through *Chem. Abstr.*, **69**, 44757 (1968).

351. K. Schwartz, *Proc. Intern. Clean Air Congr., Part I*, London, 136 (1966).
352. M. N. Magnus, *J. Air Pollution Control Assoc.*, **15**, 149 (1965).
353. H. Brandt, *Staub* (*English Transl.*), **25** (10), 23 (1965).
354. S. Maartman, *Proc. Intern. Clean Air Congr., Part I*, London, 131 (1966).
355. H. Knecht, W. W. Moore, and F. W. Schmitz, *Proc. Am. Power Conf.*, **28**, 525 (1966).
356. R. T. Mitchell, Iron Steel Inst. (London), Spec. Rept. No. 83, 80 (1963).
357. H. B. Lloyd and N. P. Bacon, Iron Steel Inst. (London), Spec. Rept. No. 83, 65 (1963).
358. W. A. Dickenson and J. L. Worth, *J. Metals*, **17**, 261 (1965).
359. W. A. Dickenson and J. L. Worth, *Open Hearth Proc.*, **47**, 214 (1965).
360. A. Jackson, Iron Steel Inst., London, Spec. Rept. 83, 61 (1963).
361. K. Guthermann, *Staub* (*Engl. Transl.*) **26** (4), 13 (1966).
362. E. R. Watkins and K. Darby, Iron Steel Inst. (London), Spec. Rept. No. 83, 24 (1963).
363. H. Schnitzler, *Staub*, **24**, 201 (1964).
364. G. Graue and R. Flossmann, *Staub*, **23**, 485 (1963).
365. R. Meldau, *Krupp Tech. Rev.*, **22** (11), 99 (1964).
366. V. S. Kulyaka, *Stal'*, **26**, 961 (1966); through *Chem. Abstr.*, **66**, 12488 (1967).
367. H. B. Boyce, Iron Steel Inst. (London), Spec. Rept. No. 83, 48 (1963).
368. K. Ussleber, *Giesserei*, **52**, 194 (1965).
369. A. Archer, *Proc. Intern. Clean Air Congr.*, Part I, London, 99 (1962).
370. T. Wada, *Netsu to Keiei*, **14**, 8 (1965); through *Air Pollution Control Assoc. Abstr.*, **11**, 7296 (1966).
371. G. Punch, *Steel Intern.*, **3** (12), 8 (1967).
372. C. M. Parker, *J. Air Pollution Control Assoc.*, **16**, 446 (1966).
373. P. S. Khomutinnikov and Iu. M. Dronevich, *Steel* (*USSR*) (*English Transl.*), 529 (1960).
374. D. H. Wheeler, *J. Air Pollution Control Assoc.*, **18**, 98 (1968).
375. K. R. Parker, *Chem. Proc. Eng.*, **44**, 505 (1963).
376. K. E. Blessing and D. Hysinger, *Chem. Eng. Progr.*, **59** (3), 60 (1963).
377. M. G. Dronsek and R. Pohle, *Giesserei*, **50**, 181 (1963).
378. W. F. Hammond and J. T. Nance, in *Air Pollution Engineering Manual*, J. A. Danielson, Ed., U.S. Public Health Service Publ. No. 999-AP-40, 1967, pp. 258–270.
379. W. F. Hammond, J. T. Nance, and K. D. Luedtke, in *Air Pollution Engineering Manual*, J. A. Danielson, Ed., U.S. Public Health Service Publ. No. 999-AP-40, 1967, pp. 241–257.
380. J. L. Sullivan and R. P. Murphy, *Proc. Intern. Clean Air Conf., London*, Part 1, 144 (1966).
381. R. A. Herrick, J. W. Olsen, and F. A. Ray, *J. Air Pollution Control Assoc.*, **16**, 7 (1966).
382. H. C. Henschen, *J. Air Pollution Control Assoc.*, **18**, 338 (1968).
383. M. Borenstein, *Ind. Heating*, **34**, 1646 (1967).
384. A. T. Lawson, Proc. Clean Air Conf., University of New South Wales, Vol. 2, Paper No. 11, 1965.
385. J. Katz, *Iron Steel Engr.*, **41** (5), 124 (1964).
386. W. Sebasta, in *Air Pollution*, Vol. 3, 2nd ed., A. C. Stern, Ed., Academic Press, New York, pp. 143–169.
387. G. Graue, R. Flossmann, and H. Schacky, *Staub* (*English Transl.*) **25** (10), 19 (1965).
388. R. Ziemendorff, *Zement-Kalk-Gips*, **19**, 171 (1966).
389. M. Tomaides, *Proc. Intern. Clean Air Congr.*, Part I, London, 125 (1966).

390. G. Funke and H. Fischer, *Zement-Kalk-Gips*, **20**, 146 (1967).

391. R. E. Doherty, U.S. Public Health Serv. Publ. 1649, 242 (1966).

392. W. Koehler, *Proc. Intern. Clean Air Congr.*, *London*, Part I, 114 (1966); *Wasser Luft Betrieb*, **11**, 155 (1967).

393. A. B. Walker, *J. Air Pollution Control Assoc.*, **13**, 622 (1963).

394. S. R. Cooper and C. F. Haskell, *Paper Trade J.*, **151** (13), 58 (1967).

395. R. W. Kittle, U.S. Public Health Serv. Publ. 1649, 232 (1966).

396. P. A. Podosinken, A. I. Postoronko, A. P. Grizodub, Z. P. Kal'na, and A. G. Lyapina, *Khim. Prom., Nauk.-Tekhn. Zb.*, **1963** (3), 82; through *Chem. Abstr.*, **60**, 5195 (1964).

397. C. H. Johnson, *Pulp Paper*, **42** (15), 31 (1968).

398. V. P. Owens, *Paper Trade J.*, **152** (33) 52 (1968).

399. R. Skrzys and A. Sliwa, *Rudy Metale Niezelazne*, **11**, 519 (1966); through *Chem. Abstr.*, **66**, 68672 (1967).

400. W. F. Hammond, J. T. Nance, and E. F. Spencer, in *Air Pollution Engineering Manual*, J. A. Danielson, Ed., U.S. Public Health Service Publ. No. 999-AP-40, 1967, pp. 270–283.

401. A. Lange and W. Trinks, *Neue Huette*, **12**(2), 81 (1967); through *Air Pollution Control Assoc. Abstr.*, **14**, 10558 (1968).

402. V. F. Denisov and K. I. Gorbenko, *Sb. Nauchn. Tr., Gos. Nauch.-Issled. Inst. Tsvet. Metal.*, **24**, 97 (1966); through *Chem. Abstr.*, **66**, 31129 (1967).

403. K. E. Savraeva and M. F. Bogatyrev, *Ref. Zh., Met.*, **1963**, Abstr. No. 2G217; through *Chem. Abstr.*, **59**, 6892 (1963).

404. E. O. Foster, *Trans. Can. Inst. Mining Met.*, **66**, 245 (1963); *Can. Mining Met. Bull.*, **56**, 469 (1963).

405. W. F. Hammond and H. Simon, in *Air Pollution Engineering Manual*, J. A. Danielson, Ed., U.S. Public Health Service Publ. No. 999-AP-40, 1967, pp. 284–292.

406. W. Jessnitz, *Wasser Luft Betrieb*, **6**, 169 (1962).

407. J. E. Tress, T. T. Campbell, and F. E. Block, U.S. Bur. Mines Rept. Invest. No. 6835 (1966).

408. K. Stopperka, *Staub*, **25**, 508 (1965).

409. R. J. MacKnight and S. T. Cuffe, in *Air Pollution Engineering Manual*, J. A. Danielson, Ed., U.S. Public Health Service Publ. No. 999-AP-40, 1967, pp. 695–701.

410. E. R. Hendrickson and J. S. Lagarias, *Proc. Intern. Clean Air Congr.* (*London*), Part I, 97 (1966).

411. J. L. Smith and H. A. Snell, *Chem. Eng. Progr.*, **64**, 60 (1968).

412. E. F. Spencer and R. M. Ingels, in *Air Pollution Engineering Manual*, J. A. Danielson, Ed., U.S. Public Health Service Publ. No. 999-AP-40, 1967, pp. 701–704.

413. P. Sutton, *Chem. Process Eng.*, **49**, 96 (1968).

414. R. Ackerman, *Rock Prod.*, **68** (10), 64 (1965).

415. M. A. Kornev and S. A. Meliksetyan, *Ref. Zh. Khim.*, **1964**, Abstr. 13P187; through *Chem. Abstr.*, **62**, 6319 (1965).

416. J. Petroll and B. Dressler, *Chem. Tech.*, **18**, 21 (1966).

417. C. R. Bruce, I. D. Johnson, and R. H. Reitsema, U.S. Patent 3,395,193 (July 30, 1968).

418. N. A. Kornev and E. V. Kirsh, *Ref. Zh. Khim.*, **1963**, Abstr. No. 21271; through *Chem. Abstr.*, **59**, 3568 (1963).

419. E. Ya. Stetsenko, *Koks i Khim.*, **1965** (5), 50; through *Chem. Abstr.*, **63**, 9470 (1965).

420. S. I. Smol'yaninov, K. K. Stramkovskaya, A. P. Smirnov, N. F. Olitskii, and S. A. Kvashnin, *Izv. Tomskogo Politekhn. Inst.*, **126**, 91 (1964); through *Chem. Abstr.*, **63**, 16092 (1965),

421. E. P. Stastny, *Chem. Eng. Progr.*, **62**, 47 (1966).
422. G. M. A. Aliev, A. E. Gonik, N. F. Bugaev, and L. M. Meitin, *Ogneupory*, **32** (9), 11 (1967); through *Chem. Abstr.*, **68**, 14352 (1968).
423. L. Barka and O. S. Privett, *Lipids*, **1**, 104 (1966).
424. C. Scheidel, *Dechema Monograph*, **52**, 229 (1964).
425. L. Raichle and G. John, *Staub* (*English Transl.*), **25** (4), 1 (1965).
426. W. F. Bixby, *Air Eng.*, **10** (8), 22 (1968).
427. R. W. Sickles, *Chem. Eng.*, **75** (22), 156 (1968).
428. H. Mortensen, *Elektroteknikeren*, **61** (9), 171 (1965); through *Elec. Electron. Abstr.*, **68**, 18714 (1965).
429. G. Schiemann, *Brennstoff-Waerme-Kraft*, **19**, 440 (1967).
430. M. Andritzky, *Brennstoff-Waerme-Kraft*, **19**, 436 (1967).
431. B. Brancato, *Fumi/Poveri*, **7** (4), 70 (1967); through *Air Pollution Control Assoc. Abstr.*, **13**, 8791 (1967).
432. C. A. Rogus, *Public Works*, **97** (6), 100 (1966).
433. E. R. Kaiser, *J. Air Pollution Control Assoc.*, **16**, 324 (1968).
434. A. B. Walker, *Proc. Natl. Incinerator Conf.*, Am. Soc. Mech. Engrs., New York, 1964.
435. J. A. Fife, U.S. Public Health Serv. Publ. 1649, 317 (1966).
436. Anon., *Modern Power and Engineering*, **61** (7), 68 (1967).
437. A. B. Walker, *Architectural Eng. News*, **9** (7), 71 (1967).
438. R. L. Bump, *J. Air Pollution Control Assoc.*, **18**, 803 (1968).
439. A. B. Walker, *Am. City*, **79** (9), 148 (1964).
440. C. A. Rogus, *Power*, **111** (12), 81 (1967).
441. F. M. Alpiser, *Air Eng.*, **10** (11), 18 (1968).
442. W. L. Polyglase, H. F. Dey, and R. T. Walsh, in *Air Pollution Engineering Manual*, J. A. Danielson, Ed., U.S. Public Health Service Publ. No. 999-AP-40, 1967, pp. 746–760.
443. R. B. O'Brien, Publ. No. TID-7677, CFSTI, 1963.
444. W. C. Bailor, in *Air Pollution Engineering Manual*, J. A. Danielson, Ed., U.S. Public Health Service Publ. No. 999-AP-40, 1967, pp. 792–798.
445. L. Silverman, *Proc. Symp. Pract. Treat. Low-Intermed.-Level Radioactive Waste, Vienna*, 89 (1965).
446. M. Kawano, in *Natural Radiation Environment*, J. A. Adams and W. M. Louber, Eds., University of Chicago Press, 1964, pp. 291–314.
447. R. Richard-Foy, *Ind. At.*, **8** (11), 117 (1964).
448. G. Thomas, in *Air Pollution Engineering Manual*, J. A. Danielson, Ed., U.S. Public Health Service Publ. No. 999-AP-40, 1967, pp. 401–410.
449. J. Goldfield and R. G. McAnlis, *Am. Ind. Hyg. Assoc. J.*, **24**, 411 (1963).
450. S. M. Weiss, in *Air Pollution Engineering Manual*, J. A. Danielson, Ed., U.S. Public Health Service Publ. No. 999-AP-40, 1967, pp. 378–383.
451. D. T. King, *Mining Eng.*, **19** (8), 64 (1967).
452. H. E. Chatfield, in *Air Pollution Engineering Manual*, J. A. Danielson, Ed., U.S. Public Health Service Publ. No. 999-AP-40, 1967, pp. 390–393.
453. G. Schiele, *Aluminium*, **43**, 171 (1967).
454. D. J. Sutton, H. A. Cloud, P. E. McNall, K. M. Nodolf, and S. H. McIver, *ASHRAE J.*, **6** (6), 55 (1964).
455. B. Ando, *Clean Air* (*Tokyo*) **4** (3), 9 (1966); through *Air Pollution Control Assoc. Abstr.*, **13**, 8449 (1967).
456. K. Wagner, *Giesserei*, **54**, 150 (1967).

457. H. Ishii, *Clean Air (Tokyo)*, **4** (3), 18 (1966); through *Air Pollution Control Assoc. Abstr.*, **13,** 8458 (1967).

458. E. Junker, *Giesserei*, **54,** 152 (1967).

459. B. Y. H. Liu and A. C. Verma, Publ. No. COO-1248-11, CFSTI, 1967.

460. B. Y. H. Liu, K. T. Whitby, and H. H. S. Yu, *Rev. Sci. Instr.*, **38,** 100 (1967).

461. P. E. Morrow and T. T. Mercer, *Am. Ind. Hyg. Assoc. J.*, **25,** 8 (1964).

462. P. C. Reist, Publ. No. CONF-660904, CFSTI, 1967.

463. R. B. Parker and C. R. Hill, Publ. No. CONF-365-4, CFSTI, 1963.

464. T. T. Mercer, M. J. Tillery, and M. A. Flores, Publ. No. LF-7, CFSTI, 1963.

465. D. A. Lundgren, *Bull. Appl. Sci. Div.*, Litton Industries, Minneapolis, Minnesota, 1965.

466. B. Binek, *Staub*, **25,** 261 (1965).

467. M. Polydorova, *Chem. Listy*, **58,** 312 (1964); through *Chem. Abstr.*, **60,** 12694 (1964).

468. G. Langer, *Intern. J. Air Water Pollution*, **8,** 167 (1964).

469. K. T. Whitby and W. E. Clark, *Tellus*, **18,** 573 (1966).

470. F. E. Adley, *Am. Ind. Hyg. Assoc. J.*, **19,** 75 (1958).

471. R. T. Allemann and U. L. Upson, Rept. HW-74210, CFSTI, 1962.

472. E. L. Collier, M. C. Gourdine, and D. H. Malcolm, *Ind. Eng. Chem.*, **58,** 26 (1966).

473. L. E. Faith, S. N. Bustany, D. N. Hanson, and C. R. Wilke, *Ind. Eng. Chem. Fundamentals*, **6,** 519 (1967).

474. C. Eyraud, J. Joubert, R. Morel, C. Henry, and B. Roumesy, *Proc. Intern. Clean Air Congr. (London)*, Part I, 129 (1966).

475. C. Henry, R. Morel, and J. Joubert, Paper 68-105, Air Pollution Control Assoc., 1968.

476. F. Molyneux, *Chem. Proc. Eng.*, **44,** 517 (1963).

477. P. Impris and J. Sencek, *Czech. Patent* 117,895, (March 15, 1966); through *Chem. Abstr.*, **66,** 12530 (1967).

478. P. Impris and J. Sencek, Czech. Patent 118,548, (May 15, 1966); through *Chem. Abstr.*, **66,** 87149 (1967).

479. P. Impris and J. Sencek, Czech. Patent 117,899, (March 15, 1966); through *Chem. Abstr.*, **60,** 12531 (1964).

480. D. A. Hedges, *Colloq. Electrostatic Precipitators*, Inst. Elec. Engrs., London, 1965.

481. R. L. Chamberlin and G. Moodie, *Proc. Intern. Clean Air Congress, London*, Part I, 133 (1966).

482. R. L. Chamberlin and P. B. Crommelin, *1st World Congr. Air Pollution*, Buenos Aires, 1965.

483. J. A. Fife and R. H. Boyer, *Proc. Natl. Incinerator Conf.*, *Am. Soc. Mech. Engrs.*, 89 (1966).

484. J. B. Opfell and W. T. Sproull, *Ind. Eng. Chem. Proc. Design Develop.*, **4,** 173 (1965).

485. R. Sickles, *Power*, **111** (11), 75 (1967).

486. S. Rabkin and P. B. Crommelin, *Paper Trade J.*, **144** (16), 48 (1960).

487. H. H. Uhlig and J. Sansone, *Mater. Protect.*, **3** (2), 21 (1964).

488. O. Guepner, Federal German Patent 1,160,827, June 9, 1964.

489. Verein Deutscher Ingenieure, Richtlinie 2260, *Technical Guarantees for Dust Collectors*, 1963.

490. Verein Deutscher Ingenieure, Richtlinie, 2264, *Operation and Maintenance of Dust-Collecting Plants*, 1966.

491. Industrial Gas Cleaning Institute, Rye, N.Y., Publ. No. EP-4, "Bid Evaluation Form for Electrostatic Precipitators," 1968.

492. Industrial Gas Cleaning Institute, Rye, N.Y., Publ. No. 3, "Criteria for Performance Guarantee Determination," 1965.
493. Industrial Gas Cleaning Institute, Rye, N.Y., Publ. No. 2, "Procedures for Determination of Velocity and Gas Flow Rate," 1965.
494. Industrial Gas Cleaning Institute, Rye, N.Y., Publ. No. EP-1, "Technology for Electrostatic Precipitators," 1967.

Symbols

a particle radius, m

A total collecting surface area, m^2

A' cumulative collecting surface area from inlet, m^2

A_c Cunningham correction coefficient, dimensionless

A_g constant of the gas, V/m

b ion mobility, $m^2/(V\text{-sec})$

b_b back-corona ion mobility, $m^2/(V\text{-sec})$

b_f forward-corona ion mobility, $m^2/(V\text{-sec})$

b_0 ion mobility at 0°C and 1 atm, $m^2/(V\text{-sec})$

b_p particle mobility, $m^2/(V\text{-sec})$

B dimensionless variable

B_g constant of the gas, $V/m^{1/2}$

c one-half wire-to-wire spacing (between centers), m

c_0 constant, m^{-1}

C capacitance, F

C_D drag coefficient, dimensionless

C_{in} inlet dust concentration, kg/m^3

C_{out} outlet dust concentration, kg/m^3

C_p particle concentration, kg/m^3

$\bar{C}_p$ cross-sectional average particle concentration, kg/m^3

C_w particle concentration at wall, kg/m^3

d dimensionless variable

E electric field, V/m

E_c corona-starting field, V/m

E'_c charging electric field, V/m

E_g radial component of electric field near a particle, V/m

E_p precipitating electric field

F force, N

j current density, A/m^2

j_{bl} linear back-corona current density, A/m

j_{fl} linear forward-corona current density, A/m

j_i free-ion current density, A/m^2

j_l linear current density, A/m

j_{lc} linear current density in clean gas, A/m

j_{ld} linear current density in dirty gas, A/m

j_p particle current density, A/m²

j_s average current density at plate, A/m²

j_{sx} current density at plate opposite a wire, A/m²

j_t total current density, A/m²

j_{tl} total linear forward and back corona current density, A/m

k Boltzmann's constant, J/°K

k_A Anderson's precipitator constant

k_p function of tube radius, m

K dimensionless variable

K_0 loss coefficient, dimensionless

K_1 constant, $\mathrm{m}^{-1/2}$

l characteristic dimension, m

L length, m

n number of wires

N ion concentration in potential field, m^{-3}

N_0 undisturbed ion concentration, m^{-3}

N_p particle number density, m^{-3}

$(N_p)_{in}$ inlet particle number concentration, m^{-3}

p dimensionless parameter, eq. 51

p_a pressure, atm

p_g pressure, N/m²

P probability of achieving efficiency greater than η_0, dimensionless

q ion charge, C

Q particle charge, C

$Q_{\max}$ limiting particle charge, C

Q_s particle charge in absence of corona quenching, C

r radius, m; radial distance from tube axis, m; radial distance from center of particle, m

r_0 wire radius, m

r_1 tube radius, m

Re Reynolds number, dimensionless

s one-half plate-to-plate spacing, m

S particle surface per unit volume of gas, m^{-1}

t time, sec

T temperature, °C

T_A absolute temperature, °K

U potential energy of ion, J

v average gas velocity, m/sec

v_i ion drift velocity, m/sec

$\bar{v}_i$ rms ion velocity, m/sec

V potential, V

V_c corona-starting potential, V

V_c' corona-starting potential in presence of dust space charge, V

V_g gas flow rate, m³/sec

V_0 potential at wire surface, V

V_{sl} sparkover voltage for system of one wire, V

V_{sn} sparkover voltage for system of n wires, V

w particle migration velocity, m/sec

$\bar{w}$ mean particle migration velocity, m/sec

w_e effective migration velocity, m/sec

w_{ev} electric-wind velocity, m/sec

w_g migration velocity for particle of the geometric mean radius, m/sec

x longitudinal distance from wire axis, m

y transverse distance from wire axis, m

α bipolar charging constant, sec⁻¹

γ dimensionless current, eq. 43

δ gas density relative to 0°C and 1 atm, dimensionless

δ' gas density relative to 25°C and 1 atm, dimensionless

δ_{cr}^+ positive critical density relative to 25°C and 1 atm, dimensionless

δ_{cr}^- negative critical density relative to 25°C and 1 atm, dimensionless

ε permittivity, F/m

ε_0 permittivity of free space, 8.86×10^{-12} F/m

η fractional efficiency, dimensionless

η_0 minimum acceptable efficiency, dimensionless

θ polar angle, rad

κ relative dielectric constant, dimensionless

κ_g relative dielectric constant of gas, dimensionless

κ_p relative dielectric constant of particle, dimensionless

λ molecular mean free path, m

λ_i mean free path of ions, m

μ viscosity, decapoise

ξ back-corona ratio, dimensionless

ρ space-charge density, C/m³

$\bar{\rho}$ average space-charge density, C/m³

ρ_b ion space charge of back corona, C/m³

ρ_d bulk particle resistivity, ohm-m

ρ_f ion space charge of forward corona, C/m³

ρ_g gas density, kg/m³

ρ_i ion space-charge density, C/m³

ρ_i' density of ions available for particle charging, C/m³

ρ_p particle space-charge density, C/m³

ρ_v volume resistivity, ohm-m

ρ_∞ constant of material, ohm-m

σ standard deviation of effective particle migration velocities, m/sec

σ_g geometric standard deviation of particle radii, dimensionless

τ charging time constant, sec

τ_d time constant of dust, sec

τ_g time constant of gas, sec

ϕ dimensionless current, eq. 17

χ $C_w/\bar{C}_p$

Collection of Particles by Fiber Filters*

F. LÖFFLER

Institute for Mechanical Process Technology, University of Karlsruhe, Germany

I. INTRODUCTION

An important technical problem from the broad area of gas cleaning is the cleaning of air used in air conditioning and ventilating. Here one is generally

* Translated by W. Strauss.

concerned with pollutants in the form of solid particles or liquid droplets present in relatively low concentrations, of the order of mg/m^3 or $gr/1000\ ft^3$. Even these low concentrations are frequently too high for the conditions required for a facility. Such concentrations can cause soiling and be a nuisance, for example in air conditioned homes, offices, and conference rooms. They can affect health, for example, in operating theatres. They can also cause damage to property and production losses, examples being computer rooms, laboratories, motor car spray painting installations, and in "clean rooms."

For reducing dust concentrations to acceptable values in these cases, fibrous filters are installed. The fibers in these filters are either natural fibers (wool, cotton, cellulose) or synthetic fibers (glass, organic polymers). The fiber layers are extremely porous and the actual fiber volume in a filter is between 1 and a maximum of 10%. The mean interfiber distances are therefore between 9 and 3 times the fiber diameters. Because the interfiber distances are considerably greater than the majority of the particles being collected, particles are not sieved from the gas. The actual gas cleaning process occurs within the fiber layer through which the gas is passing, the particles reaching the fiber surface and being held there by adhesive forces. Following saturation of the fiber layer with dust, the filters are generally discarded, although some types can be reused after cleaning by washing or similar processes. For the action of fiber filters the two processes of greatest importance are separation and adhesion.

The parameters needed to describe the filtration are the particle collection efficiency or degree of separation and the flow resistance of the filter. These parameters depend on the properties of the filter, the dust, and the carrier gas. Their determination has been the aim of numerous theoretical and experimental investigations.

A further application of fibers in woven or felted (nonwoven) fabrics is in tubular or flat filters in industrial gas cleaning. Here, gases with high dust loadings (up to $100\ g/m^3$ or $40\ gr/ft^3$) have to be cleaned. The filtration process here is different to the "air cleaning" high-volume fiber filters in that, with the exception of a brief initial period, the filtration occurs at the filter surface. A filter cake is formed, which itself acts as the filter medium. After a specified pressure loss occurs in the fiber filter, the filter cake is removed by a suitable technique. The filtration process is then recommenced and is repeated periodically. In these filters the type and magnitude of the adhesive forces between dust and filter surface are of particular interest with respect to the cleaning process. In this application, both theoretical and experimental investigations have concentrated on the time variation in pressure drop. As this is not of direct concern in this paper, the reader is referred to other publications (1–5).

This review is chiefly concerned with the deposition processes in the loose filters described earlier.

II. SEPARATION MECHANISMS

Different effects may be responsible for the transport of dust particles to the fiber surface. The development of the current theories for calculating these separation effects has been adequately described in a number of studies. These include, in particular, the work of Strauss (1), Ranz and Wong (6), Chen (7), Landt (8), Richardson (9), Fuchs (10), Dorman (11), Pich (12), and Spurný (13). In the following section only the basic relations are briefly presented, and some modifications, as a result of recent publications, have been added.

It is usual for different separation mechanisms to be first considered separately for an isolated fiber. In a fiber layer, however, the neighboring fibers can influence each other, and this must be allowed for by the use of empirical correlations. The separation efficiency of the whole filter is then obtained from the efficiency for a single fiber, and by assuming a model for the whole filter.

It is assumed in the calculation for the collection efficiency of a single fiber that the particles are spherical and evenly distributed throughout the gas stream. The particle concentration is also assumed to be so low that it does not influence the gas flow, and when particles contact the fiber surface, they are held there. This last assumption, however, as will be shown later, is frequently invalid. Furthermore the classical filtration theories only apply to clean fibers, which implies that the separation of particles is not influenced by those already held on the fiber surface. This dependence of filtration efficiency on the deposited dust, which is a time-dependent function, called the filtration kinetics of the system, has so far not been adequately covered in the literature.

The separation efficiency of a single fiber, η, is defined as the ratio of the number of particles collected by a unit length of fiber, "a," in unit time, to the total number which flow toward the projected surface of the fiber in unit time.

$$\eta = \frac{a}{D \cdot v_0 n_0} \tag{1}$$

where D is fiber diameter, v_0 is velocity of the undisturbed gas stream upstream of the fiber, and n_0 is upstream particle concentration.

The most important mechanisms for particle deposition are inertial impaction, interception, diffusion, and electrical forces. Sedimentation due to gravity can generally be neglected. Fuchs (10) has shown this is

only a secondary effect (10a). Similarly, thermal precipitation effects can
generally also be neglected.

A. Velocity Distribution around a Fiber

For calculating the collection by different mechanisms the velocity distribution around the fiber must be known. The characteristic parameter
for the flow is the Reynolds number based on fiber diameter:

$$\mathrm{Re}_c = \frac{v_0 \cdot D}{\nu}$$

where ν is the kinematic viscosity of the gas.

At high Reynolds numbers the flow at the upstream side of an infinitely
long cylinder, whose axis is perpendicular to the direction of flow, can be
approximately described with the equations for potential flow for two-
dimensional, friction-free, incompressible flow. More accurate calculations
have, of course, to allow for the boundary layer. The stream function ϕ
for potential flow is given by

$$\phi = v_0 \cdot r \cdot \sin \theta (1 - r_f^2/r^2) \tag{2}$$

Here r and θ are polar coordinates and r_f is the fiber radius; the velocity
components v_r and v_θ follow from the stream function

$$v_r = \frac{1}{r} \frac{\partial \phi}{\partial \theta} \qquad v_\theta = - \frac{\partial \phi}{\partial r} \tag{3}$$

Because of the separation of flow, these equations do not hold on the downstream side of the cylinder.

For very low Reynolds numbers ($\mathrm{Re} \ll 1$), Oseen's approximation of the
Navier-Stokes equation, as given by Lamb, is generally used. The stream
function for viscous flow is then

$$\phi = \frac{r_f \cdot v_0 \cdot \sin \theta}{2(2 - \ln \mathrm{Re})} \left(\frac{r_f}{r} - \frac{r}{r_f} + 2 \frac{r}{r_f} \ln \frac{r}{r_f} \right) \tag{4}$$

This equation is only valid at the cylinder surface. In contrast to potential flow, the stream lines in viscous flow are deflected at much greater
distances from the cylinder and much more gradually. The stream lines
are also deflected further sideways.

Equations 2 and 4 described the flow around a single cylinder. Kuwabara
(14) and Happel (15) determine a velocity field for a system of cylinders,
which, in contrast to Lamb's solution, no longer depends on the Reynolds

number. This is

$$\phi = \frac{r_f \cdot v_0 \cdot \sin \theta}{2(-\frac{1}{2} \ln \beta - C)} \left(\frac{r_f}{r} - \frac{r}{r_f} + 2 \frac{r}{r_f} \ln \frac{r}{r_f} \right) \tag{5}$$

Here β is the relative fiber volume and $\beta = 1 - \varepsilon$ where ε is the porosity. The constant C is 0.75 according to Kuwabara and 0.5 according to Happel. Equation 5 is only valid for the conditions

$$\frac{r - r_f}{r_f} \ll 1 \quad \text{and} \quad \beta \ll 1$$

that is, only close to fibers and for highly porous (or low-density) filters.

In an experimental study of the flow distribution in a system of cylinders, Kirsch and Fuchs (16) found better agreement between their results and the Kuwabara equation than the Happel equation. The range of Reynolds numbers studied was 0.01–0.05. In a further publication Kirsch and Fuchs (17) compared experimental pressure drop measurements with theoretical equations. The maximum Reynolds number here was also 0.05. The fiber volume ratio β was varied between 0.0043 and 0.27. Here, too, the Kuwabara equation gave better agreement with experimental values than the Happel equation. This paper was also able to show the extent to which the pressure drop depends on the arrangement of the cylinders. Small deviations from the parallel arrangement effect a decrease in pressure drop. It can therefore be deduced that there is a marked dependence of the velocity field on their arrangement.

When the differences between the models and real filters are considered, it is clear that the theories of deposition cannot provide exact results, but at best are able to show trends and provide approximate answers. A further difficulty is that commercial filters do not, generally, operate at either very low or very high Reynolds numbers, for which the above theories are available, but at intermediate values ranging from 1 to 10. Because of the difficulties of solving the Navier-Stokes equation in this range, only very few calculations are available (see Strauss (1a)).

The velocity distributions given above apply in the range of continuous flow, where there is no "slipping" of the gas past the fiber surface. This is realistic at ambient pressures for fibers larger than 5-μm diameter. Pich (12,18) has modified the Kuwabara-Happel equation to allow for this slipping by introducing a "slip" coefficient. This is of particular importance for the fiber filters developed in recent years where fibers are frequently less than 1-μm diameter. However this case will not be considered further here.

B. Collection by Changing Direction (Inertial Impaction)

The theoretical basis for inertial separation was first considered by Sell (19) in 1931 and developed concurrently by Albrecht (20). Later developments are numerous (1,6,7,10,12,13). It is usual to proceed from the concept that the dust particles are unable to follow fully the deviations of the gas stream lines as these move around an obstruction. This leads to particle contact and deposition of the particle on the obstruction or fiber (Fig. 1). It is assumed that every particle touching a fiber will stick there.

For calculating particle paths, the equations of motion for particles are set up. It is generally assumed that no forces except inertial and resistance forces act on the particle. Furthermore, the resistance forces are calculated by Stokes law, which is a reasonable approximation in the flow range of fiber filters.

In these equations of motion the theoretical or experimental velocity distributions given in Section II-A must be used. The differential equations which are arrived at cannot be solved analytically but must be solved by numerical methods. Various workers have obtained somewhat different answers [see Chen (7), Strauss (1b)] because of different assumptions and this in turn has affected the accuracy of the calculation.

As characteristic parameter for collection by inertial impaction, the dimensionless parameter ψ (frequently also called the Stokes number, Stk) is introduced. ψ is defined as

$$\psi = \frac{C\rho_p d_p^2 v_0}{18\mu D} \qquad (6)$$

where $C =$ Cunningham correction, $\rho_p =$ particle density, $d_p =$ particle diameter, and $\mu =$ dynamic viscosity of the gas.

In the Stokes' region, the dimensionless number ψ represents the relation of the "stopping distance" of a particle with initial velocity v_0 to a fiber, diameter D. The stopping distance is the maximum distance a particle travels if it enters still air with a velocity v_0. If it is assumed that a particle

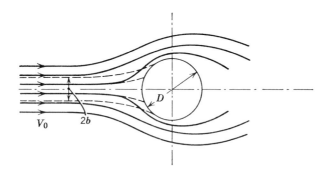

Fig. 1. Stream lines (full lines) and particle paths (broken lines) in flow around a cylinder.

as mass but no physical size, the efficiency of deposition, according to eq. 1 is given by

$$\eta_I = \frac{2b}{D} \tag{7}$$

where b = distance is the cross stream distance from the fiber center line upstream of the fiber to the particle which will tangentially graze the fiber. It follows that $\eta_I = f(\psi, \mathrm{Re})$.

In potential flow the velocity field is independent of the Reynolds number, and here

$$\eta_I = f(\psi)$$

As a result of the less abrupt deviation of the stream lines in viscous flow, the efficiency of deposition is less for small Re_c numbers than in potential flow.

A number of the theoretical curves have been presented together with experimental measurements by Strauss (1) and it is seen from these that the experimental values are in reasonable agreement with the theoretical curves of Langmuir and Blodgett (potential flow) and of Landahl and Hermann ($\mathrm{Re}_c = 10$). These two theoretical curves differ somewhat only for $\psi < 2$.

Pich (12) reports that Landahl and Hermann give the following simplified expression:

$$\eta_I = \frac{\psi^3}{\psi^3 + 0.77\psi^2 + 0.22} \tag{8}$$

Equation (8) gives a useful representation of the efficiency of inertial separation incorporating present experimental results.

May and Clifford (21) separated 20–40 μm droplets using metal wires and a Reynolds number range of 165–8500. Although this was high, they found good agreement with the earlier results of Langmuir and Blodgett.

Whitby (22) pointed out that the experimental results of Wong and Johnstone approximately followed a log-normal distribution. The characteristic parameter for such a distribution is the median value ψ_M (i.e., for the $\eta_I = 50\%$ value) and the geometric standard deviation, σ_g. Whitby found that for circular cylinders $\sigma_g = 1.65$ for $0.2 < \mathrm{Re} < 150$

$$\psi_M^{1/2} = 1.253 \, \mathrm{Re}^{-0.0685} \tag{9}$$

and for $\mathrm{Re} < 0.2$

$$\psi_M^{1/2} = 1.4 \tag{10}$$

Equation 9 covers the range of greatest importance in filtration technology, and has the additional advantage of being simple to apply.

In general, the equations and curves show that the efficiency of collection by inertial impaction approaches 100% value asymptotically. This is influenced by increasing particle diameter d_p, increasing stream velocity, v_0, or decreasing fiber diameter D. This remains the case as long as the condition of particle retention following collection is fulfilled. However, Whitby (22) has indicated that for particles larger than 5-μm diameter this is not the case in practice.

C. Interception

In the above the particles were essentially considered as point masses. In actual practice, their finite size must be taken into account, and so particles can touch the fibre surface whose centers are no more than $d_p/2$ from the fiber.

The dimensionless parameter introduced to describe this mechanism was the ratio

$$R = \frac{d_p}{D} \tag{11}$$

The efficiency of collection corresponding to this is given by η_c. According to the flow regime about the fiber, different functions for $\eta_c = f(R)$ are given in the literature (1,7,10,12). Only the equation for potential flow will be given here:

$$\eta_c = 1 + R - \frac{1}{1 + R} \tag{12}$$

Thus with increasing particle size d_p the interception effect is increased. Examples given by Fuchs (10b) and Strauss (1c) show that for $R \leqslant 0.1$ the effect of inertial impaction far outweighs that of interception. However, with increasing R and increasing ψ, η_c increases rapidly relative to η_I.

It can be noted from eq. 12 that η_c can be greater than 1. For Knudsen numbers in the range $10^{-3} < \mathrm{Kn} < 0.25$ Pich (18) calculated the interception effect with a modified Kuwabara-Happel equation for the range where slipping of the gas occurs. Here the Knudsen number is given by

$$\mathrm{Kn} = \frac{\lambda}{r_f} = \frac{2\lambda}{D}$$

where λ = mean free path of the gas molecules and r_f = fiber radius. Pich found that η_c increases with increasing R, increasing $\beta(=1-\varepsilon)$, and increasing λ. This last also implies decreasing pressure.

D. Diffusion Collection

As a result of the collisions with gas molecules, small dust particles do not follow a direct path, but a zig-zag movement around their statistical mean path, which is called Brownian movement. This can lead to deposition of particles from the gas stream surrounding the fibers, but it is only of significance for particles smaller than 0.5 μm. Because of the Brownian motion becoming more pronounced with decreasing particle size, it follows that diffusion collection also becomes more significant.

The characteristic parameter for the diffusion process is the Peclet number, which is defined as:

$$\text{Pe} = \frac{Dv_0}{\mathscr{D}} \tag{13}$$

where $\mathscr{D}$ is the diffusion coefficient or diffusivity of the particles. For Brownian movement, the diffusivity can be calculated by eq. 14:

$$\mathscr{D} = \frac{kTC}{3\pi\mu d_p} \tag{14}$$

where k is Boltzmann's constant, T is absolute temperature, and C is the Cunningham correction factor.

The methods for calculating the diffusion collection efficiency have been extensively treated by Strauss (1), Pich (12), Dorman (11), and Spurný (13). The basic assumption is that during the time the gas stream flows past the fiber, the particles from a certain layer are able to diffuse to the fiber surface. This layer, thickness x, is proportional to $(\mathscr{D} \cdot t)^{1/2}$. The time t depends on the velocity distribution. It therefore follows that in the viscous flow regime the diffusion collection efficiency $\eta_D = f(\text{Re}, \text{Pe})$ and in the potential flow regime $\eta_D = f(\text{Pe})$. Langmuir derived the following equation for a thin boundary layer in viscous flow:

$$\eta_D = \frac{1}{(2 - \ln \text{Re})^{1/3}} \cdot \frac{1}{\text{Pe}^{2/3}} \tag{15}$$

For potential flow, according to different workers (see Pich (12))

$$\eta_D = K \cdot \left(\frac{1}{\text{Pe}^{1/2}}\right) \tag{16}$$

The factor K is about 3.

Torgeson (quoted by Strauss (1) and Whitby (22)) found the following:

$$\eta_D = 0.75 \frac{(C_D \, \text{Re})^{0.1}}{2} \cdot \text{Pe}^{-0.6} \tag{17}$$

Here C_D is the friction factor for the fiber. Strauss (1) and Whitby (22) find that this equation gave good agreement with experimental values.

Thus eqs. 15–17 show that if the definition of the Peclet number is considered, the collection efficiency by diffusion decreases with increasing particle size, gas velocity, and fiber diameter.

E. Deposition by Electrical Forces

Electrical force effects between particles and fibers can occur when either the particles or the fibers have an electrostatic charge or when an external electrostatic field is imposed on the filter. These cases are therefore treated separately.

1. The Influence of Electrostatic Charging

It is possible that either the dust particles or the fibers, or both, carry an excess charge which can result in attractive or repulsive forces. For determining the particle path under these conditions the electrical forces have to be added as an additional term to the equation for particle movement. When both particle and fiber are charged, Coulombic forces operate. When either particle or fiber is charged, then image forces act. Under these conditions there is also a space charge effect surrounding charged particles. The relevant equations are fully presented by Strauss (1), Pich (12), and Spurný (13).

The degree of deposition depends on the electrostatic forces and on the dielectric constants. In practice, however, it is difficult to predict the charge which limits the use of the theory in actual cases. In the case of natural aerosols, the charge is relatively small.

On the other hand, in laboratory experiments with synthetic aerosols comparatively large electrostatic charges can be induced on the particles. Such experiments have been reported by Lundgren and Whitby (23) who charged 1-μm methylene blue particles. The Reynolds numbers in these experiments were between 0.1 and 1. Marked increases in collection efficiency followed with increasing particle charge; for example, 16% of uncharged 1-μm particles were retained by a wool felt filter, but 320 elementary charges on each particle increased collection efficiency to better than 99%.

On the basis of their measurements, Lundgren and Whitby suggested the following equation for the viscous flow regime (Re < 1):

$$\eta_E = 1.5(K_M)^{1/2} \tag{18}$$

where the dimensionless parameter K_M for the attractive force between a charged particle and its image charge on a neutral fiber is given by

$$K_M = \frac{\varepsilon_f - 1}{\varepsilon_f + 1} \cdot Q_p{}^2 \frac{1}{12\pi^2 \varepsilon D_0{}^2 \mu d_p v_0} \tag{19}$$

where ε_f = dielectric constant of the fiber, ε_0 = permittivity = $8.859.10^{-21}$ coulomb²/(dyne cm²), and Q_p = particle charge.

Both Lundgren and Whitby as well as Gillespie (24) consider that electrostatic effects are of particular importance for particles with diameters less than 1 μm($d_p < 1$ μm) and low velocities. Larger particles are largely separated by inertial impaction.

2. External Electrical Field

If an electrical field is established around the filter, both fibers and particles are polarized. This also establishes forces between them. Havliček (25) was able to show that collection efficiency can be improved, particularly when the external field is established in the direction of flow.

Walkenhorst and Zebel (26) report that considerable collection with very low pressure drop was achieved in a very loosely packed filter (fiber diameter $D = 40$ μm), particles of 4 μm diameter and electric fields of 6000–10,000 V/cm. Zebel (27,28) obtained collection efficiency equations for the case of loosely packed fiber filters in which the charge between fibers can be neglected. Inertial impaction, interception, and diffusion effects were also neglected. The theory is valid for particles from approximately 0.1 to several microns in diameter.

The critical parameters for deposition are as follows: (a) for uncharged particles

$$F = \frac{\varepsilon_p - 1}{\varepsilon_p + 2} \cdot a \cdot \frac{r_p^{\,3}}{r_f} \cdot \frac{E_0^{\,2} \cdot B}{v_0} \tag{20}$$

where $a = \varepsilon_f - 1/\varepsilon_f + 1$, ε_f = dielectric constant of the fiber, ε_p = dielectric constant of the particle, E_0 = field strength, v_0 = gas velocity, and B = mobility of the particles = $(3\pi\mu d_p)^{-1}$.

(b) For particles with charge q

$$G = \frac{E_0 q B}{v_0} \tag{21}$$

These parameters represent the relation between the velocity which the electric field gives the particles and the gas velocity v_0.

The differential equation for particle motion cannot be solved directly for uncharged particles. Zebel presents a series of paths that were calculated by an approximation method. For charged particles, the collection efficiency for both potential and viscous flow is given by

$$\eta_q = G \cdot \frac{a + 1}{G + 1} \tag{22}$$

This equation is of great importance for very small particles because the polarization is very low, and only a few elementary charges are necessary to charge the particles.

F. Combination of Different Deposition Mechanisms

In the flow around a fiber, several of the mechanisms described can contribute to particle deposition. A number of empirical relations or approximation formulas are available for the interaction of the different effects (7,10–13), which do not include electrical effects. An exact solution has not yet been established because of the mathematical problems involved. In the simplest case it is assumed that the single effects are superimposed without interaction and so can be added. Fuchs (10b) indicates that the total deposition for a fiber is larger than each of the single mechanisms, but less than their arithmetic sum. Ranz and Wong (6) assume that the deposition effect with the largest dimensionless parameter predominates, and the other effects can be neglected.

As was shown above, the diffusion effect decreases with increasing particle size while the inertial impaction effect increases. At the same time interception increases with increasing size, and it is therefore to be expected that a deposition minimum occurs which depends on particle size and also velocity. Friedlander (29) derives the following expression from an analysis of experimental results:

$$\eta = 6 \, \mathrm{Re}^{1/6} \, \mathrm{Pe}^{-2/3} + 3R^2 \, \mathrm{Re}^{1/2} \tag{23}$$

This expression incorporates diffusion and interception mechanisms and assumes that inertial impaction is negligible. On differentiation of eq. 23 Friedlander obtains an equation for the particle size with minimum collection efficiency.

Hasenclever (30) compared experimental values and calculated values based on eq. 23. He found that for synthetic fibers of 1 μm and dust particles of 0.5 μm, the Friedlander equation gave qualitative agreement. Exact quantitative comparisons were not possible. Whitby (22), in comparing his measurements with the different theories, finds that Torgeson's equations give the best fit for the calculation of single fiber combined collection (31) efficiencies. The combination is of the diffusion, the interception, and of the inertial impaction mechanisms. These equations are unfortunately not readily solved. For Re $\geqslant$ 0.5 Whitby gives a combination of the Langmuir equation for interception:

$$\eta_c = \frac{1}{2(2 - \ln \mathrm{Re})} \left[2(1 + R) \ln (1 + R) - (1 + R) + 1/(1 + R) \right] \tag{24}$$

together with eq. 9 for inertial impaction (see Sec. II-B). This gives equally satisfactory agreement with experimental values, but with simpler calculations. Whitby has determined the experimental collection efficiencies for single fibers on fiber layers whereby the "effective" fiber diameter was determined by a pressure drop method. The influence of neighboring fibers on the single fiber is included in this.

G. Interference Effect

In a fiber layer the fibers can influence one another, and here the effectiveness of an isolated fiber is different from a fiber in a bed. This is because of the change in the flow relations. Chen (7) and Pich (12) show that the presence of neighboring fibers increases the collection efficiency, and this increase is a function of the fiber volume fraction β.

Chen (7) finds that for $\beta < 0.09$ experimental results fit the empirical relation

$$\eta_\beta = \eta_0(1 + 4.5\beta) \tag{25}$$

where η_β = deposition efficiency of a single fiber and η_0 = deposition efficiency of an isolated fiber.

Van der Waal and Clarenburg (32) derive an expression for the aerosol penetration of a fiber layer, where they introduce a "structure" and a "shadow" effect. These effects occur when fibers have different diameters. In mixing coarse and fine fibers, more coarse pores are formed (structure effect). Fine fibers also tend, in part, to lie on the coarse ones, screening these (shadow effect). The effective contributions of the different fibers are corrected for by assuming a particular geometrical model. For the aerosol penetration P, the following approximate relation is given:

$$-\log P \simeq \frac{(1 - \varepsilon)^2}{\varepsilon}$$

In experiments with glass fiber filters ($\varepsilon = 0.88$–0.96) and 0.6-μm diameter droplets, the theoretical assumptions were confirmed and the structure and shadow effects contribute more to the deposition of particles than to the pressure drop.

H. Total Collection Efficiency of a Fiber Layer

In order to deduce from the collection efficiency of a single fiber the efficiency of a whole bed with many successive fibers, the following model is usually used. The fibers lie in regular layers perpendicular to the gas flow. The flow is therefore perpendicular to the direction of the fiber axis. The particles are evenly distributed before each fiber throughout the gas, and this implies that the conditions of particle deposition are the same for all fibers.

If l is the total fiber length per unit volume of filter, then in a unit area of surface, in a layer of depth dL, the fiber has length $l\,dL$. If one assumes a monodisperse fiber diameter D, and a monodisperse dust particle diameter d_p, then the dust da deposited in layer dL is given by substituting in eq. 1:

$$da = \eta \cdot l\,dL \cdot D \cdot n\,\frac{v_0}{\varepsilon}$$

$$= -dnv_0 \tag{28}$$

where n = particle concentration, η = single fiber efficiency, and v_0/ε = upstream velocity of the gas within a fiber bed with porosity ε. l is found from

$$1 - \varepsilon = \frac{\pi}{4}\,D^2 \cdot l \tag{29}$$

If eq. 29 is substituted in eq. 28 and the expression is then integrated between the limits of particle concentration N_0 before the filter and N following the filter,

$$\frac{N}{N_0} = \exp -\left(\frac{4}{\pi} \cdot \frac{L}{D} \cdot \frac{1-\varepsilon}{\varepsilon} \cdot \eta\right) \tag{30}$$

The collection efficiency of a fiber layer for particles of size d_p is

$$E(d_p) = 1 - \frac{N}{N_0} = 1 - \exp -\left(\frac{4}{\pi}\frac{L}{D} \cdot \frac{1-\varepsilon}{\varepsilon} \cdot \eta\right) \tag{31}$$

This equation has been derived by different authors (see Strauss (1), Dorman (11), and Pich (12)).

Whitby (22) gives for the collection efficiency of a fiber layer:

$$E(d_p) = 1 - \exp -(S \cdot \eta) \tag{32}$$

where $S = 4/\pi \cdot w/l_f D$, (w = weight of the filter per unit surface area,

and l_f = fiber density). $\tag{33}$

S is called the "solidarity factor" and means, in physical terms, the projected fiber surface per filter surface, which is presented to the gas stream. For the values of ε approaching unity, eq. 32 is the same as eq. 31. The eqs. 31 and 32 postulate "fractional" efficiencies. They give the efficiencies for a definite particle size d_p.

A more common practical case is that the filter consists of fibers of the same diameter, whereas there is a distribution of particle sizes. If $n(d_p)$

describes the distribution of the particles by integration the total collection over all differential quantities of particles is given by

$$E_T = \int_{d_p \min}^{d_p \max} E(d_p) \cdot n(d_p) \cdot \mathbf{d}d_p \qquad (34)$$

By substituting eq. 31 in eq. 34 and with the condition of standardization

$$\int_{d_p \min}^{d_p \max} n(d_p) \, \mathbf{d}d_p = 1$$

We find

$$E_T = 1 - \int_{d_p \min}^{d_p \max} \exp - \left(\frac{4}{\pi} \cdot \frac{L}{D} \cdot \frac{1 - \varepsilon}{\varepsilon} \cdot \eta \right) n(d_p) \, \mathbf{d}d_p \qquad (35)$$

To calculate the total collection efficiency, the size distribution function $n(d_p)$ must be known, that is, determined by measurement. In order to treat the general case, where neither fibers nor dust are monodisperse, the case where the particles are monodisperse but the fibers have a distribution of sizes is first considered.

If $f(D)$ is the distribution function of the length of the fibers between D and $D + \mathbf{d}D$ with the condition of standardization

$$\int_{D \min}^{D \max} f(D) \, \mathbf{d}D = 1$$

so, from eq. 30

$$\frac{N}{N_0} = \exp - \left[\frac{4}{\pi} L \cdot \frac{1 - \varepsilon}{\varepsilon} \int_{D \min}^{D \max} \frac{1}{D} \cdot \eta(D) f(D) \, \mathbf{d}D \right] \qquad (36)$$

where $\eta(D)$ is the deposition efficiency for particles of size d_p depending on D.

If now a particle size distribution $n(d_p)$ and fiber diameter distribution $f(D)$ is present, then for the total collection efficiency E_T, analogously with eqs. 36 and 34:

$$E_T = 1 - \int_{d_p \min}^{d_p \max} \exp - \left[\frac{4}{\pi} \frac{1 - \varepsilon}{\varepsilon} \cdot \overline{\eta(D)} \right] n(d_p) \, \mathbf{d}d_p \qquad (37)$$

where

$$\overline{\eta(D)} = \int_{D \min}^{D \max} \frac{1}{D} \eta(D) f(D) \, \mathbf{d}D \qquad (38)$$

I. Influence of Pressure

It was shown above (eqs. 6 and 14) that both the inertial impaction parameter ψ and the diffusivity $\mathscr{D}$ are directly proportional to the Cunningham correction C. If

$$C = (1 + 1 \cdot 246\lambda/D) \qquad (27)$$

then it is apparent that C decreases with increasing pressure, because the mean free path, λ, of gas molecules decreases. It may therefore be expected that both the inertial impaction efficiency as well as the diffusion collection efficiency decrease with increasing pressure (67). The converse has also been found to be the case, as was shown experimentally by Stern et al. (33) and Schuster (34).

Schuster investigated the pressure dependence of the collection efficiency of a fiber filter of PVC fibers with $D = 19.5$ μm, fiber volume fraction $\beta = 0.301$. The particles used were polystyrene spheres, 0.1–1.3 μm diameter, and paraffin oil fog, with droplets 1.08 μm diameter. The filtration velocity was between 0.5 and 4 m/sec. Electrostatic charging was carefully avoided. A strong increase on deposition efficiency with decreasing pressure was observed. Examples are that the deposition efficiency for paraffin fog (1.08-μm diameter droplets), $v_0 = 0.5$ m/sec with 650 mm Hg pressure, was approximately 55%, with 50 mm Hg approximately 82% and with 2 mm Hg approximately 98%. These values are in general agreement with those of Stern. With polystyrene spheres, Schuster found an appreciable influence of particle adhesion. His measurements for polystyrene were lower than those of Stern et al., who used sticky, impregnated filters in their work. Schuster concluded that in his work only part of the particles hitting the fibers were retained in the filter.

J. Time Dependence of Deposition

As was emphasized in Section II, the theories presented so far only represent initial conditions, when the fibers are not covered with particles. From filtration experience it is known that both the deposition efficiency and the pressure drop of the filter depend on the collected dust, and so these are a function of time. Here the structure of the filter and the type of dust play a decisive role. Dorman (11) showed with measurements that the pressure drop through a glass fiber filter increased with decreasing particle size. For many filters the (time) dependence of the filtration efficiency E and of the pressure drop follow the amount of dust deposited, shown in the patterns as indicated in Figure 2. This figure shows how the filtration efficiency changes, particularly at the beginning, while the pressure loss increases steeply with greater dust burdens. At the end of the operating cycle, the filter becomes blocked. The change in deposition efficiency can depend on the fact that the deposited particles form small chains, which themselves act as collecting surfaces. Furthermore, the retention properties also change and more particles stick to the surface. This is considered in more detail in Section III. Theoretical models for describing the filtration process under these conditions are not known. Pich (12) reports the empirical equations

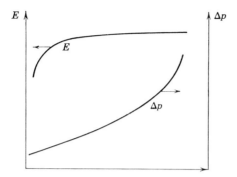

Fig. 2. Filtration efficiency E and pressure drop ΔP as a function of the dust deposited in the filter.

of Radushkevich, which have the form:

$$P = P_0 \cdot \exp -[k \cdot t] \tag{39}$$

where P_0 is the filter permeability at time $t = 0$, and k is a constant depending on experimental conditions.

So far little research has been directed at this particular problem. Experimental investigations have to be concerned with electrostatic effects, because these dictate the structure of the dust layer and the deposition behavior.

K. Comments on the Theories of Deposition

Experimental examination of the theories generally only produces qualitative agreement with marked quantitative differences. This is not surprising if one compares the assumptions with real conditions. Thus assumptions were made that the particles are all spherical and evenly distributed throughout the gas stream and all the fibers were cylindrical. The latter assumption is not realistic in the case of natural fibers. The fiber diameter influences the deposition, but is often difficult to determine. Furthermore, the velocity distribution of the gas stream is only adequately known for either very small or very large Reynolds numbers. Thus, while filter theories are restricted to these ranges, actual filters tend frequently to operate in the transition range. Doubtless of great importance is the difference between the real and model filter structure, particularly with denser filters. Whitby (22) points out that for filters with experimental efficiencies greater than 90% all theories give values which are too large, because it is the irregularities in the filter structure which determine the particle penetration. The experiments of Fuchs and Kirsch (17) clearly show that the order and orientation of fibers have a marked influence on the pressure drop and therefore, it must be presumed, on the collection efficiency.

In laboratory experiments with synthetic aerosols, charge effects can become important. The changed conditions which occur when particles are deposited on the fibers has been indicated in Section II-J. The problems of bouncing and retention will be discussed in Section III below.

III. PARTICLE ADHESION ON FIBERS

An important assumption in the calculation of theoretical collection efficiency was that the particles adhere to the fibers. Thus every particle that touches the fiber surface adheres and is not re-entrained into the gas stream. The impact between particles and fibers is nonelastic, so that there is no recoil of the particles. Furthermore, particles are not blown off the fibers.

Numerous experiments, however, show that these adhesion principles are not realistic. Thus Kaufmann (35) found that quartz particles 10–20 μm diameter were not as efficiently collected as smaller particles. Gillespie (36) considers that many "anomalous" filter efficiency measurements are due to inadequate adhesion. In his experiments with particles ranging from 2 to 3.6 μm diameter he found fewer particles on the deposition cylinder at 20.4 cm/sec than at 5.8 cm/sec. When the cylinder was treated with silicon, the number of deposited particles increased. Dorman (11) reported that particles larger than 15 μm diameter bounce off fibers, but this is also found with much smaller particles. Whitby (22,37) in his measurement of collection efficiency found that contrary to theoretical prediction particles larger than 5 μm showed a reduction in collection efficiency. Here methylene-blue particles and ragweed pollen particles were used with filters of different materials.

The poor collection efficiency of the particles greater than 5 μm was explained by Whitby by considering inadequate adhesion. The present author's own experiments with quartz particles gave similar results. Here too, the maximum collection efficiency was found at approximately 4–5 μm. By surface treatment of the fibers the fractional efficiency curves could be markedly influenced, which was similar to Gillespie's experience. With increasing face velocity the collection of +5 μm particles deteriorated. Schuster (34) in his experiments with polystyrol particles found that even with 1-μm particles, the adhesion conditions were not fulfilled in the filters used. On the other hand, in the collection of oil droplets the adhesion of droplets was more intense.

It becomes apparent that particle adhesion on fibers plays a critical part in the effectiveness of filters with particles generally greater than 5 μm, but in some cases even as small as 1-μm diameter. This is not taken into consideration in the deposition and collection theories, and may account largely for the variation between them and experimental results.

A. Review of Adhesion Mechanisms

Adhesive forces appear in technology and everyday life in a variety of ways, and can have a number of causes. The following examples may be mentioned: problems of friction and lubrication, questions concerned with adhesives, cleaning technology, electrostatic copying (xerography), wall deposits in channels and ducts and on machines, granulation, briquetting, agglomeration, mechanisms in dust collectors, and others. The literature in these areas has been reviewed in the publications of Bowden and Tabor (38), of Houwink and Salomon (39), and of Corn (40).

The following sections consider only those aspects that are important for the mechanisms of particle collection in fiber filters. It is possible to consider the deposition and adhesion of particles on fibers or on other particles as an agglomeration process. Rumpf (41) presented the most important bonding mechanisms in *pelletizing* or *granulation*. The cohesion of agglomerates can be effected by the following forces and mechanisms:

1. Attractive forces between solids: (Valence forces, Van der Waals forces, electrostatic forces as a result of excess charges, double layer forces);

2. Solid bridges: (sintering, chemical reactions, fusion, hardened binders, crystallization of dissolved materials on drying);

3. Adhesive forces in immovable binders: (viscous binders, adhesives);

4. Capillary forces on mobile liquid surfaces: (liquid bridges, capillary forces in agglomerates filled with liquid);

5. Form-closed bonds:

Of these only the following three are of importance for the adhesion of particles of fibers:

(*a*) Van der Waals forces

(*b*) Electrostatic forces as a result of excess charging

(*c*) Capillary forces in liquid bridges.

Below is a comprehensive review of the theoretical methods for calculating these forces and the most important experimental results.

B. Theories for Calculating Adhesive Forces

The "adhesive force" will be defined as the force which acts perpendicular to the bearing surface, that is, the fiber surface, and through the center of gravity of the particle, to separate the particle from the fiber surface. This strict definition is of importance, as with the measuring techniques given later, the adhesive forces in different directions have been measured and these give different results.

"Adhesion" may be strictly used to define the interaction between two solids, while "cohesion" is the attractive force between elements of the same material.

1. Van der Waals Forces

These forces are also described as secondary adhesive forces in contrast to the primary forces of valency, ionic, or metallic bonds. Van der Waals forces, unlike these specific bonds, have less dependence on the material and on impurities, and have longer range. On the other hand, they are much less powerful.

Different theories are available for the calculation of Van der Waals forces. These calculations are not yet fully developed, largely because experimental confirmation is difficult to obtain and appropriate data for the materials is not available.

The most recent comprehensive review is by Krupp (42), which contains adequate references to the literature. Krupp differentiates between a microscopic and a macroscopic model.

a. Microscopic Model

In this model the Van der Waals forces are considered as interactions between the atoms or molecules of the adhesion partners, which depend on their dipolar attraction. In the general quantum theoretical model (dispersion theory) this attraction is based on the electromagnetic vibrations of the load centers of the atoms. It is assumed that the interactions between molecules enable the forces to overlap and so are additive.

The first studies in this area are by Tomlinson (43) in 1928; but extensive calculations of the Van der Waals forces were carried out by Hamaker in 1937 (44). He considered these on the basis of an interaction potential U, which had the form

$$U = -\frac{\lambda}{r^6} \tag{40}$$

Here r = distance between molecules and λ = Van der Waals constant (see eq. 45).

Integrating over all the centers of attraction (i.e., molecules) gives the combined interaction energy U_{12} between the partners, denoted by 1 and 2. The force is obtained by differentiating the energy following the distance d between the adherents:

$$K = \frac{\partial U}{\partial d} \tag{41}$$

Hamaker calculated, for example, the force per unit surface between two planes

$$K^- = \frac{A}{6\pi d^3} \qquad (42)$$

and between a sphere, with diameter D, and a surface:

$$K^0 = \frac{A \cdot D}{12d^2} \qquad (43)$$

Here d = the free distance between the interacting bodies,

$$A = \pi^2 \lambda \cdot N_1 \cdot N_2 \qquad (44)$$

N_1 and N_2 = number of molecules per unit volume,

$$\lambda = \frac{3\hbar}{2} \alpha_1 \cdot \alpha_2 \cdot \frac{\omega_1 \cdot \omega_2}{\omega_1 + \omega_2} \qquad (45)$$

α = polarization, $2\pi\hbar$ = Planck's constant, and ω = angular frequency.

Equations 42 and 43 are only applicable for distances between surfaces d less than 1000 Å.

The difficulties in applying eqs. 42 and 43 are due to the fact that the distances "d" are barely measurable and so are not accurately known. Nevertheless in the calculations they are raised to the second or third power. In addition the Van der Waals constant is difficult to calculate. Experimental values are also not adequate. The measured and calculated value of A for glass against quartz was 10^{-11} to approximately 10^{-13} erg.

b. Macroscopic Theory

This theory was first developed by Lifshitz (1956) and extended by Krupp (42). In contrast to the microscopic theory, these workers proceed from the macroscopic interaction of the two bodies.

As a result of the statistical motion of electrons and nuclei, fast changes with time occur in the polarization with the time aggregate O. These vibration effects radiate like black radiation from the solid surface to the surrounding medium.

The energy dissipation is characterized by the imaginary part of $\varepsilon''(\omega)$ of the complex dielectric constant:

$$\varepsilon(\omega) = \varepsilon'(\omega) + i\varepsilon''(\omega)$$

Here the far UV parts of $\varepsilon(\omega)$ are of importance.

According to Lifshitz, the Van der Waals force per unit surface between two half spaces (planes)

$$P_{\text{vdW}}^- = \frac{\hbar\bar{\omega}}{8\pi^2 Z_0{}^3} \qquad (46)$$

where $\hbar\bar\omega$ is the Lifshitz-Van der Waals constant, and Z_0 is the distance between the bodies. P_{vdW}^- is also described as the Van der Waals pressure (i.e., force per unit surface). $\hbar\bar\omega$ lies between 0.6 and 11 eV, where 0.6 represents synthetic fibers. Metals and semiconductors have $\hbar\bar\omega$ between 2 and 11 eV, whereas for materials such as quartz estimates have still to be made. For adhesion between two different materials, 1 and 2, Kottler et al. (45) estimated the Van der Waals energy $\hbar\bar\omega_{12}$ as the geometric mean of the cohesion energies:

$$\hbar\bar\omega_{12} \simeq \sqrt{\hbar\omega_{11} \cdot \hbar\omega_{22}} \tag{47}$$

Z_0 in eq. 46 is the distance between the adherents where the largest attractive force occurs. Krupp estimates Z_0 at 4 Å, which corresponds approximately to the lattice constant for crystals with Van der Waals bonds. This estimate is confirmed by experiment.

Krupp (42) extended the Lifshitz calculations to the case of adhesion between sphere and half space (representing sphere and fiber) similar to the microscopic theory. This gives the force F_{vdW}° as

$$F_{\mathrm{vdW}}^\circ = \frac{\hbar\bar\omega}{8\pi Z_0{}^2} \cdot R \tag{48}$$

where R is the radius of the contact point.

When the case does not consist of ideal smooth spheres, Krupp and Sperling (46) found that R is no longer the "macroscopic radius" but the radius of curvature of the surface roughnesses, because the two contacting bodies contact the elevated sections of the rough surface. This can lead to multipoint contact. Sperling (47) gives a statistical model for these based on electron micrographs.

The order of magnitude of the Van der Waals forces for dust particles adhering to fibers will be given later and compared to electrostatic forces.

Equation 48 has been found of values in the calculation of Van der Waals forces. The major uncertainty is in the radius R of the elevated surface roughnesses, and here one depends either on estimates or on electron micrographic studies of the surface.

2. Electrostatic Forces

Only the case of excess electrostatic forces, which are unbalanced, is considered here and not effects such as the electrostatic double layer forces.

For a metallic sphere separated by a space Z_0 from a conducting half space the attractive force due to charge Q is given by:

$$K_{\mathrm{el}}^\circ = \frac{Q^2}{16\pi\varepsilon_0[\gamma + \frac{1}{2}\ln 2r_p/Z_0]^2 r_p Z_0} \tag{49}$$

where r_p is particle radius, ε_0 is influence constant $= 8.86 \cdot 10^{-14}$ As/(V cm), and γ is Eulers constant $= 0.5772$.

In filtration the case of nonmetallic, that is, insulated spheres is more important than metallic spheres with a charge. On a metallic sphere the charge is on the surface of the sphere, but on a nonmetallic sphere the charge is distributed over a certain depth, called the space charge depth. For this case the attractive force, according to Krupp (42,45), is

$$F_{el}^\circ = \frac{q^2}{16\pi\varepsilon_0 r_p \delta} \frac{\ln(1 + \delta/Z_0)}{\left(\gamma + \frac{1}{2}\ln\frac{2r_p}{Z_0}\right)\left(\gamma + \frac{1}{2}\ln\frac{2r_p}{Z_0 + \delta}\right)} \tag{50}$$

Here $q =$ effective particle charge, that is, that part of the total charge Q which is noticed by the adhesion partner (of the order of $0.1 \simeq 0.3Q$); and $\delta =$ depth of penetration of the charge (at depth δ, the charge is $1/e \times$ surface charge).

In contrast to eq. 48 the radius r_p here is the total radius; ε_0, γ have been defined for eq. 49.

Kottler et al. (45) investigated the adhesive forces between different highly charged polymer particles, 5–30 μm diameter and selenium surfaces. They found that charging the selenium surface to 2.10^{11} e_0/cm² ($e_0 =$ elementary charge) did not influence particle adhesion. When particles were charged to less than 10^4 e_0/particle, then the measured adhesive force corresponded to the Van der Waals force calculated by eq. 48. With charges greater than 10^4 e_0/particle, the influence of the electrostatic forces increases more rapidly and corresponds, at least approximately, to the values calculated by eq. 50. It should, however, be noted that this concerns charges and conditions which are not likely to occur in filter technology.

Kunkel (48,49) measured maximum charges of 500–1000 elementary charges on different dusts, such as quartz dusts (5–10 μm) after blowing these around. With the value of $r_p = 5$ μm, $Z_0 = 4$ Å $\delta = 5$–10Å, eq. 50 becomes $F_{el}^\circ \simeq 0.05$ to 0.06 mdyne (with increasing δ, F_{el} becomes smaller).

In contrast to this the Van der Waals force F_{vdW}° uses $\hbar\bar\omega_{11} = 0.6$ eV for synthetic fibers, and $\hbar\bar\omega_{22} = 3.5$–6.5 eV for quartz or limestone particles (estimated): eq. 47 gives $\hbar\bar\omega \simeq 1.5$–2 eV.

For radii of curvature at the contact points, $R'' = 0.1$ μm and $R' = 0.5$ μm is used in the calculations.

Then, for $R'' = 0.1$ μm, eq. 48 with $Z_0 = 4$ Å gives $F_{vdW}^\circ = 0.6$–0.8 mdyne; for $R' = 0.5$ μm, $F_{vdW}^\circ = 3$–4 mdyne. According to Sperling (47), with small radii of curvature ($R'' = 0.1$ μm) multipoint contact is probable. If there is three-point contact, $F_{vdW}^\circ = 1.8$–2.4 mdyne. The electrostatic forces F_{el} are also much smaller than Van der Waals forces. Only when the charge Q is 10 times as large as estimated will $F_{el}^\circ \simeq F_{vdW}^\circ$.

However, these charges are not realistic with normal filtration:

1. In the estimation it was assumed that $q = Q$, which according to Krupp is not correct.

2. Harper (50) was only able to measure higher charges than those obtained by Kunkel with extremely clean quartz surfaces.

3. Boehme et al. (51) report that in moist air, even with quartz and polyamide surfaces, the excess charge largely discharges within a few minutes.

4. Löffler's measurements (52) indicated that the adhesive force increased somewhat in damp air, which also argues against a marked influence of electrostatic forces. The measured adhesion for quartz particles on poly-amide fibers was between 1 and 5 mdyne and so were almost of the order of the values for F°_{vdW}.

It can therefore be assumed that generally the conditions in fiber filters do not favor particle adhesion by electrostatic forces. On the other hand, for separating particles from the gas stream, electrostatic attractive forces, because of their longer range than Van der Waals forces, can play a decisive role. This will be the case particularly when the excess charge is unevenly distributed over the particle and is concentrated over a limited area. In the presence of moisture the charges are soon discharged and no longer influence the adhesion.

3. Separation Force

Under the influence of the adhesive forces, Van der Waals and electro-static, the contact points are deformed (Fig. 3).

The first to calculate the deformation with the aid of Hertz's theory of elastic deformation was Deryagin (53). Krupp (42) now pointed out that nonelastic deformations also appear at the points of contact because Van der Waals forces of the order of $4 \cdot 10^8$ dyne/cm^2 can occur between the adherents. Exact calculation of the deformation is not possible. In an estimate that assumes that one adherent is markedly harder than the other, it follows that for radii less than 10 μm nonelastic deformation predominates at the point of contact.

When separating the adhering partner for measuring the adhesive forces, the attractive forces between the undeformed areas and the forces between the quasiparallel contact surfaces have to be overcome. It should be noted that the measurement always requires destroying the adhesive bond.

According to Krupp, the separation force F is

$$F = F^{\circ} \cdot \left[1 + \frac{P^-_{vdW}}{H(t)} \right] \tag{51}$$

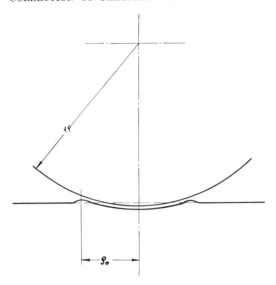

Fig. 3. Deformation of plane on particle contact.

and $H(t) = F^\circ/\pi\rho_0{}^2 =$ measure of the deformation resistance of the softer of the two adherents; a time-dependent function ("creeping").

P_{vdW}^- from eq. 46 gives the value for the force between parallel contact surfaces. (Krupp shows that $P_{\mathrm{el}}^- \ll P_{\mathrm{vdW}}^-$.) F_0 is the sum of the attractive forces beyond the contacting surfaces which cause the deformation.   Because, as was shown above, $F_{\mathrm{el}}^\circ \ll F_{\mathrm{vdW}}^\circ$ for fiber filters, it is possible to write

$$F = F_{\mathrm{vdW}}^\circ \left[1 + \frac{P_{\mathrm{vdW}}^-}{H(t)} \right] \tag{52a}$$

that is,

$$F = \frac{\hbar\bar{\omega}}{8\pi Z_0{}^2} r_p \left[1 + \frac{\hbar\bar{\omega}}{8\pi^2 Z_0{}^3 H(t)} \right] \tag{52b}$$

With $H(t) = 3.10^8$ dyne/cm² (for polymers), for the numerical example given earlier, the value of the square brackets is approximately 2–3. This means that the adhesive force is strengthened by this factor as a result of the deformation. The deformation is particularly marked when the two adherents are of very different hardness.

4. Capillary Forces

With mobile sorption layers on the two adherents, liquid bridges can be formed at the contact points between particles and fibers. In these wedges there is a low-pressure region that can be calculated using Laplace equations.

From the reduced pressure and the contact surface, the adhesive force for a sphere with radius r_p on a plane surface can be calculated [Bowden and Tabor (38) and Corn (40)].

$$F_{cap} = 4\pi\gamma \cdot r_p \tag{53}$$

γ is surface tension of the fluid, and r_p is the radius of the contact roughness.

Equation 53 is effective for small quantities of liquid at the contact points and complete wetting. When surfaces are not smooth, the radii of the ridges of the roughened surfaces must be substituted for r_p. If one considers the radii, as above, of R″ = 0.1 μm and a value of γ = 72 dyne/cm (water), then using eq. 53 the value of F_{cap} = 9 mdyne, whereas for R′ = 0.5 μm, F_{cap} = 45 mdyne.

Löffler (54) calculated the capillary force for the case of particles contacting the plane surface with a tapered point instead of a spherical shape. For this case the equation is

$$F_{cap} = \pi\gamma b\left[\frac{\sin\alpha}{tg\beta} - \frac{\sin\alpha}{1-\sin\beta}\right] \tag{54}$$

The significance of the symbols in this equation is given in Figure 4.

When $b \simeq c$, which is approximately the case for α between 30 and 60°, then eq. 54 can be simplified:

$$F_{cap} \simeq \pi\gamma \cdot c\left[\frac{\sin\alpha}{tg(45° - \alpha/2)} - 1\right] \tag{55}$$

F_{cap} becomes zero when $\sin\alpha/[tg(45° - \alpha/2)] = 1$, which is the case when $\alpha \simeq 33°$. This means that only tapers with $\alpha > 33°$ can be held by liquid bridges. If $c \simeq 500$ Å U, then for α between 45 and 60° and water as liquid (γ = 72 dyne/cm) gives $F_{cap} \simeq 0.8$–2.5 mdyne.

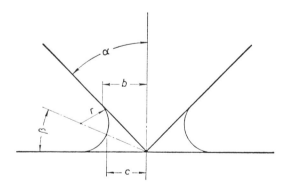

Fig. 4. Liquid bridge between a tapered point and a plane surface.

It is apparent that capillary forces can become considerably greater than Van der Waals forces, when liquid bridges are formed. That this actually occurs depends on the particular system. The most important adhesive force is considered in connection with experimental results.

C. Methods of Measurement

Different methods of measurement with different techniques have been used over the past 40 years for measuring adhesive forces. In most cases idealized systems were used such as fresh fibers with smooth surfaces and spheres. The aim of the measurements was to determine the material constants for the theories of adhesion. Boehme et al. (56) and Corn (57) review the methods used. For filtration technology two methods are of most interest—the centrifuge and the aerodynamic methods.

1. The Centrifuge Method

With particles several microns in diameter, the adhesive forces are larger than the particle weight by a factor of 10^3 or 10^5. Still, the absolute force is only a few millidynes. The measurements of such small forces are therefore best carried out in a centrifugal field because very high accelerations of known value can be achieved. For example, Polke (58) used an electronically driven ultracentrifuge where accelerations of $3 \cdot 10^6 g$ (g = acceleration of gravity) were achieved.

For determining the adhesive (i.e., separating) force, the centrifuge speed (number of revolutions per unit time) for detaching particles has to be determined. From the number of revolutions per unit time (i.e., angular velocity) and the particle mass, the separating force can be found from

$$F = m \cdot r \cdot \omega^2 \tag{56}$$

where m is particle mass, r is the distance between particle and rotor axis, and ω is angular velocity.

The centrifuge method has been used by Larsen (55), Boehme et al. (56), Kordecki and Orr (59), Corn and Stein (60), and others. In general, these workers carried out their measurements by scattering the particles on a surface which was then placed in the rotor of a centrifuge so that the centrifugal force acted either in the normal or parallel direction to the surface. The centrifuge speed was increased in a stepwise manner and the remaining particles counted after each speed increases.

An added advantage of this method is due to the fact that a number of particles can be investigated simultaneously. This is important because even apparently identical particles with identical conditions have large variations in adhesive forces. This is probably a result of microinhomogeneities at the points of contact.

Because the method of distributing the particles is an important parameter, Löffler (53) filtered the particles from an air stream on to a grating of
stretched parallel fibers. The particles were then centrifuged off the grating.
The advantage of this technique is that it resembles the actual filtration
process in the way the particles are deposited on the fibers.

2. *Aerodynamic Method*

In this method, particles are separated from their support as a result of the
stream line forces caused by blowing toward the supporting body. Gillespie
(36) and Larsen (55) worked with a single fiber; Corn and Silverman used a
woven screen (61), while Löffler (52) used screens of parallel fibers for the
blowing-off experiments. There are also a number of experiments in which
particles have been blown off plane surfaces [see Corn (40)].

In the first instance this method only leads to the re-entrainment velocities
which is, of course, of particular importance in filter design. On the other
hand, it is not yet possible to use this method to determine the adhesive
forces because the actual conditions pertaining to the particles are not known.
However, Corn (61) and Löffler (52) estimated approximate values of the
adhesive forces from their experiments by this method.

The model in this case is a sphere sitting on the surface of a cylinder.
Buoyancy forces are neglected. The resistance force acting on the particle
tangential to the cylinder surface can be found from

$$W = \tfrac{1}{2}\rho v^2 F c_w (\text{Re}) \tag{57}$$

where ρ is the density of the fluid, v is the effective gas velocity at the particle,
F is the projected surface area of the particle perpendicular to the fluid
stream, c_w is the friction factor, and $\text{Re} = v\, d_p \rho/\mu$ is the Reynolds number
for the particle.

The effective velocity is that acting on the center of mass of the particle.
This is calculated with the aid of the boundary layer equations depending on
the location, assuming that the boundary flow is not disturbed by other particles located upstream on the cylinder. The value used for c_w is that for
spheres in an infinite fluid. These last two are not strictly valid assumptions,
but other more realistic values are not available, and the results thus obtained
are at best only a crude approximation. Nonetheless Löffler (52), in contrast
to Corn (61), obtained good agreement between the resistance forces estimated
by eq. 57 and the separation forces measured in the centrifuge.

D. Experimental Results

Because of the large number of experimental results published by different
workers, which are often difficult to compare, only some that influence
particle adhesion and have particular relevance to filtration are discussed.

Corn (57) found that in measuring the adhesive forces for individual particles, different values were obtained for the same particles in multiple contacts.   He used the molten rounded end of a fiber as a "particle."   It may be expected that for irregularly shaped particles a wide variation of adhesive forces are obtained.   This is confirmed by the measurements of Boehme et al. (51), Sperling (47), Kordecki and Orr (59), and Corn and Stein (60).

Figure 5 shows the distribution of adhesive forces obtained by Löffler using the centrifuge method (52).   The measurements were done with closely sized fractions ($\sigma_g \simeq 1.15$) of quartz particles which were filtered on to the polyamide fibers.   Experiments with other fibers (polyester, glass) and other particles (limestone) gave a similar qualitative pattern.   The presentation of the results on log-probability scales shows that up to 80% of the values plotted follow a log-normal distribution.   The scatter, as shown by the slopes of the straight lines, was approximately the same for all the systems studied, indicating that the median value F_{50} may be used to characterize the adhesive force.   The figure also demonstrates the wide scatter of adhesive forces.   This is caused by the irregularities in particle shape, surface irregularities, nonhomogeneous surface properties, and the differences in particle position.   Because all the particles do not sit in the stagnation line on the fibers, the separation force during centrifuging may not act normally to the point of contact.   Measurements by Corn and Stein (60) and Unterforsthuber (62) showed that the separating force is strongly dependent on the direction relative to the supporting surface.   The forces required for separation tangentially are much smaller than those in the normal direction.

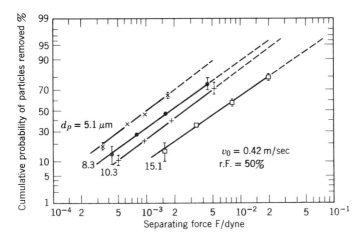

Fig. 5. Distribution of adhesive forces of quartz particles on polyamide fibers (Log-probability scales).

1. Particle Size and Filtration Velocity

It was found by Boehme et al. (56), Corn (60,63), and Larsen (55) that the adhesive force increases proportionally with particle size. Similarly the experiments by Löffler gave increases in adhesive force with particle size for all the systems investigated. Figure 6 shows the median values for the distribution of adhesive forces for particles of quartz on polyamide fibers. The parameter here is the filtration velocity.

The filtration velocity has a marked effect on the adhesive force, and at the same time the number of particles which adhered to the fibers was observed to decrease with increasing filtration velocity. This indicates the great importance which must be attached to the way the particles are deposited on the filters, when adhesive forces are being measured.

The influence of velocity is a statistical sorting-out process during filtration. The recoil momentum and flow separation forces increase with increasing filtration velocity. Thus only the particles with favorable adhesion conditions, that is, larger adhesive forces, remain on the fibers. It follows that the separating force for removing the particles is also greater. Corn and Silverman (61) observed an analogous effect. Similarly, Rumpf (65) found in his investigation of deposits in ducts that higher velocities give rise to thinner, but more adhesive deposits.

Obviously the most important processes occur during the initial impact between particle and fiber. Löffler and Umhauer (66) report an optical

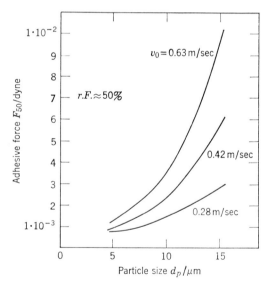

Fig. 6. Dependence of adhesive force on particle size and filtration velocity ($v_0 =$ upstream velocity).

method for observing particle paths in the immediate vicinity of fibers.   In analyzing the films it was observed that a growing number of the impacting particles were reflected.   This confirms the above-mentioned hypothesis of a sorting-out effect.   This work is continuing.

2. Influence of Different Materials

Boehme et al. (51) found in their measurement of the adhesion of gold spheres on different substrates that the adhesion of the particle depended on the material and surface texture of the substrate.   The adhesion was firmer with soft substrates, which is due to the greater deformation at the points of contact.   The median values for the adhesion for gold spheres of mean diameter 5 μm were between 70 and 100 mdyne, whereas on coarse quartz the value was 10 mdyne.

Löffler investigated the adhesion of quartz and limestone particles on polyamide, polyester, and glass fibers approximately 50 μm in diameter. The polyamide fibers had two cross sections: round and Y-shaped.   Table I presents measured values for the adhesive force for 10-μm particles in an atmosphere with a relative humidity of 50–60%.   In two cases the glass fibers showed somewhat higher adhesive forces, although these are not significant when considering their application.   This effect can be explained by the fact that surface properties in normal atmospheric conditions are largely affected by adsorbed materials, chiefly water.

In order to investigate the changes in adhesion with increasing particle deposition on the fiber, a primary layer of particles was stuck on to fibers (64). Further particles were then deposited on the fiber by filtration and then centrifuged off.   In this way the adhesive forces between particles in successive layers were measured.   No significant differences were observed compared to the particles on "clean" fibers.   The median values were in the range of 1–10 mdyne.   Krupp's hypothesis (42) on the influence of hardness

TABLE I

Median Values of Centrifugal Force F_{50} (mdyne or 10^{-3} dyne)

	Polyamide			
	Round	Y-shaped	Polyester	Glass
Quartz 10 μm				
0.28 m/s	1.4	1.5	1.7	2.0
0.42 m/s	2.4	2.4	1.8	2.4
Limestone 10 μm				
0.28 m/s	1.5	1.4	2.0	2.5
0.42 m/s	2.3	2.0	—	—

would lead toward the expectation that adhesion on hard quartz particles would give a reduction in adhesive forces, but the results indicate that the effect of the changed surface geometry, because of the particle layer, is of greater importance. In the fissured structure of the adhesive primary layer more contact points can be created between the adhesion partners. A significant difference compared to the measurements with clean fibers was observed in the number of deposited particles. When a primary layer was present, the number of particles deposited did not decrease with gas stream velocity, but increased slightly. Furthermore, 3–4 times as many particles were deposited on the primary layer than on the clean fibers. It can therefore be concluded that when particles impact on a particle layer, the damping effect is greater than on the clean fiber.

Most median values for adhesion, with and without the presence of a primary layer, were in the range of 1–10 mdyne. The numerical example given above indicated that this was in the range of Van der Waals forces calculated by eq. 48. The theoretical electrostatic forces were smaller by a factor of 100; capillary forces were markedly higher. The conclusion that may be reached from this evidence is that in the systems investigated Van der Waals forces predominate the adhesive forces.

3. Influence of Moist Air

Larsen (55) has reported that when air has a relative humidity of 40% much higher velocities were needed to blow glass spheres off a glass fiber than with conditions of 22% relative humidity. Corn (63) found that with particle diameter of 88 μm there was a greater increase in adhesion with increased air moisture than with particles only 25-μm diameter. Corn (40) refers to work by Zimon, who obtained steep increases in adhesion for glass spheres on quartz surfaces with a relative humidity of approximately 60%.

Boehme et al. (51) suggested that the reason for their observed increase in adhesion with moisture in air was that this gave a softer substrate surface. This has been confirmed by Vicker's hardness measurements. As a result of Van der Waals forces this gives larger contact surfaces which in turn lead to greater separation forces.

Figure 7 represents Löffler's measurements. The increase in adhesion with synthetic fibers can be attributed to a softening of the fiber surface. The strongest changes in adhesion were found with glass fibers when the relative humidity exceeded 60%, which leads to the conclusion that at certain contact points liquid bridges had been formed. However, if capillary forces were to become predominant, then the measured adhesion should be increased even further, as the theoretical approximation (Sec. III-B-4) indicates.

The increase in adhesive forces with humidity is a further indication that electrostatic forces do not play a decisive role in particle adhesion, because

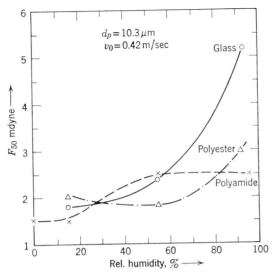

Fig. 7. Dependence of adhesion on relative atmospheric humidity.

increasing humidity tends to reduce electrostatic forces, because they are rapidly discharged.

4. Results of the Aerodynamics Method

Gillespie (36) found that to blow completely Lycopodium spores from a 150-μm steel cylinder, the blow-off velocities had to be three times as large as the filtration velocity. Larsen (55) reports that with a relative humidity of 40% a blast velocity of 20 m/sec was necessary to remove the first 16-μm glass spheres from glass fibers. When velocities had reached 65 m/sec about 90% of the spheres had been removed. When fibers had been wetted, the velocities required were much greater. Corn and Silverman (61) investigated the blowing of small quartz particles (1.3 μm) out of a screen (Tyler 100 mesh). The velocity required for blowing the particles off the screen increased with the initial filtration velocity. Thus dust filtered at 1.16 m/sec required a blow-off velocity of 12.20 m/sec in order to gain the initial pressure drop of the clean screen.

The investigations listed here show how much higher velocities are needed for removing particles than were used during the initial filtration. This is confirmed by Löffler (52,64), who obtained appreciable separation of particles from fibers at velocities which were noticeably greater than the filtration velocity.

The median values U_{50} entered on the ordinate in Figure 8 is the velocity which removed 50% of the particles. It is apparent that U_{50} is in the range of 10–20 m/sec.   In the investigations where a primary layer of particles had been stuck to the fiber the gain velocity required to remove the secondary particles was 1.5 times as large as in the experiments without the primary layer.

Corn and Silverman (61) calculated the resistance by the method given in Section III-C-2 and found that these calculated values were a factor of ten below the measured values.   These adhesive force measurements were, however, carried out using a microbalance, where rounded molten fiber ends were used as "particles."   The present author believes, in contrast to Corn and Silverman (61), that this discrepancy does not rest on uncertainties in the calculation, but on the different surface roughness on the particles used for the different experiments, because the resistance forces calculated by Corn and Silverman agree well with Löffler's measurements (52) with irregular particles.

Löffler also estimated the stream line separation force from the measured blow off velocities.   These are compared to the results from the centrifuge experiments in Figure 9.

The median values of the distributions differ by a maximum of 50% when related to the centrifuge values.   With 15-μm particles the differences were as low as 30%, which is very good agreement if the approximate nature of

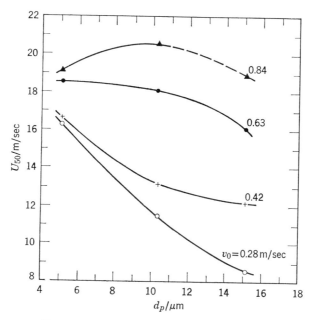

Fig. 8. Median values of the blow off velocity U_{50}.

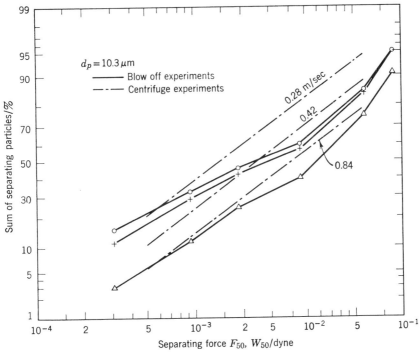

Fig. 9. Distribution of separating forces in blow off and centrifuge experiments.

the calculations are considered.   An explanation of the different slopes of the curves is given by the different directions of the stress when the particles are detached.   Thus in the centrifuge the particles are spun off in the opposite direction to their arrival during the filtration process, that is, in a direction normal to the contact surface.   On the other hand, the particle separates during blowing off in the same direction as the filtration gas stream, and so the adhesion is broken by a shear stress.

E. Conclusions

From the comparison of theoretical with experimental adhesive forces it appears that in fiber filters electrostatic forces act in the particle deposition process, that is, the transport of the particles to the surface, but are not of major significance in particle adhesion.   The extent to which Van der Waals forces and capillary forces play a part in particle adhesion depends on conditions; thus high humidities and wetted fibers will significantly influence the capillary forces.

In many cases, however, the adhesive force depends on Van der Waals forces.   Their qualitative values, particularly in the macroscopic case, are best estimated by the theory of Lifshitz. The main difficulty is determining the appropriate roughness factors. If certain assumptions about this are made, then good agreement between measured and calculated values can be obtained.

The experiments in blowing off particles from fibers show that with the velocities common in filtration technology, particles are not generally blown off even from dry fibers because the velocities required for this are not even approached. (It should, however, be noted that with very high filter loadings conditions are different.) Once particles are deposited their adhesion is very firm. It is not possible, however, to conclude from this that filters can without hesitation be operated at higher velocities. The reason is that with increasing velocities a much larger fraction of particles rebounds on impact with fibers. This is also the basis for the experimentally found decrease in collection observed with particles greater than 4–5 μm.

The loosening of deposited particles by bombardment by new particles in the gas stream is also more probable with increasing velocity. This is shown in cine-shots. The explanation of these processes, however, will require further investigation.

References

1. W. Strauss, *Industrial Gas Cleaning*, Pergamon Press, Oxford, 1966; (a) ch. 7; (b) Figure 7.2; (c) p. 222.
2. D. G. Stephan, G. W. Walsh, and R. A. Herrick, *Ind. Hyg. J.*, **21** 1–4 (1960).
3. M. W. First and L. Silverman, *J. Air Pollution Control Assoc.*, **13**, 581 (1963).
4. J. W. Robinson, R. E. Harrington, and P. W. Spaite, *Atmosph. Environ.*, **1**, 499 (1967).
5. F. H. H. Valentin, *Brit. Chem. Eng.*, **7**, 268 (1962).
6. W. E. Ranz and J. B. Wong, *Ind. Eng. Chem.*, **44**, 1371 (1952).
7. C. Y. Chen, *Chem. Rev.*, **55**, 595 (1955).
8. E. Landt, *Gesundh. Ingr.*, **77**, 139 (1956).
9. E. G. Richardson, *Aerodynamic Capture of Particles*, Pergamon Press, Oxford, 1960.
10. N. A. Fuchs, *The Mechanics of Aerosols*, Pergamon Press, Oxford, 1964; (a) p. 216; (b) p. 217.
11. R. G. Dorman, "Filtration," in *Aerosol Science*, C. N. Davies, Ed., Academic Press, London, 1966.
12. J. Pich, "Theory of Aerosol Filtration by Fibrous and Membrane Filters," in *Aerosol Science*, C. N. Davies, Ed., Academic Press, London, 1966.
13. K. Spurný, Fortschr.-Ber. VDI-Z. Reihe 3 Nr. 17, 1967.
14. S. Kuwabara, *J. Phys. Soc. Japan*, **14**, 527 (1959).
15. J. Happel, *Am. Inst. Chem. Eng. J.*, **5**, 174 (1959).
16. A. A. Kirsch and N. A. Fuchs, *J. Phys. Soc. Japan*, **22**, 1251 (1967).
17. A. A. Kirsch and N. A. Fuchs, *Ann. Occup. Hyg.*, **10**, 23 (1967).
18. J. Pich, *Staub-Reinhalt. Luft*, **26**, 267 (1966).
19. W. Sell, *VDI* (*Ver. Deut. Ingr.*) *Forschungsh.*, 347 (1931).
20. F. Albrecht, *Physik, Z.*, **32**, 48 (1931).

21. K. R. May and R. Clifford, *Ann. Occup. Hyg.*, **10**, 83 (1967).
22. K. T. Whitby, *ASHRAE (Am. Soc. Heat. Refrig. Air-cond. Engrs.) J.*, **7** (9), 56–65 (1965).
23. D. A. Lundgren and K. T. Whitby, *Ind. Eng. Chem., Process Design and Development*, **4**, 345 (1965).
24. T. Gillespie, *J. Colloid Sci.*, **10**, 299 (1955).
25. V. Havlicek, *Intern. J. Air Water Pollution*, **4**, 225 (1961).
26. W. Walkenhorst and G. Zebel, *Staub*, **24**, 444 (1964).
27. G. Zebel, *J. Colloid Sci.*, **20**, 522 (1965).
28. G. Zebel, *Staub-Reinhalt. Luft*, **26**, 281 (1966).
29. S. K. Friedlander, *Ind. Eng. Chem.*, **50**, 1161 (1958).
30. D. Hasenclever, *Staub-Reinhalt. Luft*, **26**, 288 (1966).
31. W. L. Torgeson, Paper No. J 1057, Appl. Sci. Div., Litton Systems Inc. St. Paul-Minnesota, quoted by Whitby (22).
32. J. F. Van der Waal and L. A. Clarenburg, *Ind. Eng. Chem. Proc., Design Develop.*, **5**, 110 (1966).
33. S. C. Stern, H. W. Zeller, and A. I. Schekman, *J. Colloid Sci.*, **15**, 546 (1960).
34. H. Schuster, Deutsche Luft- und Raumfahrt, Forschungsbericht 67-35, 1967.
35. A. Kaufmann, VDI-Z, **80**, 593 (1936).
36. T. Gillespie, *J. Colloid Sci.*, **10**, 266 (1955).
37. K. T. Whitby, A. B. Algren, and R. C. Jordan, *ASHRAE (Am. Soc. Heat. Refrig. Air-cond. Eng.) J.*, **4** (9), 79–88 (1962).
38. F. P. Bowden and T. Tabor *Reibung und Schmierung fester Körper* (translated by E. H. Freitag), Springer-Verlag, 1959. (Original English edition—*Friction and Lubrication of Solids*, 2nd ed., 1954).
39. R. Houwink and G. Salomon, *Adhesion and Adhesives*, Elsevier, Amsterdam, 1965.
40. M. Corn, "Adhesion of Particles," in *Aerosol Science*, C. N. Davies, Ed., Academic Press, London, 1966.
41. H. Rumpf, *Chemie-Ing.-Technik*, **30**, 144 (1958).
42. H. Krupp, *Advan. Colloid Interface Sci.*, **1**, 111 (1967).
43. G. A. Tomlinson, *Phil. Mag.*, **6**, 695 (1928); **10**, 541 (1930).
44. H. C. Hamaker, *Physica*, **4**, 1058 (1937).
45. W. Kottler, H. Krupp, and H. Rabenhorst, *Z. Angew. Physik*, **24**, 219 (1968).
46. H. Krupp and G. Sperling, *Z. Angew. Physik*, **19**, 259 (1965).
47. G. Sperling, Dr.-Ing.-Dissertation Universität Karlsruhe, 1964.
48. W. B. Kunkel, *J. Appl. Phys.*, **21**, 820 (1950).
49. L. B. Loeb, *Static Electrification*, Springer-Verlag, 1958.
50. W. R. Harper, *Advan. Phys.*, **6**, 365 (1957).
51. G. Boehme, W. Kling, H. Krupp, H. Lange, and G. Sandstede, *Z. Angew. Physik*, **16**, 486 (1964).
52. F. Löffler, *Staub-Reinhalt. Luft*, **26**, 274 (1966).
53. B. V. Deryagin, *Kolloid-Z.*, **69**, 155 (1934), quoted by Krupp (42).
54. F. Löffler, Dr.-Ing.-Dissertation, Universität Karlsruhe, 1965.
55. R. J. Larsen, *Am. Ind. Hyg. Assoc. J.*, **19**, 265 (1958).
56. G. Boehme, H. Krupp, H. Rabenhorst, and G. Sandstede, *Trans. Inst. Chem. Engrs.*, **40**, 252 (1962).
57. M. Corn, *J. Air Pollution Control Assoc.*, **11**, 523 (1961).
58. R. Polke, *Chemie.-Ing.-Techn.*, **40**, Heft 21/22 (1968).
59. M. C. Kordecki and C. Orr Jr., *Arch. Environ. Health*, **1**, 13 (1960).
60. M. Corn and F. Stein, *Am. Ind. Hyg. Assoc. J.*, **26**, 325 (1965).

61. M. Corn and L. Silverman, *Am. Ind. Hyg. Assoc. J.*, **22**, 337 (1961).
62. K. Unterforsthuber, unpublished measurements from the Institut für Mech. Verfahrenstechnik, Universtität Karlsruhe.
63. M. Corn, *J. Air Pollution Control Assoc.*, **11**, 566 (1961).
64. F. Löffler, *Staub-Reinhalt. Luft*, **28**, Heft 11 (1968).
65. H. Rumpf, *Chemie-Ing.-Technik*, **25**, 317 (1953).
66. F. Löffler and H. Umhauer, lecture to the Schwebstofftechnischen Arbeitstagung, Universität Mainz, October 1967; (publication in preparation).
67. W. Strauss and B. W. Lancaster, *Atm. Environ.*, **2**, 135 (1968).

Symbols

a Number of particles collected per unit length of fiber

B Mobility $= (3\pi\mu d_p)^{-1}$

C Cunningham correction factor

C_D Friction factor of a cylinder

C_w Friction factor for a sphere

$\mathscr{D}$ Diffusivity

D Fiber diameter

D Diameter of adhering sphere

d Separation between adhering particle and cylinder

d_p Particle diameter

$E_\circ$ Field strength

$e_\circ$ Elementary charge

E_T Total efficiency of a fiber layer

F° Attractive force between sphere and plane

F Separating force

F_{50} Median value of the distribution of adhesive forces

g Gravity acceleration

$\hbar\bar{\omega}$ Lifshitz-Van der Waals constant

$2\pi\hbar_e$ Planck's constant

Kn Knudsen number $= \lambda/r_f$

k Boltzmann's constant

L Depth of fiber layer

l Fiber length per unit volume of filter

m Mass of particle

N_0 Particle concentration before filtration

N Particle concentration after filtration

n_0 Particle concentration before single fiber

Pe Peclet number

P_{vdW}^- Van der Waals attraction per unit surface area

Q Total excess electrostatic charge

q Effective particle charge

R Radius of contact point

R', R'' Radius of curvature of the surface roughness
R Interception parameter
Re Reynolds number
r Distance
r_p Radius of sphere
T Absolute temperature
t Time
U Interaction potential
v_0 Undisturbed upstream velocity
v Flow velocity
w Weight per unit surface area of filter
Z_0 Separation for maximum attractive force
α Polarization
β Fiber volume fraction
γ Surface tension
γ Eulers constant $= 0.5772$
δ Depth of space charge
ε Porosity
ε_f Dielectric constant of fiber
ε_p Dielectric constant of particle
ε_0 Influence constant $= 8.86 \cdot 10^{-14}$ A.s/V.cm
η Collection efficiency of a single fiber
λ Mean free path of gas molecules, van der Waals constant
μ Dynamic viscosity of gas
ν Kinematic viscosity of gas
ρ Density of gas
ρ_p Density of particle
ϕ Stream function
ψ Inertial impaction parameter (Stokes number)
ω Angular velocity

Condensation Effects in Scrubbers

BRIAN W. LANCASTER* AND WERNER STRAUSS

University of Melbourne, Victoria, Australia

I. INTRODUCTION

It is some 60 years since a new principle has been successfully applied to the collection of dusts and mists. This was the principle of electrostatic precipitation and even this had been suggested as early as 1824. Over the last half century the major developments in gas cleaning have been restricted to improvements in design and operating efficiency of existing types of collection equipment.

A large variety of wet scrubbers were in use at the turn of the century, and at least one was as efficient as any in common use today (58). Major developments in this field have been associated with simplification of designs, reduction of operating costs, and reduction of maintenance requirements.

Little study was made of the fundamentals of wet scrubbing until the 1930's when Kleinschmidt and Anthony analyzed the performance of cyclonic spray scrubbers (Pease-Anthony type) (59) which was followed by the Pease-Anthony-Venturi scrubber. Although the Venturi scrubber concept was first patented in 1925, the modern version was not developed and put into commercial practice for another 20 years.

* Presently at Westinghouse R & D Center, Pittsburgh.

The Venturi scrubber has been subjected to rigorous scientific investigation during the past two decades and numerous attempts have been made to correlate scrubber performance with particle parameters and scrubber energy consumption (31,34,53). A great deal of work was carried out in the postwar years by various research groups in the United States, although more recently the literature on scrubber investigation has declined sharply.

Two basic processes occur in equipment designed to bring particles in contact with a scrubbing liquid. The first is one of particle conditioning, in which the effective size of the particle is increased, making it less difficult to capture. The second critical process is that of deposition of the particles on the surface of the scrubbing liquid.

Particle conditioning in scrubbers may be achieved by a number of mechanisms which fall into two groups: (a) agglomeration; and (b) condensation of vapor on the particles.

The various methods of achieving agglomeration have been studied extensively and some use has been made of these techniques in gas cleaning equipment   However, little work has been carried out on particle conditioning by condensation, although the data which is available suggests that this technique could find economic commercial application.

It has been suggested that condensation in gas cleaning plants may substantially improve particle capture not only by particle buildup, but also by enhancing the diffusion capture mechanisms. Consequently the study of condensation as a method of gas cleaning requires an assessment of the relative importance of both the particle buildup and the process of diffusion capture. These two aspects will be discussed separately and the available data on condensation scrubbers will be reviewed in detail.

II. PARTICLE CONDITIONING

Although there is little fundamental information available on the effect of aerosol particle buildup by a condensing vapor on the efficiency of wet scrubbers, a considerable amount of work has been carried out on the condensation process in various other applications. From this it is possible to deduce the effect of condensation on particle buildup in scrubber systems.

The initial consideration is the method by which the condensable vapor is introduced into the aerosol and condensed on the particles. To ensure maximum effective usage of vapor in the buildup of the particles it would be desirable to distribute the condensed vapor equally among the particles. This could be achieved if the vapor and the aerosol were intimately mixed before condensation was initiated. As the adiabatic expansion of saturated air produces condensation of vapor, it would be feasible to thoroughly mix vapor with an aerosol under pressure, and produce uniform condensation by subsequent adiabatic expansion. This system has been employed by Schauer (1), but unfortunately for most practical applications the energy requirements of such a system would be prohibitive.

An alternative method would be to inject a jet of vapor into the aerosol at ambient pressure, relying on the quenching of the vapor by the aerosol to promote condensation. It is this process which will be discussed in detail. In this situation the degree of mixing achieved prior to nucleation and condensation of the vapor would determine the number of particles on which condensation occurred. The kinetics of the nucleation and growth of droplets on the particles would also influence this and would determine the ultimate particle size distribution.

Consequently it is necessary to look at condensation in jets and the associated mixing and nucleation phenomena if particle buildup by this mechanism is to be understood.

Particle buildup may also occur by agglomeration of the aerosol particles in the turbulent mixing zone of the jet and the possibility of particle collisions in the turbulent field must also be considered.

A. Nucleation

1. Homogeneous Nucleation

When vapor is condensed in a turbulent jet, the possibility of condensation on homogeneously produced nuclei must be considered in conjunction with heterogeneous nucleation (i.e., on foreign nuclei). Homogeneous nucleation has been widely treated both theoretically and experimentally. The Becker-Doring theory (2) is generally accepted although it expresses nucleation rate in terms of liquid parameters which cannot be strictly applied to the clusters of 70 or 80 molecules which constitute nuclei. The theory predicts nucleation to be inappreciable below a certain critical level of supersaturation, which is about five for water vapor in air at 0°C. More rigorous treatments on a statistical mechanics basis are also available (3), but as the Becker-Doring theory accurately predicts experimental results it is adequate for most purposes.

Amelin (4) and Amelin and Belakov (5) have studied nucleation in a jet of hot vapor exhausting into a cool atmosphere. The proposed vapor structure of the jet is shown diagramatically in Figure 1. Here the zone A is a zone of unmixed vapor, B and C are boundaries of the mixing zone, while S_{crit} is the surface of critical supersaturation.

From simple heat and material balances on the system it can be shown that the supersaturation (S) at any point in the mixing zone for a two component vapor/gas system is given by:

$$S = (p_1 + p_2 Y)/(1 + Y)p_S \tag{1}$$

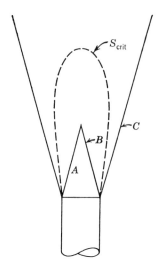

Fig. 1. Condensation in a free jet (Amelin (4)). *A*,
Undiluted vapor; *B* and *C*, boundaries of mixing
zone; S_{crit}, surface of critical supersaturation.

and the temperature at this point is:

$$T = \frac{T_\infty - T_0 b Y}{1 + b Y} \tag{2}$$

where p_1 and p_2 are the partial vapor pressures of the two components, p_s
is the vapor pressure of the saturated vapor, Y the mole ratio of vapor in the
mixture, and b the ratio of the molar specific heats of the components.
T_0 and T_∞ are the temperatures at the surfaces B and C of the jet (°C).

Thus depending on the conditions of the vapor and the atmosphere,
degrees of supersaturation high enough to produce condensation could be
realized in the mixing zone of such a jet.   In the absence of heterogeneous
nuclei, nucleation and subsequent condensation would occur after the critical
supersaturation level was achieved (represented by S_{crit}, Fig. 1).   The
presence of heterogeneous nuclei would allow nucleation and condensation
to take place at much lower values of supersaturation.   By studying
homogeneous condensation in a jet, Amelin (4) showed that the value of
S_{crit} in such a system compared favorably with values found in cloud
chamber experiments.   Further work by Amelin and Belakov (5) considered the possibility of heterogeneous nucleation on submicroscopic nuclei
introduced by the mixing components, and it was shown that homogeneous
nucleation predominated over heterogeneous nucleation in the system under
study.

The simplified analysis of the jet condensation employed by Amelin assumed that all nucleation occurred at a supersaturation level corresponding to

the point of observed cloud formation in the jet. In an extensive study of condensation in a turbulent jet, Higuchi and O'Konski (6) criticized this assumption. They assumed a solution to the equation of motion in a turbulent jet and by combining this with the Becker–Doring theory, produced an expression of the nucleation rate in a jet. The expression was tested experimentally by rapidly quenching a vapor jet with a cool gas stream and electrically counting the resulting aerosol particles. The derived expression was shown to predict experimental nucleation rates with fair accuracy. The approximations of Amelin (4) were also applied to the system under study and showed good agreement with the measured results.

A theory of condensation in the mixing zone of a jet has been developed more recently by Levine and Friedlander (7) for systems in which the Lewis number (the ratio of the Schmidt and Prandtl numbers) for the vapor is one. For the steam–air system (the system of most interest in scrubber applications) the Lewis number approximates to one and this theory may be applied. In the absence of condensation, the equations of conservation of mass and energy in a jet can be expressed in the form

$$w \cdot \frac{dy'}{d\theta} + wv \cdot \nabla y' = \nabla \cdot wD' \nabla y' \tag{3}$$

$$w \cdot \frac{dT'}{d\theta} + wv \cdot \nabla T' = \nabla \cdot (\kappa/C_p) \nabla T' \tag{4}$$

where D' is the diffusion coefficient, w is the density, v is the velocity, C_p is the specific heat of the gas at constant pressure, κ is the thermal conductivity, and where y' and T' are the dimensionless parameters defined by

$$y' = \frac{y - y_\infty}{y_0 - y_\infty} \qquad T' = \frac{T - T_\infty}{T_0 - T_\infty} \tag{5}$$

y_0, y_∞, T_0, and T_∞ are the conditions of composition and temperature at surfaces B and C of the jet (Fig. 2).

$$\mathrm{Le} \simeq 1 \simeq \kappa/wD'C_p$$

therefore

$$wD' = \kappa/C_p$$

where Le is the Lewis number and hence eqs. 3 and 4 are of the same form and have identical boundary conditions. Thus, the solutions of these two equations are identical, and y' and T' are the same. It follows that

$$\frac{y - y_\infty}{y_0 - y_\infty} = \frac{T - T_\infty}{T_0 - T_\infty} \tag{6}$$

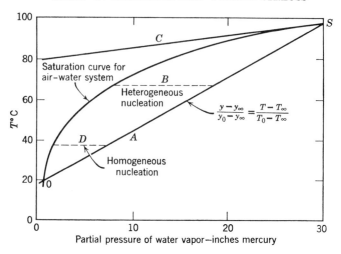

Fig. 2. Saturation curves and condensation curves for the air–water system. (The nucleation paths *B*, *C*, and *D* are explained in the text.)

If eq. 6 is plotted on the same axis as the saturation curve for the air–water system, a straight line plot is obtained (Fig. 2) and this represents the path followed in the mixing zone of the jet without condensation. If the homogeneous nucleation occurs at a supersaturation level represented by point *A*, the path followed is represented by curve *D* (neglecting latent heat effects). Heterogeneous nucleation would be expected to begin at an earlier stage and the curve *B* would be followed in this case. If the plot of eq. 6 falls entirely in the one-phase region (curve *C*) condensation is not possible.

The above analysis was tested experimentally by Levine and Friedlander (7) on a turbulent jet containing glycerol vapor. The experimental program investigated the limits of vapor concentration and temperature which led to detectable condensation. They concluded that very high supersaturation must be attained in a rapid mixing process for detectable homogeneous condensation to occur. The mass concentration of fog in a jet was found to increase by three or four orders of magnitude when ions were introduced to promote nucleation.

The time available for condensation in the system was of the order of 0.02 sec, which indicates that condensation in a jet is a very rapid process.

2. Heterogeneous Nucleation

The simplest explanation of heterogeneous nucleation is that the particles adsorb a thin film of liquid on the surface and then act as a droplet of liquid of an equivalent size. The supersaturation necessary is then given by the well-known Thompson-Gibbs equation

$$\ln S = \frac{2\gamma M}{RTwr} \tag{7}$$

where S is the supersaturation, the liquid surface tension is γ, M is the liquid molecular weight, R the gas constant, T the absolute temperature, w the liquid density, and r the droplet radius.

This explanation has not been found adequate, for it has been shown that a supersaturated vapor will not condense on a plane surface which has adsorbed a liquid film thick enough to show interference colors (8). Bradley (9), in a review of nucleation, shows that surface molecules of an adsorbed film have an energy which is dependent on the attractive forces of the solid nucleus and the energy of the composite drop differs from that of a normal drop. The change in surface energy of the water, due to the influence of the solid, does not necessarily favor condensation.

The Volmer theory of condensation on a flat surface (10) predicts that the critical supersaturation for a given nucleation rate should rise with increasing contact angle between the liquid and the solid. Twomey (11) has verified this theory experimentally over a wide range of contact angles. The Volmer theory has been extended by Fletcher (12) who shows that to be an efficient nucleus a particle must be large and have a small contact angle with the condensing liquid.

Qualitative results of a number of workers (4,13–15) show that condensation on a wetted solid begins at the dew point, but for an unwetted solid, super cooling of 0.015–0.02°C has been found necessary. This is equivalent to supersaturation of about 101%. Recent work (16) has found the degree of supersaturation necessary for condensation by a dynamic method and results show that for 0.2-μm NaCl particles nucleation began at less than 100% relative humidity.

Fletcher (12) gives a theoretical treatment of the effect of the size and surface properties of heterogeneous nuclei on the rate of nucleation. He predicted that nucleation efficiency is independent of size for particles larger than 0.1 μm. Time lags which have been noted in heterogeneous nucleation studies are attributed to size effects. It is necessary for subcritical particles to grow to the critical size prior to spontaneous condensation. This is a chance occurrence which results in a time lag.

B. Droplet Growth Rate

Numerous experimental and theoretical studies have been made of the rate of mass transfer to drops immersed in a gas stream. These studies have been critically reviewed by Fuchs (17). In the case of nucleation and growth on submicron aerosol particles, it may be assumed that the particles are without inertia and will move with the same velocity as the gas stream. That is, they may be considered motionless relative to the medium in all dimensions.

If Stefan flow and temperature effects are ignored,

$$\frac{\partial(cp)}{\partial\theta} = D' \frac{\partial^2(cp)}{\partial p^2} \tag{8}$$

where c is concentration of vapor, p is the distance from the droplet center, θ is time, and D' is the diffusion coefficient, or in spherical coordinates for no angular variation

$$\frac{\partial(c)}{\partial\theta} = D' \left(\frac{\partial^2 c}{\partial p^2} + \frac{2}{p} \cdot \frac{\partial c}{\partial p} \right) \tag{9}$$

For the initial condition

$$c = c_\infty \quad \text{at} \quad \theta = 0 \quad p > r$$

and boundary condition

$$c = c_0 \quad \text{at} \quad \theta > 0 \quad p = r$$

The solution is given by

$$\frac{dm}{d\theta} = I_0 \left(1 + \frac{r}{\sqrt{(\pi D'\theta)}} \right) \tag{10}$$

(Ref. 17)

where I_0 is the quasi-stationary droplet growth rate. This gives the growth rate of an isolated droplet in an infinite medium.

By introducing a factor to allow for the distortion of the concentration profile around a droplet due to droplet growth, Fuchs deduces:

$$\frac{dm}{d\theta} = I_0 \left[1 + \left(\frac{c_\infty - c_0}{w} \right)^{1/2} \right] \tag{11}$$

The solution assumes constant boundary conditions and is thus not very useful for practical applications.

Luchak and Langstroth (18) arrived at a similar solution for the evaporation of a droplet in a spherical vessel, with constant boundary conditions.

In a system of droplets in which vapor is condensing, each droplet may be considered as being centrally located in a spherical container whose diameter is equal to the average distance between droplets. If the distance between droplets is very large compared with the droplet diameter, the boundary condition $\delta c/\delta p = 0$ at $r = b$ may be assumed. This condition has been used by Reiss and LaMer (19) to obtain expressions for the growth rate of mono-disperse assemblages of competing particles. Their analysis was simplified by assuming that $\delta c/\delta \theta \simeq 0$ and reducing Fick's law to the form

$$\nabla^2 c = 0$$

for static boundaries, and

$$\nabla^2 c = \frac{F(\theta)}{D'}$$

for a growing droplet. The resulting expressions are differential equations which require numerical solution. They show that the rate of increase of the surface area of the droplet is independent of the original drop radius but is not a linear function of time. This analysis has been established for more general conditions (20) and extended to polydisperse systems.

Frish and Collins (21) eliminated the simplifying assumptions made by Reiss and also rejected the assumption that the concentration of vapor at the droplet surface was zero. They substituted the boundary condition at the droplet surface:

$$\frac{\partial c}{\partial \rho} = \frac{c\alpha}{\lambda} \quad \text{for} \quad \theta > 0 \tag{12}$$

This assumption is based on the Fuchs "Δ" concept which considers that in a layer of thickness Δ, approximately equivalent to the mean free path, and adjacent to the droplet surface, the interchange of vapor molecules proceeds unobstructed as in a vacuum. Thus the Fick equation and those derived from it are applicable only at distances greater than Δ from the drop surface.

The solution for a single droplet in an infinite medium is given by

$$\frac{\Delta}{\alpha} \cdot (r - r_0) + \tfrac{1}{2}(r^2 - r_0) = \frac{c_\infty D' \theta}{w} \tag{13}$$

This shows that for $r \gg \Delta$ the droplet growth proceeds by a diffusional mechanism, and when $r \ll \Delta$ the growth is controlled by a kinematic mechanism as in a vacuum. Expressions were also developed for situations with competing particles and time-dependent source functions were included in the formulations. However, the resulting equations are not in a form which is readily soluble.

Houghton (22) has shown experimentally that droplet growth by diffusion is governed by the equation

$$\frac{d(r)^2}{d\theta} = 2D'(w - w_0) \tag{14}$$

which is the same form as eq. 13 when $r \gg \Delta$.

The following simplified analysis has been used in conjunction with the wet scrubber investigation by Lancaster (23) and follows from the analysis of Fuchs (17):

Consider a system of droplets which are motionless relative to the medium and on which water vapor is condensing. Each droplet may be considered as being centrally located in a spherical container of radius b such that if the particle concentration is n,

$$\tfrac{4}{3}\pi b^3 = \frac{1}{n} \tag{15}$$

If $b \gg r$ (the radius of the droplet), and if the vapor is assumed to have constant composition, then the amount of vapor in the sphere of radius b at time θ will approximate to:

$$B \simeq \tfrac{4}{3}\pi b^3 c_\theta \tag{16}$$

where c_θ is the vapor composition at the spherical boundary at time θ.

From the Maxwell equation (16) the rate of condensation at time θ is given by:

$$\frac{dB}{d\theta} = -4\pi r D'(c_\theta - c_0) \tag{17}$$

where c_0 is the vapor composition at the droplet surface.

Differentiating eq. 16 with respect to time,

$$\frac{dB}{d\theta} = \tfrac{4}{3}\pi b^3 \frac{dc_\theta}{d\theta} \tag{18}$$

Hence, by combining eqs. 17 and 18

$$-rD'(c_\theta - c_0) = \tfrac{1}{3}b^3 \frac{dc_\theta}{d\theta} \tag{19}$$

This may be integrated with boundary conditions $c_\theta = c_i$ at $\theta = 0$:

$$-\frac{3\pi r D'\theta}{b^3} = \ln \frac{c_\theta - c_0}{c_i - c_0} \tag{20}$$

or

$$c_\theta - c_0 = (c_i - c_0)e^{-A\theta} \tag{21}$$

where $A = 3rD'/b^3$. Combining eqs. 17 and 21:

$$-\frac{dB}{d\theta} = 4rD'(c_i - c_0)e^{-A\theta} \tag{22}$$

If constant r is assumed, eq. 22 may be integrated between the limits $B = B_1$ and $B = B_2$ at $\theta = 0$ and $\theta = \theta_1$:

$$B_2 - B_1 = \tfrac{4}{3}\pi b^3 (c_i - c_0)(e^{-4\pi n r D'\theta} - 1) \tag{23}$$

substituting from eq. 15.

$$B_2 - B_1 = \frac{c_i - c_0}{n} (e^{-4\pi r n D'\theta} - 1) \tag{24}$$

Thus if r is assumed constant over the range of droplet growth, the time θ for a given increase in particle mass may be estimated.

C. Coagulation

The current theory of Brownian coagulation was first considered by Smoluchowski (24) and his basic equation has been modified to make it more general in application. The general equation for the rate of coagulation of an aerosol is

$$-\frac{dn}{d\theta} = \frac{RTs}{6\mu N} \cdot \frac{(r_1 + r_2)^2}{r_1 r_2} \cdot C n^2 \tag{25}$$

where r_1 and r_2 are the radii of the coagulating particles and C is the Cunningham correction factor at a mean value of r_1 and r_2, n is the number of particles, s is the influence sphere ratio and μ is the gas viscosity.

If the coagulation of aerosols in a scrubber is considered, a residence time of the order of one second would be a reasonable basis for an estimate of the magnitude of the effect. For an aerosol of $1\ \mu$ particles at a concentration of $10^6/cm^3$, the coagulation is negligible in this time interval. 10^6 particles/cm^3 corresponds approximately to a dust burden of 3 gr/ft^3 for a dust of specific gravity 3. This is of the same order of magnitude as encountered in industrial gas cleaning problems. For Brownian coagulation to become significant in a scrubber, burdens of the order of 10^9 particles/cm^3 or 3000 gr/ft^3 would be necessary. Such burdens would rarely be encountered in any practical application.

Smoluchowski has also developed an expression for coagulation in a velocity gradient (25). For $1\ \mu$ particles this gives the relationship between Brownian coagulation ϕ_{Br}, and gradient coagulation ϕ_{gr} as

$$\frac{\phi_{Br}}{\phi_{gr}} = \frac{\Gamma}{60} \tag{26}$$

where Γ is the velocity gradient (sec^{-1}). Hence velocity coagulation will only be of importance in very steep velocity gradients: for example, in the boundary layers.

Coagulation due to the spectrum of velocity fluctuations in turbulent flow must also be considered.

On the assumption that the transfer of energy from large eddies to smaller scale fluctuations is not accompanied by appreciable dissipation of heat, the size of the eddies must be large compared with internal scale of turbulence (λ). When the size of the eddies (λ) < (λ_0), the decrease in fluctuation energy decreases rapidly with λ. It has been shown experimentally by Obukhov and Yaglom (26) that for an average gas velocity in a duct ($\bar{u}$) of 12.2 m/sec, λ is 1.1 cm and for $\bar{u} = 24.4$ m/sec, $\lambda_0 = 0.7$ cm. Considering the case of $\lambda_0 \simeq 1$ cm, if ε is the rate of dissipation of turbulent energy per gram of the medium, then (26)

$$\lambda_0 = \left(\frac{\nu^3}{\varepsilon}\right)^{1/4} \tag{27}$$

where ν is kinematic viscosity and ε is rate of energy dissipation.

Obukhov and Yaglom (26) derive the following expression for the gradient of the fluctuating velocity:

$$\frac{d(u\lambda)}{d(\lambda)} \simeq \left(\frac{2\varepsilon}{15\nu}\right)^{1/2} \tag{28}$$

Therefore

$$\frac{\phi_{gr}}{\phi_{Br}} = \frac{\sqrt{2\varepsilon/15\nu}}{60}$$

$$\simeq 150\left(\frac{\varepsilon}{\nu}\right)^{1/2} \tag{29}$$

Hence, for ϕ_{gr} to be significant compared with ϕ_{Br}, $\sqrt{\varepsilon/\nu}$ must be greater than about 5. Now for $\lambda_0 = 1$ cm

$$\left(\frac{\nu^3}{\varepsilon}\right)^{1/4} = 1 \tag{30a}$$

therefore

$$\nu = \left(\frac{\varepsilon}{\nu}\right)^{1/2} \tag{30b}$$

and

$$\frac{\phi_{gr}}{\phi_{Br}} \simeq 150\nu \tag{31}$$

Now ν for air is very much less than 5 and thus turbulent coagulation is not significant compared with Brownian coagulation, and so is negligible in scrubber applications.

The experimental study of turbulent coagulation is limited. The only reliable results are due to Langstroth and Gillespie (27). The interpretation

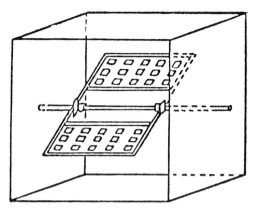

Fig. 3. Smoke chamber and stirring paddle (Langstroth and Gillespie (27)).

of these results is seriously limited by the unfortunate geometric complexity of the apparatus used (Fig. 3). This consisted of a cubic container in which was mounted a stirrer consisting of two cardboard vanes fixed to a frame and with a gap between them. The stirrer oscillated through 180°. This was claimed to have "well defined characteristics as a stirrer." Attempts have been made to analyze the results obtained from this experimental work (28), but the uncertain nature of the turbulent field produced by the apparatus makes the application of the results to flow systems impossible.

III. PARTICLE COLLECTION

The second stage of the scrubbing process is that of particle collection. In conventional scrubbers the predominant collection mechanisms are impaction and coagulation, and these are more effective if particle buildup has preceeded. However, other mechanisms of collection must be considered when the collection of particles from a hot, saturated gas stream by a cool scrubbing liquid is envisaged.

Under such circumstances appreciable thermal gradients would exist in the vicinity of the collecting surfaces. Consequently, thermal diffusion processes should be considered.

Also, condensation of vapor from the air stream on the scrubbing liquid produces a hydrodynamic flow of the noncondensing carrier gas directed toward the collecting surfaces. It has been shown that this bulk flow or "Stephan flow" could sweep with it sufficient aerosol to appreciably alter the efficiency of the gas cleaning plant. This sweep diffusion mechanism must also be considered in conjunction with the thermal diffusion processes (thermophoresis).

Finally, particle collection on duct surfaces due to inertial impaction through the boundary layer may be significant, and if the inertia of the particles were increased by buildup, the rate of particle capture by this mechanism would be increased.

A. Impaction and Coagulation

Johnstone and Roberts (29) studied gas absorption and particle deposition in a 1.5 × 8-in. rectangular Venturi throat at air velocities of up to 240 ft/sec. Their data was correlated on the basis that the efficiency of the unit

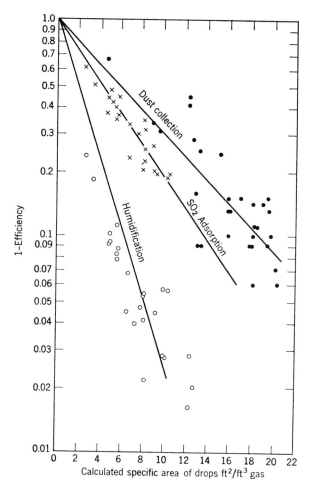

Fig. 4. Efficiency of Venturi scrubber for absorption, humidification, and dust collection (29).

was a function of the surface area produced by the atomization of the scrubbing liquid. The specific surface was estimated from the following equations due to Nukiyama and Tanasawa (30): (a) average droplet diameter

$$D_0 = \frac{16,000}{V_t} + 1.4L^{1.5} \tag{32}$$

(b) specific surface of droplets

$$S = \frac{245L}{D_0} \tag{33}$$

where D_0 = average droplet diameter (μm), V_t = relative velocity between air and liquid stream (m/sec), L = gallons of water/1000 ft³ air, and S = ft²/ft³ gas. The correlations (Fig. 4) were surprisingly good.

Further work by Johnstone, Field, and Tassler (31) considered in detail the possible mechanisms of particle capture in the initial stages of gas–liquid contacting in a Venturi scrubber. They established that impaction collection was the predominant mechanism and correlated their results by the expression:

$$\eta = 1 - \exp(-KL\psi^{1/2}) \tag{34}$$

where K is a constant, L is the liquid flow rate, and ψ is an inertial impaction parameter defined by

$$\psi = \frac{2Cur^2}{9\mu D} \tag{35}$$

where D is the diameter of the collecting droplet and u is the particle velocity. This parameter is the ratio of the force necessary to stop a particle in a distance $D/2$ to the fluid resistance. The results are shown graphically in Figure 5.

The relationship derived from these results (eq. 34) is if the same form as the Ranz and Wong expression (eq. 32) for the rate of change of dust concentration when dust is removed from air by a spray. The general Ranz and Wong equation has been used by Gieseke (33) to predict dust removal in an extended Venturi throat. This treatment is more rigorous than Johnstone's and accounts for heat, mass and momentum transfer through the contact zone. The expression derived relates the number of transfer units for a scrubber to the pressure drop thus:

$$N_t = \frac{C\eta \, \mathrm{Re}^{1/2} \, \Delta P}{\rho v C_p{}^2} \tag{36}$$

where C is a constant, η is the target efficiency, Re is the Reynolds number, ΔP is the pressure drop, ρ is the gas density, v is the gas velocity, and C_p is the liquid heat capacity.

This has the same general form as the equation of Semrau (34) for the performance of wet collectors

$$N_t = \alpha(P_T)^\gamma \tag{37}$$

where P_T is a constant multiplied by ΔP.

Figure 6 shows some of Semrau's results plotted on this basis. Semrau reports that γ should be related to aerosol particle size distribution and α should be related to both particle size and size distribution. The Gieseke (33) analysis (eq. 36) suggests that Semrau's comments are an oversimplification of the real situation.

Brownian coagulation (or deposition) in the contact zone of the Venturi scrubber has been shown to be negligible (31,33). The mechanisms of turbulent coagulation and deposition have been summarized by Levich (35). Coagulation of aerosol and water droplets for most mechanisms can be estimated on the basis of the average gas and droplet velocities. In turbulent flow the local variations between actual and average velocities can be high and should be considered. Levich (35) estimates that the distortion of stream lines due to intense turbulence dominates the distortion due to inertial effects and concludes that each encounter based on straight line trajectories would be realized. However, no method for estimating the capture by "straight line" trajectories in a turbulent field is available.

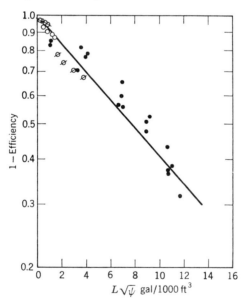

Fig. 5. Correlation of Venturi scrubber efficiency on the basis of inertial impaction (29). L is the liquid flow rate (gal/1000 ft³), ψ is inertial impaction parameter, ($\bullet$) 10-μ dibutyl phthalate (Eckman), ($\varnothing$) 1.22-μ ammonium sulfite, ($\bigcirc$) 0.58-μ dibutyl phthalate, and ($\otimes$) is 0.27-μ ammonium chloride.

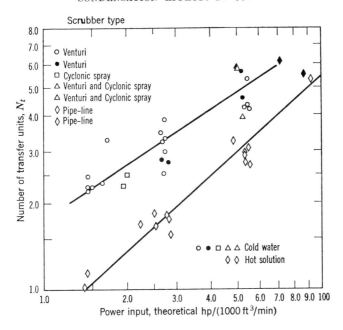

Fig. 6. A correlation of scrubber performance showing $N_t = (P_t)^\gamma$, decreased efficiency with hot scrubbing liquid was considered due to evaporation effects (34).

B. Thermophoresis

The forces which act on a particle in a thermal gradient depend on the relationship between the particle size and the mean free path of the carrier gas molecules. In gas cleaning applications the particles are generally larger than the mean free path and discussion will be limited to this case.

Consider a particle suspended in a gas in which a temperature gradient exists. It can be considered that the particle surface and the adjacent gas layer will have approximately the same tangential temperature gradient. The molecules of gas in the latter regions have a higher average velocity than those in the cooler regions and thus the molecules bombarding the particle surface give the particle a net impulse toward the cooler end of the thermal gradient. An equal impulse in the direction of the hotter zone is imparted to the gas causing a radiometric flow of gas around the nonuniformly heated particle (thermal creep). The thermal creep is influenced by the thermal gradient in the particle which is a function of heat conduction.

From these considerations Epstein (36) derived an expression for the thermal force on a particle, and this shows excellent agreement with experimental observations for particles of low thermal conductivities. However for particles of high thermal conductivity the Epstein equation is seriously in error (37). Brock (38) has shown that the Epstein analysis used boundary conditions which were not appropriate for thermal creep and that Epstein's solution of the continuum equation neglected corrective terms. Considering these points, Brock developed an equation for the thermal force which holds well over a full range of thermal conductivities. Further work (39) indicates that this analysis is limited to cases where the ratio of the particle parameter to the mean free path of gas molecules is greater than 4 (i.e., $r/\lambda > 4$). The equation given by Epstein is

$$F_t = \frac{9\pi D_0 \mu^2}{2wT\left(2 + \dfrac{\kappa_p}{\kappa_g}\right)} \frac{dT}{dx} \tag{38}$$

where F_t is thermal force on the droplet, D_0 is droplet diameter, κ_p is thermal conductivity of the droplet, κ_g is thermal conductivity of the gas, w is the density of the liquid or solid and dT/dx is thermal gradient across the space.

If the resistance of the gaseous medium to the droplets is given by Stokes' law ($F = 3\pi\mu D_0 u$) then eq. 38 can be rewritten to give the velocity, u_t, of the particle in the thermal gradient:

$$u_t = \frac{3\mu}{2wT\left(2 + \dfrac{\kappa_p}{\kappa_g}\right)} \frac{dT}{dx} \tag{38a}$$

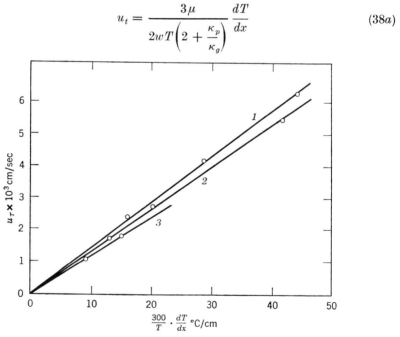

Fig. 7. Dependence of thermal velocity upon temperature gradient and particle size (24). *1*, 0.28 μm; *2*, 0.68 μm; *3*, 1.3 μm. Pressure: 76 cm Hg.

For a given aerosol material of constant particle radius and at constant pressure, this equation may be simplified to

$$u_t = \frac{K}{T} \cdot \frac{dT}{dx} \tag{39}$$

where K is a constant ($K/T \simeq 10^{-4}$ for 1-μ oil drops at $300°K$). K decreases with increase in thermal conductivity of the particles under consideration.

Figure 7 shows the experimental results of Rosenblatt and LaMer (24). The plots exhibit the straight line form predicted by eq. 39.

As steep thermal gradients may exist in the vicinity of liquid droplets in a scrubber, the particle velocity due to these gradients would be appreciable and this could contribute to scrubber efficiency.

C. Stephan Flow, Sweep Diffusion, and Diffusiophoresis

Near a surface on which a vapor is condensing, a hydrodynamic flow of the gaseous medium must be directed toward the surface. The rate of flow is given by

$$u = \frac{D'}{c} \cdot \frac{dc}{dp} \tag{40}$$

where u is the velocity, c is the vapor concentration, D' is the diffusion coefficient, and p is the distance from the droplet center, which for a spherical drop of radius r becomes

$$u = \frac{D'(c_s - c_\infty)rM_g}{cp^2M_v} \tag{41}$$

where c_s is the concentration at the drop surface, and c_∞ the concentration as $p = \infty$.

If the gas through which the vapor is diffusing contains small particles, these particles may be expected to move with the hydrodynamic flow and at about the same velocity. Derjaguin and Bakanov (40) and Waldmann (41) have treated this mechanism for the case of spherical particles for which $r < \lambda$ on the basis of the Chapman-Enskog kinetic theory. The Waldmann expression is

$$u_p = -\frac{\sqrt{M_v}}{\{y_v\sqrt{(M_v)} + y_g\sqrt{(M_g)}\}} \cdot \frac{D'}{y_g} \cdot \frac{dy_v}{dx} \tag{42}$$

where M_v and M_g are the molecular weights of vapor and gas, respectively, y_v and y_g are the concentrations of vapor and gas, respectively, D' is the diffusion coefficient, and u_p the particle velocity; it can be shown that when $M_v = M_g$ this expression reverts to the expression 40. For the case of steady-state, counter-current diffusion of a binary gas mixture without Stephan flow, the Waldmann expression becomes

$$u_p = \frac{\sqrt{(M_1)} - \sqrt{(M_2)}}{y_1\sqrt{(M_1)} - y_2\sqrt{(M_2)}} \cdot D' \operatorname{grad} y_1 \qquad (43)$$

(the two components being referred to by suffixes 1 and 2), which indicates that in such a system the particles will move in the direction of the heavier gas molecules. This motion is termed diffusiophoresis. When the expression for Stephan flow (40) is added to eq. 43, eq. 42 results.

The Waldmann equation has been developed for the case of $r < \lambda$, but according to Waldmann and Schmitt (42) it can also be used in the case $r > \lambda$ within an accuracy of 9%, Schmitt (43) verified the Waldmann equation by measuring directly the velocities of large oil droplets in a diffusing vapor. The experiments were carried out in a Millikan-type cell and low pressures were used to increase λ. For these experiments r/λ ranged from 1 to 6.

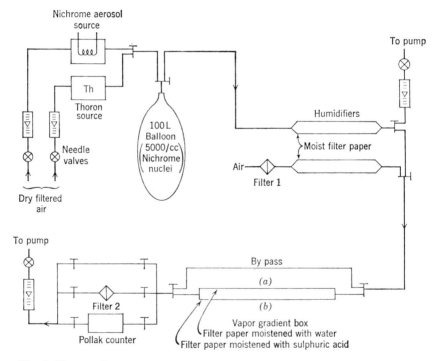

Fig. 8. The experimental arrangement used in the measurement of velocities imposed on thoron marked nichrome particles along gradients of water vapor pressure. Deposition of the particles was ascertained by measuring radioactivity of papers in the gradient box (44).

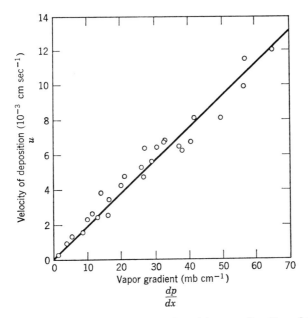

Fig. 9. The velocity of submicron particle deposition as a function of the water vapor pressure gradient.

Goldsmith, Delafield, and Cox (44) showed experimentally that the velocity of submicron particles in a water vapor gradient in air is in excellent agreement with that predicted from the eq. 42 and is very closely approximated by the Stephan flow velocity (eq. 40).

They passed a radioactively marked nichrome aerosol between two parallel plates across which was established a controlled water vapor gradient. The apparatus is shown schematically in Figure 8. Deposition of the aerosol ascertained by metering the activity of the plates and from this the deposition velocity of the particles was deduced. Their results are shown in Figure 9 and establish the linear relationship

$$u = -(1.89 \times 10^{-4}) \frac{dp}{dx} \tag{44}$$

where u is the particle velocity and dp/dx is the water vapor partial pressure gradient. Equation 40 predicts

$$u = -(2.4 \times 10^{-4}) \frac{dp}{dx} \tag{45}$$

while eq. 42 predicts

$$u = -(1.9 \times 10^{-4}) \frac{dp}{dx} \tag{46}$$

These workers deduced an expression for the efficiency of aerosol collection on stationary droplets growing by condensation. The analysis assumed that all particle transfer to the droplet was due to Stephan flow, and for an isolated droplet which has grown by mass m, the expression has the form

$$\frac{n}{n_0} = \exp\left(-\frac{km}{D'}\right) \tag{47}$$

where n/n_0 is the efficiency of particle collection, D' the diffusivity, and for dilute systems k is a constant. This expression was used to predict the efficiency of deposition of aerosol particles by diffusion mechanisms in clouds, and it was concluded that this mechanism would be secondary to Brownian diffusion, condensation, and evaporation mechanisms. However, the expression has not been thoroughly tested experimentally. In any case some kinetic data for condensation must be available before the expression can be usefully applied to gas cleaning problems.

Proceeding from eq. 42, Fuchs and Kirsch (45) have produced an analysis of aerosol deposition by condensation in a granular bed. Their analysis also neglects the effects of Brownian diffusion and assumes that particle concentration is constant from the fluid bulk to the condensation surface. For the case of complete absorption of the vapor in the bed the predicted ratio of aerosol concentrations before and after passage through the bed was given by the simple expression

$$\frac{n}{n_0} = \frac{y}{(y + y_0/\beta)} \tag{48}$$

where y is the total molar concentration of vapor and gas, y_0 is the initial molar concentration of vapor, and β is given by $\beta = (M_1^{1/2} - M_2^{1/2})/M_2^{1/2}$. Thus for $M_1 > M_2$ the aerosol concentration decreases, but for $M_2 > M_1$ the aerosol concentration increases after passing through the bed. On a vapor-free basis the expression becomes

$$\frac{n}{n_0} = \left(1 - \frac{y_0}{y}\right)\left(1 + \frac{y_0\beta}{y}\right) \tag{49}$$

As the minimum value of β is -1, n/n_0 is always less than 1.

For the case of incomplete absorption the expression becomes

$$\frac{n}{n_0} = \frac{y_1 + y/\beta}{y_0 + y/\beta} \tag{50}$$

Experimental work on absorption and evaporation of ethyl alcohol vapor in a bed of silica gel and with selenium aerosols of $r = 0.1\ \mu$ showed fair agreement with the predicted results (Fig. 10). The experimental data

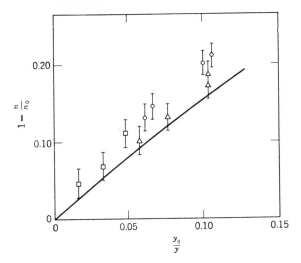

Fig. 10. Effect of sorption and of evaporation of ether on the deposition of selenium aerosol in a granular bed. ($\bigcirc$) 0.1 μm, sorption. ($\triangle$) 0.2 μm, sorption. ($\square$) 0.1 μm, sorption. n is particle concentration, n_0 is the initial particle concentration, y is the molar concentration of vapor, and y_0 is the initial molar concentration of vapor.

shows consistently higher collection efficiencies than predicted by eq. 49. Considering the difficulties of studying one collection mechanism exclusively, the results are within the accuracy which could be expected from the experiment.

Applications of the above must be restricted to cases of low droplet Reynolds number as the expressions assume the collecting surface to be substantially motionless relative to the carrier gas. Consequently they cannot be expected to yield accurate results in a situation in which the droplets are being accelerated from rest by a turbulent gas stream, as in a wet scrubber. In this case it is necessary to analyze the situation with reference to a film theory model.

Gieseke (33) has considered deposition on a moving evaporating droplet assuming constant boundary conditions. This analysis leads to the following conclusion—that evaporation–condensation mechanisms are of minor importance in conventional scrubbers, but the conditions considered were not favorable to condensation. Diffusiophoresis was not considered and as the development contains a number of errors the conclusions are doubtful.

Lancaster (23), using the same initial concepts as Gieseke, has produced the following analysis:

Consider a droplet on which vapor is condensing from a dusty atmosphere. Consider also that a temperature gradient exists between the atmosphere and the droplet.   The equation of continuity for the dust component is:

$$\frac{\partial n}{\partial \theta} + \nabla \cdot N = 0 \tag{51}$$

To evaluate the flux N, Brownian diffusion, Stephan flow, and thermophoresis must be considered.

If the coefficient of Brownian diffusion is D', the flux due to Brownian diffusion (N_{Br}) is

$$N_{\mathrm{Br}} = wD'\nabla n \tag{52}$$

For dilute systems of water vapor in air it has been shown (44) that the velocity of aerosol particles due to Stephan flow is given by $K_1 \times$ (velocity of Stephan flow), where K_1 is a constant determined by the molecular weight and diffusivity of the gases involved.   Consequently the particle flux due to Stephan flow (N_{S}) is given by

$$N_{\mathrm{S}} = K_1 n(N_A + N_B) \tag{53}$$

(This assumes that the dust concentration is such that its contribution to Stephan flow is negligible.)

The aerosol particle velocity due to thermophoresis can be expressed as (46)

$$u = K_g \cdot \frac{dT}{dx} \tag{54}$$

where K_g is a constant.   Assuming a film model and considering heat transfer through the film to be by conduction only

$$\frac{Q}{K} = -\frac{dT}{dx} \tag{55}$$

where Q is the heat flux; thus the particle flux due to thermophoresis (N_{T}) is given by

$$N_{\mathrm{T}} = nK_2 Q \tag{56}$$

where $K_2 = K_g/k$.

These expressions can be combined to give an equation equivalent to Fick's first law:

$$N = wD' \nabla \cdot n + K_1 n(N_A + N_B) + K_2 nQ \tag{57}$$

Combining eqs. 51 and 57

$$\frac{\partial n}{\partial \theta} + \nabla \cdot wD' \nabla n + \nabla \cdot n\, K_1 v + \nabla \cdot nK_2 Q = 0 \tag{58}$$

For constant w and D' this becomes

$$\frac{\partial n}{\partial \theta} + v \cdot \nabla K_1 n + Q \cdot \nabla K_2 n + wD' \nabla^2 n = 0 \tag{59}$$

Considering a film theory model at steady state

$$\frac{\partial n}{\partial \theta} = 0$$

Thus in spherical coordinates, for no angular variation

$$K_1 v_r \frac{dn}{dp} + K_2 Q_r \frac{dn}{dp} + D' \frac{1}{p^2}\left[\frac{\partial}{\partial p}\left(p^2 \frac{\partial n}{\partial p}\right)\right] = 0 \tag{60}$$

where

$$v_r = \left(\frac{\Delta m}{4\pi p^2 w_F}\right) \quad \text{and} \quad Q_r = \frac{\Delta q}{4\pi p^2} \tag{61}$$

Δm and Δq represent the heat and mass transfer rates per droplet and w_F is the average film density.

Rearranging with

$$A = \frac{K_1 \Delta m/w_F + K_2 \Delta q}{4\pi D'} \tag{62}$$

$$p^2 \frac{d^2 n}{dp^2} + (2p + A)\frac{dn}{dp} = 0 \tag{63}$$

Solving for the boundary conditions $n = n_0$ at $p = r_2$ (radius of the boundary layer) and $n = n$ at $p = r_1$ (radius of the droplet) gives

$$\frac{n}{n_0} = \frac{(\exp A/p - \exp A/r_1)}{(\exp A/r_2 - \exp A/r_1)} \tag{64}$$

If the particle concentration at r_1 is zero, the particle deposition rate at the surface is given by

$$N = D'\left(\frac{\partial n}{\partial p}\right) \tag{65}$$

From eq. 64

$$\left(\frac{dn}{dp}\right) = \frac{n_0 A/r_1{}^2 \exp(A/r_1)}{\exp A/r_1 - \exp A/r_2} \tag{66}$$

Therefore

$$N = \frac{D' n_0 (A/r_1{}^2)}{1 - \exp(A/r_2 - A/r_1)} \tag{67}$$

Applying this to the collection of dust in a scrubber with residence time θ and droplet concentration of S',

$$\frac{n_0 - n}{n_0} = \frac{4\pi\theta S' D' A}{1 - \exp{(A/r_2 - A/r_1)}} \tag{68}$$

The values of A and r_2 can be determined from the correlation of Ranz and Marshall for heat and mass transfer to a sphere (47):

$$\mathrm{Nu} = 2 + 0.6(\mathrm{Re})^{\frac{1}{2}}(\mathrm{Pr})^{\frac{1}{3}} \tag{69}$$

$$\mathrm{Sh} = 2 + 0.6(\mathrm{Re})^{\frac{1}{2}}(\mathrm{Sc})^{\frac{1}{3}} \tag{70}$$

where Nu is the Nusselt number, Pr is the Prandtl number, Re is the Reynolds number, Sh is the Sherwood number, and Sc is the Schmidt number.

For particles of the order of 1-μm diameter, the coefficient of Brownian diffusion is of the order of 10^{-7} and at reasonable condensation rates the value of A becomes quite large. In this case the exponential term in eq. 68 tends to zero. Thus

$$\frac{n_0 - n}{n_0} = 4\pi\theta S' D' A \tag{71}$$

If v' and q' indicate the amount of mass and heat transfer per unit volume for the system in time θ, then:

$$\frac{n_0 - n}{n_0} = k_1' v' + k_2' q' \tag{72}$$

where n_0 is the initial particle concentration and n the final particle concentration. This suggests that the particle capture efficiency is dependent on the amount of condensation occurring in the system and not on the rate at which the condensation occurs. This is similar to the result obtained by Fuchs and Kirsch (45) for particle capture in a packed bed.

The semiquantitative results of Cichelli (48) on the removal of smoke from air by "sweep diffusion" have been widely quoted. Tobacco smoke was introduced into an annular column as shown in Figure 11. Steam introduced through the central porous cylindrical section diffused across the annular section and condensed on a water curtain flowing down the inner surface of the cylindrical case. The diffusing steam swept with it the smoke particles and this was demonstrated by sampling the gas in the annular column at points near to and remote from the condensation surface. The results indicated gas-phase segregation of smoke particles but gave no indication of particle deposition rates. The evidence of this work suggests only low efficiency for particle removal in spite of steam consumption rates of up to 1 lb/ft^3 of gas treated. However, the information available does not allow any quantitative assessment of the work.

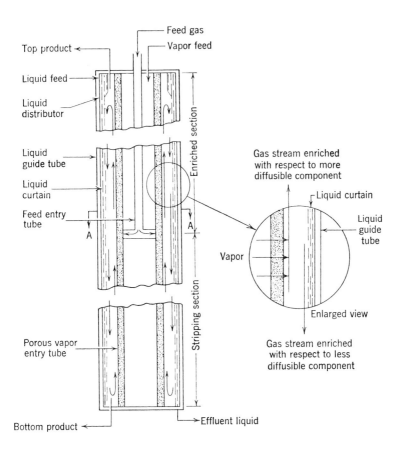

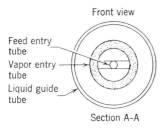

Fig. 11. Sketch of the sweep diffusion apparatus of Cichelli et al. (48). Vapor diffuses from an internal porous cylinder to the external liquid curtain while feed gas is introduced at the center of the column and particles are swept by diffusion to the liquid curtain zone. This section of the gas is carried down by the motion of the liquid and emerges as a bottom product in which the particles are concentrated. The remaining cleaned gas is taken off overhead.

All available theoretical and experimental evidence is relevant only to a simplified situation and cannot be directly applied to scrubber theory without some reservations. The balance of evidence does strongly suggest that Stephan flow and diffusiophoresis will not contribute appreciably to improvements in efficiency in conventional scrubbers. The work of Fuchs and Kirsch (45) shows that at high condensation rates the improvement may be considerable.

D. Surface Loss

As particle buildup by condensation may be expected to have some influence on the rate of wall deposition, this mechanism of collection will be considered briefly.

When an aerosol flows in turbulent motion over a surface, particles may be deposited by diffusional and inertial mechanisms. The deposition by diffusion through the boundary layer is controlled by the nature of the boundary layer. The boundary layer theory of Prandtl predicts that the "laminar" boundary layers for mass and momentum transfer coincide and have a thickness of the order of $10\nu/\sqrt{\tau/\rho}$ where ν is the kinematic viscosity, τ is shear stress, and w is density. If these layers coincide, then $\nu_t/\nu = D_t/D'$ and since $\nu_t/D_t \simeq 1$, the Schmidt number must also approximate to unity. However for aerosols the Schmidt number is very much greater than one. Thus for aerosol deposition the boundary layer in which particle transfer is purely by Brownian diffusion is very much thinner than that predicted from the Prandtl theory.

Landau and Levich (49) have developed an argument for turbulent motion penetrating the boundary layer. Their analysis assumes that turbulence completely disappears only at the surface and there is a boundary layer in which the Brownian diffusion of particles predominates over the turbulent diffusion. This layer has a thickness

$$\delta = 0.57 \times \frac{10\nu}{\sqrt{\tau/\rho}} \times \frac{1}{(\text{Sc})^{1/4}} \tag{73}$$

and the aerosol deposition rate I' in a pipe of radius R_P is given by

$$I' = \frac{n_0 D'^{3/4}(\text{Re})^{7/8}}{90R_P} \cdot \nu^{1/4} \tag{74}$$

where n_0 is the particle concentration, D' the diffusivity, and Re the Reynolds number.

Fuchs has shown (25) that the work of Deissler (50) on the prediction of velocity profiles in a turbulent stream (which show excellent agreement with experiment) can be combined with experimental data for the friction velocity in smooth pipes to produce the same equation as given by Landau and Levich (49).

Alexander and Coldren (51) have studied the problem from a semiempirical approach using the concept of an overall film coefficient for mass transfer, assuming that mass transfer was film controlled. They developed two expressions which were tested experimentally. It was found that once the flow had become fully developed a film-controlled expression described closely the deposition rate. The overall coefficient resulting from the experiments allowed both turbulent and laminar deposition to occur. The fact that the process was film controlled was deduced from the shape of concentration profiles.

Fuchs (45) doubts the accuracy of this work on the basis that the results depended entirely on the droplet distribution from the particle generator employed. The generator was a water spray mounted axially in a straight 6-in. diameter duct. The experimental results showed that in the initial zone of duct work, the deposition was related to the spray distributor. Experiments carried out downstream from this section were not dependent on the spray distribution, and Fuchs' objections do not appear to be valid. The droplets were of the order of 20 μm and this is very much larger than particles encountered in gas scrubbers.

Friedlander and Johnstone (52) have studied the deposition of particles ranging from 0.8 to 2.63 μm on the walls of vertical tubes at Reynolds numbers of up to 50,000. In conjunction with this work they developed semi-empirical expressions to predict the rate of transfer of particles through a boundary layer by inertial forces. This correlated well with experimental determinations of particle deposition rates from turbulent flow. The results are two orders of magnitude greater than predicted on the basis of the Levich and Landau theory, which indicates that inertial deposition is the major mechanism in the particle range considered (0.8–2.4 μm). As the problem area in gas cleaning is in the submicron range, the correlation of Friedlander and Johnstone is the most useful available for scrubber application. They expressed their correlation in terms of a transfer coefficient k_g, viz

$$\frac{k_g}{U} = \frac{f/2}{1 + \sqrt{f/2}(1525/S^{*2} - 50.6)} \tag{75}$$

where f is the friction factor and S^* is a function of the particle stopping distance.

Figure 12 shows this expression plotted (dotted curves) on the same axis as some experimental results. The scatter is rather large, especially considering the plot is on log–log axis, but these results represent the best experimental work available. The expression thus developed predicts the

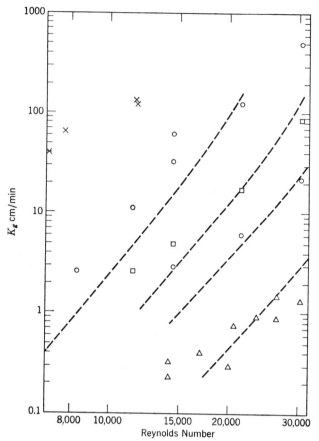

Fig. 12. Deposition in 2.5-cm glass and brass tubes plotted as a transfer coefficient versus Reynolds number (52). (✕) Lycopodium particles, 2.5-cm brass tube, glycerol jelly adhesive; (△) 0.8-μm iron particles, 2.5-cm brass tube, pressure-sensitive tape; (▲) 0.8-μm iron particles, 2.5-cm glass tube; (●) 1.32-μm iron particles, 2.5-cm brass tube, pressure-sensitive tape; (□) 1.81-μm iron particles, 2.5-cm brass tube, pressure-sensitive tape; (○) 2.63-μm iron particles, 2.5-cm brass tube, pressure-sensitive tape; (—) eq. 75.

results within an order of magnitude, which is somewhat better than the other method quoted.

IV. CONDENSATION SCRUBBERS

The experimental investigations of the effects of condensation on the performance of wet scrubbers are fragmented through the literature. There have been no sustained studies and consequently the available data are incomplete and unsatisfactory.

Schauer (1) has carried out extensive work on the use of convergent–divergent steam nozzles to promote condensation on aerosol particles before scrubbing. In this work polluted flue gas from the combustion of radio-active wastes was drawn into a steam exhauster and the resulting aerosol–steam mixture was expanded through a nozzle, causing condensation on the aerosol particles. The nozzle system was followed by a Pease-Anthony cyclonic spray scrubber.

Figure 13 shows one of the experimental layouts employed in the work, while Figure 14 indicates the nozzle types which were tested. Aerosols were produced by a dioctyl-phthalate (DOP) smoke machine and by the combustion of radioactive waste. The efficiency of the various nozzles in removing the DOP smoke (which was of controlled particle size and loading) are listed in Table I.

From these initial experiments, Schauer designed an improved steam nozzle (Fig. 15) which was used in conjunction with the pilot plant of Figure 13.

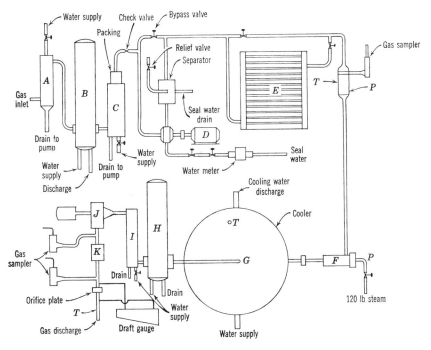

Fig. 13. Gas process equipment of Schauer's pilot plant (1). *A*, rough washer; *B*, heat exchanger (water cooled); *C*, Pease-Anthony type scrubber; *D*, gas pump (Nash); *E*, heat exchanger (water cooled); *F*, steam nozzle; *G*, expansion chamber (water cooled); *H*, heat exchanger (water cooled); *I*, Pease-Anthony type scrubber; *J*, blower; *K*, CWS filter; *P*, pressure gauges; *T*, thermometers.

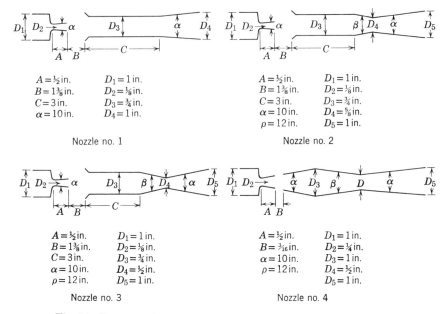

$A = \frac{1}{2}$ in. $D_1 = 1$ in.
$B = 1\frac{3}{8}$ in. $D_2 = \frac{1}{8}$ in.
$C = 3$ in. $D_3 = \frac{3}{4}$ in.
$\alpha = 10$ in. $D_4 = 1$ in.

Nozzle no. 1

$A = \frac{1}{2}$ in. $D_1 = 1$ in.
$B = 1\frac{3}{8}$ in. $D_2 = \frac{1}{8}$ in.
$C = 3$ in. $D_3 = \frac{3}{4}$ in.
$\alpha = 10$ in. $D_4 = \frac{5}{8}$ in.
$\rho = 12$ in. $D_5 = 1$ in.

Nozzle no. 2

$A = \frac{1}{2}$ in. $D_1 = 1$ in.
$B = 1\frac{3}{8}$ in. $D_2 = \frac{1}{8}$ in.
$C = 3$ in. $D_3 = \frac{3}{4}$ in.
$\alpha = 10$ in. $D_4 = \frac{1}{2}$ in.
$\rho = 12$ in. $D_5 = 1$ in.

Nozzle no. 3

$A = \frac{1}{2}$ in. $D_1 = 1$ in.
$B = \frac{3}{16}$ in. $D_2 = \frac{1}{4}$ in.
$\alpha = 10$ in. $D_3 = 1$ in.
$\rho = 12$ in. $D_4 = \frac{1}{2}$ in.
    $D_5 = 1$ in.

Nozzle no. 4

Fig. 14. Steam nozzle types tested with dioctyl phthalate smoke (1).

This system gave efficiencies of up to 99.9% for the collection of 0.3 μm particles of DOP whereas the same equipment when operated without steam gave zero collection efficiency for the same fume. The equipment was operated at low gas treatment rates (approximately 30 scfm) and steam consumption was high (0.2 lb. steam/ft³ gas). No attempt was made to investigate the particle collection mechanisms and the efficiency improvement was ascribed solely to particle buildup.

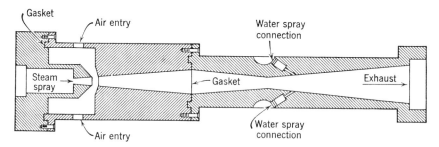

Fig. 15. Schauer's improved steam nozzle. Dimensions of nozzle: first orifice, 0.283-in. diameter; second orifice, 0.400-in. diameter; angles of divergence 11°; angles of convergence 11°; steam entry, 1-in. diameter; exhaust, 1½-in. diameter; overall length, 15⅝ in.; water spray openings (four), 0.0135-in. diameter; air entry openings (four) ⅛-in. diameter.

TABLE I

Results Obtained when Steam Nozzles
Shown in Figure 14 Were Used in Conjunction
with a Pease-Anthony Cyclonic Spray Scrubber[a]

Steam	Nozzle	Smoke flow rate, ft³/min	Particle size, μm	Particle conc./ft³	Penetration %
Wet	1	15	0.3	5.9×10^{11}	8.0
Wet	2	13	0.3	5.9×10^{11}	4.0
Wet	3	10	0.3	5.9×10^{11}	3.0
Wet	4	15	0.3	5.9×10^{11}	0.15
Wet	5 (not shown)	13	0.3	5.8×10^{11}	6.8
Superheated	1	15	0.3	5.9×10^{11}	29.2

[a] From Ref. 1.

In an investigation of the performance of various wet scrubbers, Lapple and Kamack (53) found that injection of steam into a dust-laden gas prior to its passage through a Venturi scrubber increased the collection efficiency of the unit. Injection of two to three times the amount of steam necessary for saturation was shown to reduce the amount of dust passing through the scrubber to one-fifth of its previous value for a given pressure drop. A number of alternative mechanisms were advanced to account for this phenomenon:

1. Wetting characteristics of the particles could be altered by steam injection, making capture less difficult.

2. Bulk flow and diffusion mechanisms could sweep particles toward condensation surfaces.

3. Particle buildup by condensation could make particles more readily captured.

4. Electrostatic effects could assist collection.

5. Sonic agglomeration due to supersonic steam velocities could flocculate particles.

6. Thermal gradients set up by steam injection could result in thermal precipitation.

None of these mechanisms were investigated.

Fahnoe et al. (54) studied the buildup of aerosol particles by condensing steam in a duct through which an aerosol was passed. They employed adiabatic steam and ethylene glycol vapor injection to promote condensation, as well as studying various other methods of particle buildup. The layout of their experiment is shown in Figure 16. Sodium chloride aerosols were employed and a complex system of sampling devices was used to

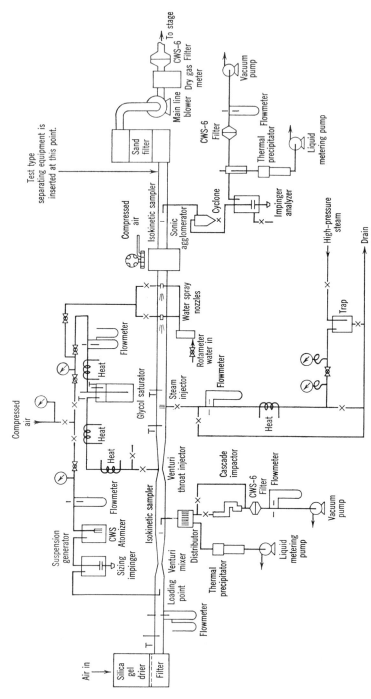

Fig. 16. Experimental equipment for investigations of particle buildup mechanisms (54).

determine the particle size distribution of the aerosol before and after buildup. The results of the investigations are summarized in Table II and Figure 17.

Their results showed that 97% of 1-μm particles could be grown to greater than 2 μm by steam injection. The dust loadings used were very low and the steam condensation rates high. Less than 10% of the steam condensed on the particles, while over 90% either condensed on the duct walls or was homogeneously nucleated (self-nucleated). No data were given on the kinetics of the process and neither condensation, nor the effects of other mechanisms were investigated.

The work was projected to two industrial types of scrubber—a cyclone representing a single unit of a multiclone designed for 5-μm particle collection, and an impingement scrubber of the Peabody type. Results of this projection are shown in Table III. The results indicate an increase in overall efficiency from 40 to 90% for the injection of 0.01 lb. steam/ft^3 of air treated.

The first experimental evidence of mechanisms other than particle buildup being introduced by the condensation process arises from the work of Browning (55). He found that, in the collection of a mixture of 0.002 and 0.007 μm particles on a dry filter without steam injection, the 0.002-μm particles were filtered with higher efficiency than the 0.007-μm particles. He explained this by suggesting that the smaller particles were more readily captured by diffusion. With steam injection the efficiencies of collection of both size fractions were increased, with the 0.002-μm particles still being more easily

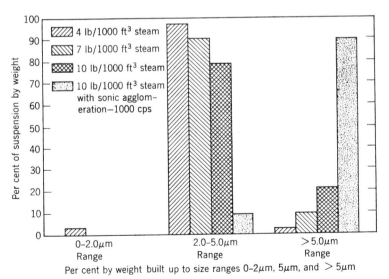

Fig. 17. Particle buildup by steam injection (54).

TABLE II

Particle Buildup Achieved by Direct Steam Injection into an Aerosol Stream[a]

Original mean particle diameter, μm	Loading, gr/1000 ft³	Steam flow, lb/1000 ft³	Steam condensed, gr/1000 ft³	Air temp. (after steam injection), °F	Particle size analysis after buildup, %		
					Overall (>2.0 μm)	Jet impinger (2.0–5.0 μm)	Cyclone (>5.0 μm)
1.0	0.85	0	0	—	37.5	31.0	6.5
1.0	0.93	10.0	4800	131	98.94	74.5	24.4
1.0	0.68	10.0	4800	125	99.51	81.0	18.5
1.0	0.60	7.0	3800	118	99.19	91.0	8.2
1.0	0.93	7.0	3800	119	98.75	88.7	10.0
1.0	0.60	4.0	2200	101	96.58	101.0	0.0
1.0	0.85	4.0	2200	101	97.30	93.0	4.3
0.6	0.45	0	0	—	55.5	42.0	13.5
0.6	0.45	10.0	4800	125	98.72	87.0	11.7

[a] From Fahnoe, Lindroos, and Abelson (54).

Table III

Evaluation of Removal Efficiencies of Representative Process Equipment
with and without Steam Injection[a,b]

Conditioning	Impinger–analyzer analysis of suspension, %		Amount removed, %	
	2–5 μm	>5 μm	Cyclone	Peabody scrubber
Dry suspension	6.6	—	36.2	40.0
Steam, 10 lb/1000 ft³, 4800 gr condensed/1000 ft³	32.0	58.2	94.3	90.5

[a] From Ref. 54.

[b] Suspension (sodium chloride), gr/1000 ft³, 0.6–0.8; maximum particle size, μm, 1.2; main air flow, ft³/min at 80°F, 10; relative humidity, %, <20.

captured. The possible significance of this was not apparent to Browning and other workers (55).

If particle buildup had been the only active mechanism in these experiments, two possibilities would have arisen:

1. Both size fractions could have been built up to such a level that impaction collection on the filter would have predominated. In this case the larger particles would have been captured with greater efficiency than the smaller ones.

2. Buildup could have been limited such that the particles were still small enough for diffusion collection to predominate. In this case the smaller particles would have been collected with greater efficiency than the larger. As both particle fractions would have been built up, the efficiency of their capture by diffusion would be expected to be lower than that without steam injection.

As experimental evidence supports neither of these possibilities, some other mechanism must account for the results. The possible nature of this mechanism can be deduced from the work of Lapple and Kamack (50) and Semrau et al. (34), who indicated that the bulk flow of a carrier gas which accompanies diffusion would sweep with it sufficient aerosol to appreciably alter the efficiency of the gas cleaning plant. The existence of this sweep diffusion process was first recognized by Russian workers (25).

The basis by Semrau (34) for considering this mechanism in relation to gas-cleaning equipment was the large difference in efficiency he noted between scrubbers operating with hot- and cold-water sprays. He reasoned that the evaporation from the hot water sprays and the resulting bulk flow away from the surface opposed the collection mechanisms, decreasing the collection efficiency. The experimental evidence used to support this conclusion was produced from scrubbers operating on fumes of uncontrolled and unknown particle size distributions. On this basis there must be some doubt in the conclusions drawn. Further, the comparison of scrubber efficiency with hot and cold spray systems was made at conditions of equivalent power dissipation in the scrubber. The power dissipation was based in part on the pressure drop across the scrubber. Other workers have shown by experiment (56) and by calculation (57) that the pressure drop across a scrubber decreases with decreasing spray temperature. Thus for equivalent conditions of pressure drop through a scrubber, a higher gas velocity would be necessary at low liquid temperatures. It has been established in a number of experiments that particle collection efficiency in a wet scrubber increases with increase in gas velocity when impaction is the major collection process (31,32,42). On this basis it should be possible to relate the observations of Semrau et al. to pressure variations as well as condensation effects. The data published is insufficient to allow any such qualitative assessment of the experiments.

Recent work by Lancaster (23) has investigated the relative importance of particle buildup and sweep diffusion mechanisms in a pipe line scrubber into which steam was injected. The significance of the major series of experiments can be best explained by reference to Figure 18. It can be seen that the steam injected was split into two streams; the first (A) injected before and the second (B) injected after the point of particle injection. The steam available for condensation in stream A was homogeneously nucleated or condensed on duct walls and lost from the system, while the condensing

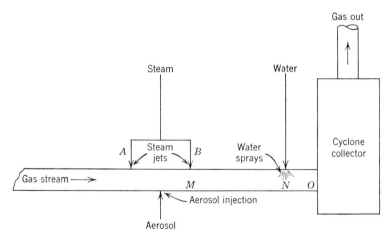

Fig. 18. Diagrammatic layout of experimental condensation scrubber (23).

fraction of stream B was available for condensation on the particles. By adjusting the ratio of these two streams for any end condition (i.e., temperature at point 0) the amount of steam available for particle buildup could be controlled. As condensation on the water curtain (and the resulting diffusion mechanisms) depends only on the end condition, the variation in particle buildup could be studied independently of the diffusion collection process.

An increase in particle loading under any particular set of operating conditions increased the number of nuclei available for condensation and this reduced the ultimate diameter of the buildup of particles. This allowed some estimate of the relative importance of the buildup-impaction process and the diffusion collection mechanism.

The position of the water spray relative to the steam injection port controlled the residence time between steam and water injection. As the residence time for buildup was increased by increasing $(M - N)$, the residence time in the water contact zone $(N - O)$ was reduced and hence interaction between the buildup and diffusion mechanisms could be studied and some indication of the kinetics of the process could be inferred.

To further distinguish between the effects of particle buildup and Stephan flow, a series of experiments were run in which the efficiency of the scrubber was found under conditions which favored Stephan flow and compared with results obtained when operating under conditions which precluded Stephan flow.

To promote Stephan flow the scrubber was operated under conditions which favored condensation on the water sprays, although to preclude Stephan flow and still maintain a strictly comparable set of operating conditions, hot water sprays at the temperature of the gas stream were substituted for the cold water sprays. This prevented condensation and thus eliminated Stephan flow.

A second series was carried out in which particle buildup was examined under conditions of constant Stephan flow. Steam was injected after particle injection and the efficiency of the unit compared with that achieved when an identical quantity of steam was injected prior to aerosol injection. Cold water scrubbing was employed in each case. As the terminal conditions of the air stream were identical for both treatments, the extent of Stephan flow capture was also identical. However in the first treatment condensation on the particles and consequent buildup was possible, but in the second, condensation occurred prior to aerosol injection and no buildup was possible.

The work was extended to examine the importance of other collection and buildup mechanisms. The mechanisms considered were turbulent coagulation, wall deposition, and particle interception by high-velocity condensate droplets.

To test the importance of the first two effects and to distinguish between them and the particle buildup due to condensation, a noncondensing gas (air) was injected through the steam injection port at rates which were comparable with the steam rates employed. The efficiency of the scrubber with air injection was compared with the efficiency achieved with an equivalent amount of steam injection. It was reasoned that turbulent coagulation and wall deposition should be about the same in each case, and any difference in efficiency could be attributed to one of the original two mechanisms.

A number of points are inadequately covered by the above techniques. It is known that wall deposition increases when walls are wetted. As the walls are wetted in the case of steam injection and are not wetted with air injection, wall deposition may be higher for the former case. Also, if particle buildup took place in the steam jet due to condensation, wall losses by inertial impaction of the built up particles would be expected to be much higher than in the case of an air jet in which no buildup is possible. Turbulent agglomeration might also be enhanced by the wetting of particles due to condensation, and this is not allowed for.

These experiments were supplemented by direct measurement of the wall deposition by determining the dust loading of the gas stream at the scrubber outlet without operating the scrubber sprays. The loading at this point was first determined without steam injection, and then measured at various steam injection rates. The drop in particle loading was attributed to wall deposition either by impaction from the steam jet or by inertial deposition through the boundary layer from the turbulent flow field.

To further investigate the extent to which steam injection contributes to turbulent coagulation and deposition, a series of steam jets were used which allowed the same total quantity of steam to be introduced over a range of Reynolds numbers. These experiments allowed some estimate of the effects of jet turbulence, independent of other factors such as wall irrigation and diffusion collection.

The possibility that the gas stream was being scrubbed by the introduction of high-velocity water droplets in the steam jet was also investigated. A series of experiments were run in which the dryness fraction of the steam was varied from 60 to 100%.

Results from the initial series showed an increase in scrubber efficiency which was comparable with the results Fahnoe et al. (54) achieved with commercial-type scrubbers. However, residence time was found to have no significant effect on scrubber efficiency and the results could be correlated with the steam jet Reynolds number by the equation:

$$\mathrm{Re} = -1.15 \times 10^5 \left(\frac{1 - \eta}{1 - \eta_0} \right) + 1 \tag{76}$$

where η is the scrubber efficiency with steam injection and η_0 is the efficiency without steam injection.   Results plotted on this basis are shown in Figure 19.

The direct comparison of aerosol penetration when utilizing hot water sprays to that obtained when cold water sprays were employed showed little change in the overall scrubber performance.

The second experiment in this series showed that although a small improvement in scrubber efficiency could be achieved by injecting steam prior to aerosol injection, the efficiency improvement was very much less than that achieved by injecting the steam after the aerosol had been introduced into the air stream.   Comparative results are shown graphically in Figure 20. This supports the conclusion that diffusion collection on the cold water curtain by Brownian diffusion, thermal diffusion, and Stephan flow are of only secondary importance in this system.

Referring to the analysis of diffusion collection, it has been shown that for steady-state conditions the efficiency of dust collection, can be estimated by eq. 72.

From the experimental results, the heat and mass transfer to the spray droplets can be estimated and eq. 72 used to predict the efficiency of aerosol

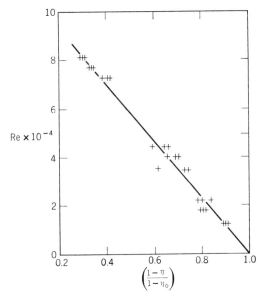

Fig. 19. Ratio of scrubber penetration with steam injection to penetration without steam injection correlated to steam jet Reynolds number in a condensation scrubber. Penetrations determined from regression equations at particle loadings of 0.1 gr/ft³, 1.0 gr/ft³, and 4.0 gr/ft³.

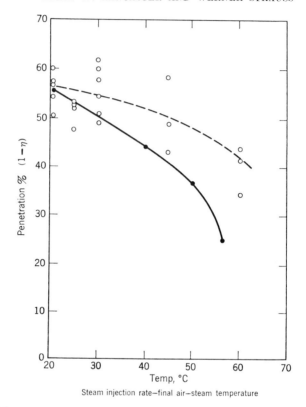

Fig. 20. Comparison of scrubber performance when steam is injected before (○) and after (●) particle injection. The scrubber performance is expressed as penetration %.

collection by diffusion mechanisms. Considering the runs which produced conditions most favorable to diffusion collection, the aerosol penetration has been estimated at about 94%.

When this was assessed in relation to the impaction collection mechanism, it was shown that the predicted efficiency compared favorably with the experimental results.

Experiments on inertial mechanisms showed wall deposition to be considerable, but suggested that particles deposited on the walls would be captured by the washing liquid in the absence of wall effects. Other inertial mechanisms were shown to be insignificant.

Data on the power consumption in the scrubber showed poor overall efficiency compared with conventional scrubbers of equivalent power consumption (Fig. 21). However, on a cost basis the two systems could become comparable for high-efficiency units.

From the results of the experimental program it is clear that the dominant mechanisms which influenced particle collection in the scrubber under test were particle buildup and subsequent impaction collection.   The correlations of scrubber efficiency with steam jet parameters indicate a relationship between the efficiency of the unit and the mixing characteristics of the system at steam injection.   The particle buildup has been shown to be substantially independent of the steam jet configuration over a limited range, and also it can be assumed, on the balance of evidence available, that substantially all condensation took place within the mixing zone of the steam jet.

Considering the course of condensation in a mixing system a number of possibilities arise.   It has been indicated by Levine and Friedlander (7) and by Amelin and Belakov (5) that with rapid mixing of a condensable vapor with a noncondensing gas in the presence of foreign nuclei, high supersaturation values are obtained and homogeneous nucleation may dominate heterogeneous nucleation.   In the system under discussion, supersaturation values could never exceed the theoretical maximum of about 3 and hence homogeneous nucleation can be neglected.   However, it must be recognized that supersaturation values approaching this maximum may have occurred in the mixing zone of the steam jet and it is likely that the majority of condensation occurred at supersaturation values near to the maximum theoretically obtainable.

A possible course for the condensation process is shown in Figure 22. It is reasonable to assume that heterogeneous nucleation would begin at relatively low values of supersaturation at a point A.   As the rate of steam removal by condensation is governed by the driving force and by the number of nuclei available, the removal in the initial phases would be low and a path similar to AB would be followed.   At some point B the condensation rate would begin to dominate the process and from B to C the majority of condensation would occur.   As this proceeds, heat release due to the latent heat of condensation would raise the temperature of the mixture above the minimum value.   Paths $AB'C$ and $AB''C$ are equally feasible, depending on

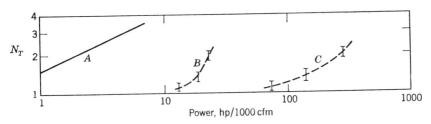

Fig. 21. Power input to scrubber compared to scrubber performance (expressed as transfer units). A, typical curve from literature (66); B, data from present study based on condensed steam; C, data from this study based on total steam injected.

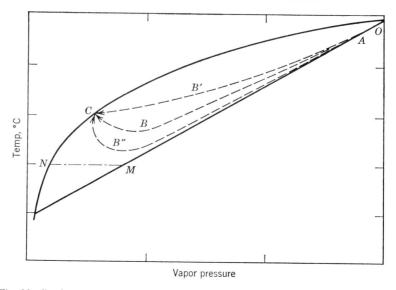

Fig. 22. Condensation curves for a vapor jet exhausting into a cooling atmosphere. (Temperature versus vapor pressure.) ABC, actual condensation curve; OCN, equilibrium curve; $AMNC$, model curve.

the system under study. From this it is obvious that a large percentage of the total amount of condensable steam could condense prior to total mixing of the steam and air stream. In this case only a small percentage of the total number of particles in the gas stream would be in a situation to nucleate this condensation and an uneven distribution of the condensed steam on the aerosol particles would result.

In an attempt to predict scrubber efficiency from the above considerations the following simplified model of the condensation process was developed:

It was considered that all the steam available for condensation was distributed equally between all the particles which had mixed with the steam jet at the maximum value of supersaturation. To give this concept some physical significance it may be assumed that the condensation curve $AMNC$ is followed. At point M, $S = S_{max}$ and condensation takes place instantaneously on all particles associated with the mixture at this point, resulting in the metastable condition N. The heat release from the latent heat of condensation returns the system to condition C on completion of the mixing process. This involves some evaporation of the condensate which takes place uniformly from all the droplets.

From this model the percentage of particles which built up was estimated and their ultimate size determined. The Johnstone, Field, and Tassler correlation was used in the following manner.

From eq. 34, the penetration $(1 - \eta)$

$$1 - \eta = e^{-KL\sqrt{\bar{\psi}}} \tag{77}$$

where K is a constant, L is the liquid rate, and ψ is the impaction parameter.

Using the experimental data for scrubber penetration at any given particle loading and with zero steam injection (P_0')

$$P_0' = e^{-KL\sqrt{\bar{\psi}}} \tag{78}$$

For the system under consideration at constant air velocity the parameter $\sqrt{\bar{\psi}}$ varies directly as the particle radius. If the initial radius is r, and the built up radius is Ar_1, then the penetration of built up particles is given by

$$(1 - \eta_{\text{B.U.}}) = (P_0')^A \tag{79}$$

A can be calculated from the model proposed and the penetration evaluated. By combining the penetration of the built up particles with the penetration of the remainder an overall penetration for the system can be determined as follows.

Example. All condensation is assumed to occur at the maximum supersaturation achieved in the jet $(S_{\max})$ as outlined. The value of $S_{\max}$ is found from eq. 1 and the corresponding mole ratio of vapor in the mixture at $S_{\max}$ determined.

Consider air at 20°C and saturated mixing with a jet of steam at 100°C and saturated: $S_{\max}$ occurs at 40°C. At $S_{\max}$ the steam/air ratio is 0.298 mole/mole.

Consider now a flow system into which sufficient steam is injected to achieve a final condition of 40°C and saturated. In such a system the steam rate–air rate ratio is 0.0675 mole/mole.

Hence the percentage of air mixed with steam in the jet at $S_{\max}$ is given by

$$\frac{0.0675}{0.298} \times 100 = 22.6 \ \%$$

Thus at a particle loading of 0.1 gr/ft³, the particles on which condensation occurs account for 0.0226 gr/ft³ (i.e., 22.6% of the total).

The amount of steam actually condensed in the jet is 3.95 gr/ft³ and the particle buildup for zinc oxide particles is given by

$$r_2 = r_1 \sqrt[3]{\frac{3.95 + 0.0226/5}{0.0226/5}} = 9.6 r_1$$

(ZnO has a specific gravity of 5) where r_1 is the initial radius and r_2 the built up radius. From the experimental results and the Johnstone, Field and

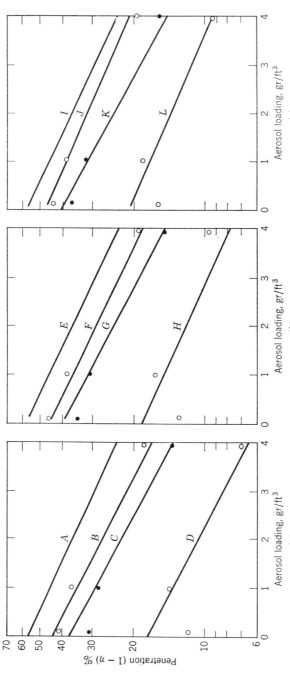

Fig. 23. Aerosol loading versus penetration for steam injection for scrubber operating conditions. Zinc oxide fume, air rate 0.45 ft/sec, water at 1.5 gal/min; inlet air saturated at (a) 20°C, (b) 25°C, and (c) 30°C. Predicted points: (○) Steam injected to 40°C; (●) steam injected to 50°C; (○) steam injected to 60°C; A–L, regression lines based on experimental data (Lancaster); A–L correspond to regression lines G–L. A, E, I, no steam injection. B, F, J, steam injected to 40°C; C, G, K, steam injected to 50°C; D, H, L, steam injected to 60°C.

Tassler correlation

$$(1 - \eta_0) = 0.55 = e^{-KL\sqrt{\bar{\psi}}}$$

Thus the penetration of the built up particles is given as

$$(1 - \eta) = (1 - \eta_0)^{9.6} = (0.55)^{9.6} \rightarrow 10^{-4}$$

(which is negligible). Therefore the predicted penetration is

$$(1 - \eta)_p\% = (22.6 \times 0) + (77.4 \times 0.55)$$

$$= 41.7\%$$

Similarly, at a loading of 4 gr/ft³ the particles on which condensation occurs account for 0.904 gr/ft³; thus

$$r_2 = r_1 \sqrt[3]{\frac{3.95 + 0.904/5}{0.904/5}}$$

or

$$r_2 = 2.84 r_1$$

$$(1 - \eta_0) = 0.23 \qquad 1 - \eta = (0.23)^{2.84} = 0.017$$

thus

$$(1 - \eta)_p\% = (22.6 \times 0.017) + (77.4 + 0.23) = 18.1\%$$

The predicted scrubber performance shows fair agreement with the experimental data over the range of the experiments. The largest deviation is at high steam injection rates with low particle loadings. Under these conditions the penetration through the scrubber was 30% higher than predicted (see Fig. 23).

Over the rest of the operating range the predicted results were within 10% of the values obtained from experimental regression equations. Considering the simplicity of the model and its arbitrary basis the agreement is good and indicates that the assumption of an uneven distribution of condensation between the particles is correct.

V. CONCLUSIONS

Condensation phenomena offer a means of improving considerably the operational efficiency of existing wet scrubber units, but the technique of direct steam injection is inefficient and could only be considered in situations where low pressure waste steam was available.

It has been shown that the poor usage of available steam is due to inadequate distribution of the steam between the potential condensation nuclei and a crude model for predicting scrubber efficiency has been proposed. This model has only been tested on small laboratory scale scrubbers and additional work is necessary before this can be extended with confidence to large industrial units.

If the efficiency of steam usage can be increased by the design of special condensation scrubbers which adequately distribute the condensed steam, among aerosol particles, the system could be competitive with conventional scrubbers in normal applications. For the removal of ultrafine particles the condensation scrubber may prove to be superior.

References

1. P. J. Schauer, *Ind. Eng. Chem.*, **43**, 1532 (1951).
2. H. L. Green and W. R. Lane, *Particulate Clouds—Dusts, Smokes and Mists*, Spon, London, 1964, p. 11.
3. H. Reiss, *J. Phys. Chem.*, **20**, 1216 (1952).
4. A. G. Amelin, *Kolloidn. Zhur.*, **10**, 169 (1948).
5. A. G. Amelin and M. I. Belakov, *Kolloidn. Zhur.*, **17**, 10 (1951).
6. W. I. Higuchi and C. T. O'Konski, *J. Colloid Sci.*, **15**, 14 (1960).
7. D. C. Levine and S. K. Friedlander, *Chem. Eng. Sci.*, **113**, 49 (1960).
8. D. H. Banghali and Z. Saweris, *Trans. Faraday Soc.*, **34**, 554 (1938).
9. R. S. Bradley, *Quart. Rev. Chem.*, **5**, 315 (1951).
10. M. Volmer, *Kinetic der Phasenbildung*, Steinkopff, Leipzig, 1939.
11. S. Twomey, *J. Phys. Chem.*, **30**, 941 (1959).
12. N. H. Fletcher, *J. Chem. Phys.*, **29**, 572 (1958).
13. N. A. Fuchs, *J. Phys. Chem. U.S.S.R.*, **6**, 410 (1935).
14. W. I. Higuchi and C. T. O'Konski, *J. Colloid Sci.*, **15**, 14 (1960).
15. W. Weilan, *Angew. Math-Phys.*, **7**, 428 (1956).
16. W. Rathse and I. N. Stranski, *Proc. Conf. Interfacial Phenomena Nucleation, Boston, 1951* (1955).
17. N. A. Fuchs, *Evaporation and Droplet Growth in a Gaseous Media*, Pergamon, London, 1959.
18. G. Luchak and G. O. Langstroth, *Can. J. Res.*, **28A**, 574 (1960).
19. H. Reiss and V. K. LaMer, *J. Phys. Chem.*, **18**, 1 (1950).
20. H. Reiss, *J. Phys. Chem.*, **19**, 482 (1951).
21. H. L. Frish and F. C. Collins, *J. Chem. Phys.*, **20**, 1797 (1952).
22. H. G. Houghton, *Pap. Phys. Ocean Meteor*, **6**, 32 (1938).
23. B. W. Lancaster, Ph.D. thesis, University of Melbourne, December 1967; see also B. W. Lancaster and W. Strauss, to be published.
24. H. L. Green and W. R. Lane, *Particulate Clouds—Smokes and Mists*, Spon, London, 1964, p. 140.
25. N. A. Fuchs, *Mechanics of Aerosols*, Pergamon, London, 1964.
26. A. Obukhov and A. Yaglom, quoted in Ref. 25, p. 334.
27. G. O. Langstroth and T. Gillespie, *Can. J. Res.*, **25B**, 455 (1947).
28. J. Corner and E. D. Pendlebury, *Proc. Phys. Soc. (London)*, **B64**, 645 (1951).
29. H. F. Johnstone and M. H. Roberts, *Ind. Eng. Chem.*, **41**, 2417 (1949).
30. S. Nukiyama and Y. Tanasawa, *Trans. Soc. Mech. Eng. (Japan)*, **4** (14), 86 (1938).
31. H. F. Johnstone, R. S. Field, and M. C. Tassler, *Ind. Eng. Chem.*, **46**, 1601 (1954).
32. W. E. Ranz and J. B. Wong, *Ind. Eng. Chem.*, **44**, 1371 (1952).
33. J. A. Gieseke, Ph.D. thesis, University of Washington, 1964.

34. K. T. Semrau, K. Marynowski, E. Lunde, and C. E. Lapple, *Ind. Eng. Chem.*, **50**, 1615 (1958).
35. V. Levich, *Physiochemical Hydrodynamics*, Prentice-Hall, New York, 1962.
36. P. Epstein, *Z. Phys.*, **54**, 537 1929.
37. C. F. Schadt and R. D. Cadle, *J. Phys. Chem.*, **65**, 1689 (1961).
38. J. R. Brock, *J. Colloid Sci.*, **10**, 768 (1962).
39. J. R. Brock, *J. Phys. Chem.*, **66**, 1763 (1962).
40. B. V. Deryaguin and S. Bakanov, *Dokl. Akad. Nauk. S.S.S.R.*, **111**, 613 (1956).
41. L. Waldmann, *Z. Naturforsch*, **14A**, 589 (1959).
42. L. Waldmann and K. Schmitt, *Z. Naturforsch*, **15A**, 843 (1960).
43. K. Schmitt, *Z. Naturforsch*, **16A**, 144 (1961).
44. P. Goldsmith, H. S. Delafield, and L. C. Cox, *Quart. J. Roy. Meteorol. Soc.*, **89**, 43 (1963).
45. N. A. Fuchs and A. A. Kirsch, *Chem. Eng. Sci.*, **20**, 181 (1965).
46. H. L. Green and W. R. Lane, *Particulate Clouds, Dusts, Mists and Fogs*, Spon, London, 1964, p. 205.
47. N. A. Fuchs, *Evaporation and Droplet Growth in a Gaseous Media*, Pergamon London, 1959.
48. M. T. Cichelli, W. D. Weatherford, and J. R. Bowman, *Chem. Eng. Progr.*, **47**, 63, 123 (1951).
49. V. Levich, *Physiochemical Hydrodynamics*, Prentice-Hall, New York, 1962, ch. 3.
50. R. Deissler, National Advisory Commission on Aeronautics, Rept. No. 1210, 1955.
51. L. Alexander and C. Coldren, *Ind. Eng. Chem.*, **43**, 1325 (1951).
52. S. K. Friedlander and H. Johnstone, *Ind. Eng. Chem.*, **49**, 1151 (1957).
53. C. E. Lapple and H. J. Kamack, *Chem. Eng. Progr.*, **51**, 110 (1955).
54. F. Fahnoe, A. E. Lindroos, and R. J. Ableson, *Ind. Eng. Chem.*, **43**, 1336 (1951).
55. W. E. Browning, U.S. At. Energy Comm. T.I.D., **7641**, 130 (1962).
56. T. T. Collins, *Paper Ind. Paper World*, **29**, 680 830 (1947).
57. J. A. Gieseke, M.Sc. thesis, University of Washington, 1964.
58. K. E. Lunde and C. E. Lapple, *Chem. Eng. Proc.*, **53**, 385 (1957).
59. R. V. Kleinschmidt and A. W. Anthony. *Am. Soc. Mech. Eng.*, **63** 349 (1941).
60. K. T. Semrau, *J. Air Poll. Control Assoc.*, **10**, 200 (1960).

Symbols

b Radius

     Ratio of specific heats

B Mass of water condensed

c Concentration of vapor, g/cm^3

C Cunningham correction factor constant

C_p Specific heat at constant pressure, $cal/g°C$

D Diameter of collecting droplet, cm

D_0 Average drop diameter, cm

D' Diffusion coefficient, cm^2/sec

D'_t Turbulent diffusion coefficient, cm^2/sec

e Correction factor $\simeq 1$

f Friction force

F_t Thermal force, dyne

g_c Gravitational constant

I Condensation rate, g/sec

I' Aerosol deposition rate, particles/sec

I_0 Quasi-stationary droplet growth rate, g/sec

k_g Film transfer coefficient

K Constant

L Liquid flow rate

Le   Lewis number

m Mass, g

M Molecular weight

n Particle concentration, particles/cm^3

N Avogadro's number

N' Particle flux, particles/cm^2 sec

Nu   Nusselt Number

N_t Number of transfer units

p Partial pressure

     distance from droplet center, cm

P' Penetration

Pe   Peclet number

P_G Power dissipation in gas

P_L Power dissipation in liquid

Pr   Prandtl number

q' Heat transfer per unit volume, cal/cm^3

Q Heat flux, cal/cm^2 sec

r Particle radius, cm

R Gas constant

Re   Reynolds number

s Ratio of particle radius to the radius of its sphere of influence

S Super saturation

     Specific surface of drop

S' Droplet concentration in gas

S^* Function of particle stopping distance

Sc   Schmidt number

Sh   Sherwood number

T Absolute temperature

u Velocity, cm/sec

U Velocity, cm/sec

v Molar velocity

     Velocity vector

v' Mass transfer/unit volume, cm^3/cm^3

V_t Relative velocity between liquid and gas

w Density of liquid or solid, g/cm^3

x Length,

y Mole fraction in vapor phase

Y Mole ratio in vapor phase

α Probability of absorption

β Parameter

γ Surface tension; dynes cm

  Parameter

Γ Velocity gradient, $\sec^{-1}$

δ Thickness of layer, cm

ε Rate of dissipation of turbulent energy

η Efficiency of particle collection

θ Time, sec

κ_g Thermal conductivity of a gas, cal/cm °C sec

κ_p Thermal conductivity of a liquid, cal/cm °C sec

λ Mean free path of gas molecules, cm

μ Gas viscosity

ν Kinematic viscosity, cm²/sec

ν_t Turbulent kinematic viscosity, cm²/sec

ρ Gas density

  Shear stress

ϕ Coagulation rate, particles/cm³ sec

χ Scale of turbulent fluctuations, cm

ψ Impaction parameter

AUTHOR INDEX

429

SUBJECT INDEX